The editors have transformed migration stuc[...] [...]sive conceptualization of the emerging field [...] [...] with thoughtful contributions from leading scholars that draw on global case studies, the complexities of humanitarian crises, moving beyond a series of isolated events into issues of global governance and structural inequality, are discussed. This is a timely and much needed book that should be required reading for all policy actors and practitioners responsible for the protection of people affected by humanitarian crises, whether forced to move or trapped in place.

Professor Susan McGrath, *Centre for Refugee Studies, York University,*
Toronto, Canada

The highly innovative and comprehensive study of migration in the context of humanitarian crises is most welcome and relevant not only for researchers but also policy-makers. It contains a wealth of information, including on little-known situations, that greatly enhance our understanding of the complex causes and dynamics of population movements.

Professor Walter Kälin, *Institute of Public Law, University of Bern, Switzerland*

This important work provides a substantial and valuable contribution to the fields of international migration and humanitarian response. The relationships between these two sets of issues have enormous implications for human security and the analyses and recommendations in this study provide invaluable insights for policy-makers and practitioners.

Professor Eric P. Schwartz, *Humphrey School of Public Affairs,*
University of Minnesota, USA, and former US Assistant Secretary of State
for Population, Refugees and Migration

HUMANITARIAN CRISES AND MIGRATION

Whether it is the stranding of tens of thousands of migrant workers at the Libyan–Tunisian border, or the large-scale displacement triggered by floods in Pakistan and Colombia, hardly a week goes by in which humanitarian crises have not precipitated human movement. While some people move internally, others internationally, some temporarily and others permanently, there are also those who become "trapped" in a place, unable to move to greater safety. Responses to these "crisis migrations" are varied and inadequate. Only a fraction of "crisis migrants" are protected by existing international, regional or national law. Even where law exists, practice does not necessarily guarantee safety and security for those who are forced to move or remain trapped. Improvements are desperately needed to ensure more consistent and effective responses.

This timely book brings together leading experts from multi-disciplinary backgrounds to reflect on diverse humanitarian crises and to shed light on a series of exploratory questions: In what ways do people move in the face of crisis situations? Why do some people move, while others do not? Where do people move? When do people move, and for how long? What are the challenges and opportunities in providing protection to crisis migrants? How might we formulate appropriate responses and sustainable solutions, and upon what factors should these depend? This volume is divided into four parts, with an introductory section outlining the parameters of "crisis migration," conceptualizing the term and evaluating its utility. This section also explores the legal, policy and institutional architecture upon which current responses are based. Part II presents a diverse set of case studies, from the earthquake in Haiti and the widespread violence in Mexico, to the ongoing exodus from Somalia, and environmental degradation in Alaska and the Carteret Islands, among others. Part III focuses on populations that may be at particular risk, including non-citizens, migrants at sea, those displaced to urban areas, and trapped populations. The concluding section maps the global governance of crisis migration, and highlights gaps in current provisions for crisis-related movement across multiple levels.

This valuable book brings together previously diffuse research and policy issues under the analytical umbrella of "crisis migration." It lays the foundations for assessing and addressing real challenges to the status quo, and will be of interest to scholars, policy makers and practitioners committed to seeking out improved responses and ensuring the dignity and safety of millions who move in the context of humanitarian crises.

Susan F. Martin, Institute for the Study of International Migration, School of Foreign Service, Georgetown University, USA.

Sanjula Weerasinghe, Institute for the Study of International Migration, School of Foreign Service, Georgetown University, USA.

Abbie Taylor, Institute for the Study of International Migration, School of Foreign Service, Georgetown University, USA.

HUMANITARIAN CRISES AND MIGRATION

Causes, consequences and responses

Edited by Susan F. Martin,
Sanjula Weerasinghe and Abbie Taylor

Routledge
Taylor & Francis Group

LONDON AND NEW YORK

First published 2014
by Routledge
2 Park Square, Milton Park, Abingdon, Oxon OX14 4RN

and by Routledge
711 Third Avenue, New York, NY 10017

Routledge is an imprint of the Taylor & Francis Group, an informa business

© 2014 Selection and editorial matter: Susan F. Martin, Sanjula Weerasinghe and
Abbie Taylor; individual chapters: the contributors

British Library Cataloguing in Publication Data
A catalogue record for this book is available from the British Library

Library of Congress Cataloging in Publication Data
Humanitarian crises and migration : causes, consequences and responses / edited by
Susan Martin, Sanjula S. Weerasinghe and Abbie Taylor.
pages cm
Includes bibliographical references and index.
1. Emigration and immigration—Case studies. 2. Migration, Internal—Case studies.
3. Forced migration—Case studies. 4. Population geography—Case studies.
5. Humanitarian assistance—Case studies. 6. Disasters—Social aspects—Case studies.
7. Natural disasters—Social aspects—Case studies. I. Martin, Susan Forbes, editor of
compilation. II. Weerasinghe, Sanjula S., editor of compilation. III. Taylor, Abbie,
1988–, editor of compilation.
JV6035.H86 2014
304.8—dc23
2013039182

ISBN: 978-0-415-85731-4 (hbk)
ISBN: 978-0-415-85732-1 (pbk)
ISBN: 978-0-203-79786-0 (ebk)

Typeset in Bembo
by Swales & Willis Ltd, Exeter, Devon

Printed and bound in the United States of America
by Edwards Brothers Malloy

CONTENTS

LIST OF FIGURES

LIST OF TABLES

NOTES ON CONTRIBUTORS

Tamer Afifi is Associate Academic Officer in the Environmental Migration, Social Vulnerability and Adaptation (EMSVA) section of the United Nations University Institute for Environment and Human Security (UNU-EHS). He is research director of the Where the Rain Falls Project, and is doing research work on environmentally induced migration in the Middle East, Sub-Saharan Africa, Asia and Latin America; he has conducted field research in all these regions. He holds a PhD in Economics from the University of Erlangen-Nürnberg, Germany, and a Master of Arts degree in Economics of Development from the Institute of Social Studies (ISS), The Hague, Netherlands.

Sebastián Albuja is the Head of the Africa and Americas Department at the Norwegian Refugee Council's Internal Displacement Monitoring Centre, based in Geneva, Switzerland. He has researched and published extensively on forced displacement in the Americas. An Ecuadorian national, he holds a JD (Universidad San Francisco de Quito) and a PhD (Northeastern University).

Alexander Betts is Associate Professor in Refugee Studies and Forced Migration at the University of Oxford. His research focuses on the international politics of refugees, migration and humanitarianism, with a focus on Sub-Saharan Africa. His recent books include *Protection by Persuasion: International Cooperation in the Refugee Regime* (Cornell University Press 2009), *Refugees in International Relations* (with Gil Loescher, Oxford University Press 2010), *Global Migration Governance* (Oxford University Press 2011), and *Survival Migration: Failed Governance and the Crisis of Displacement* (Cornell University Press 2013). He has worked as a consultant to UNHCR, UNDP, IOM, UNICEF and the Council of Europe. He has also held teaching and research positions at Stanford University and the University of Texas at Austin.

Richard Black is Pro-Director (Research and Enterprise) and Professor of Development Studies at SOAS, University of London, as well as Visiting Professor in Geography at the University of Sussex. His research focuses on the relationship between migration and poverty, forced migration, and most recently on migration in the context of climate change. From 2009-2011, he was Chair of the Lead Expert Group for a Foresight project on Migration and Global Environmental Change within the UK's Government Office for Science. He served as Co-Editor of the Journal of Refugee Studies from 1994-2009, and was elected as an Academician of the Academy of Social Sciences in 2012. He has field experience across a wide range of countries in sub-Saharan Africa and beyond.

Robin Bronen works as a human rights attorney. She received her PhD in 2012 and has been researching the climate-induced relocation of Alaska Native communities since 2007. She has published her research in the *Proceedings of the National Academy of Sciences*, the *New York Review of Law and Social* and the *Guardian*. She currently works as the executive director of the Alaska Institute for Justice. The Alaska Bar Association awarded her the 2012 International Human Rights award. The Federal Bureau of Investigation awarded the Alaska Institute for Justice the 2012 FBI Director's Community Service award, and the International Soroptomists awarded her the 2012 Advancing the Rights of Women award.

Michael Collyer is Senior Lecturer in Geography at the University of Sussex. He has held visiting fellowships at universities in Morocco, Egypt and Sri Lanka, and in 2012/13 was Fulbright Scholar at the University of Washington in Seattle. He has worked as a consultant for various international organizations, including UNHCR and the European Commission, and is currently a member of the Independent Advisory Group on Country of Origin Information, advising the UK Home Office.

Alex De Sherbinin is a senior researcher at the Center for International Earth Science Information Network (CIESIN), an interdisciplinary data and analysis center within the Earth Institute at Columbia University. He has published widely on the human aspects of environmental change at local, national and global scales, including on climate change impacts, vulnerability and adaptation. Prior to CIESIN he worked at the International Union for Conservation of Nature (Switzerland) and the Population Reference Bureau (Washington, DC), and also served as a Peace Corps Volunteer (Mauritania, West Africa). Alex has degrees in geography from Dartmouth College and Syracuse University.

Michael Edelstein is an infectious diseases epidemiologist, currently on a European field epidemiology fellowship at the Public Health Agency of Sweden. In 2013, Dr Edelstein was part of the WHO emergency response to Typhoon Haiyan in the Philippines. In 2012, he worked with WHO on polio eradication in Burkina Faso. Since 2009 he has also been a resident in public health medicine, working on outbreak response and disease surveillance for the Health Protection Agency (now Public Health England). In the past, he worked in Nepal and Argentina advising NGOs on public health program implementation. He received his medical degree from Birmingham University (UK) and his MPH from London School of Hygiene and Tropical Medicine (UK).

Patricia Weiss Fagen is a Senior Fellow at the Institute for the Study of International Migration. From 1987–96 she was a UNHCR official, serving in the United States, El Salvador and Geneva, and subsequently spent two years in the World Bank. Dr. Fagen's work has focused on post-conflict reconstruction, refugees, refugee/returnee integration, remittances and disaster management. Through ISIM and independently, she has undertaken research assignments for USAID, US Peace Corps, Christian Michelsen Institute (Norway), the Inter-American Development Bank, the Brookings Institution, the IDRC (Canada) and the Overseas Development Institute (UK). She was awarded a resident fellowship at the US Institute for Peace 2010–11.

Elizabeth Ferris is the co-director of the Brookings–LSE Project on Internal Displacement and a senior fellow in Foreign Policy, where her work encompasses a wide range of issues related to internal displacement, humanitarian action, natural disasters and climate change. Her book, *The Politics of Protection*, examines the challenges – and limitations – of protecting vulnerable populations from the ravages of war and natural disasters.

David L. Heymann is Professor of Infectious Disease Epidemiology at the London School of Hygiene and Tropical Medicine, Head of the Centre on Global Health Security at Chatham House, and Chairman of Public Health England. Previously he was the World Health Organization's Assistant Director-General for Health Security and Environment, and the representative of the Director-General for polio eradication. From 1998–2003 he was Executive Director of the WHO Communicable Diseases Cluster, during which time he headed the global response to SARS. In 2009 he was appointed an honorary Commander of the Most Excellent Order of the British Empire (CBE) for service to global public health. He is the editor of the 20th edition of the *Control of Communicable Diseases Manual*.

Khalid Koser is Deputy Director and Academic Dean at the Geneva Centre for Security Policy. He is also Non-Resident Senior Fellow in Foreign Policy Studies at the Brookings Institution, Associate Fellow at Chatham House, Research Associate at the Graduate Institute of International and Development Studies in Geneva, and Non-Resident Fellow at the Lowy Institute in Sydney. He is Extraordinary Professor in Conflict, Peace, and Security at the University of Maastricht. Dr. Koser is chair of the World Economic Forum Global Agenda Council on Migration, and editor of the *Journal of Refugee Studies*.

Judith Kumin worked for the UN refugee agency UNHCR for 32 years in Southeast Asia, Europe and North America. Her last assignment was as the agency's Director for Europe. Ms. Kumin authored the sixth edition of *The State of the World's Refugees*, published by Oxford University Press in 2012. She holds a PhD from the Fletcher School of Law and Diplomacy, and has taught international human rights law and policy at Carleton University in Ontario, at the Free University of Brussels, Belgium, and at the University of New Hampshire.

Anna Lindley is a lecturer in Development Studies at the School of Oriental and African Studies, University of London. She works on migration and livelihoods issues, with particular interests in displacement and refugee situations; transnationalism and remittances; and mobility, education and skills. Several years of research on Somali mobilities resulted in the publication of *The Early Morning Phone Call: Somali Refugees' Remittances* (Berghahn 2010), as well as various articles on displacement experiences, politics and policy.

Susan Martin is the Donald G. Herzberg Professor of International Migration and serves as the Director of the Institute for the Study of International Migration in the School of Foreign Service at Georgetown University. Her recent books include A Nation of Immigrants; The Migration-Displacement Nexus: Patterns, Processes and Policies (ed.); and Managing Migration: The Promise of Cooperation. Her forthcoming book, International Migration: Evolving Trends from the Early Twentieth Century to the Present, examines modes of international cooperation in addressing cross border movements of people from 1919 to the present. Dr. Martin received her MA and PhD in the History of American Civilization from the University of Pennsylvania.

Jane McAdam (BA (Hons), LLB (Hons) (Sydney), DPhil (Oxford)) is Scientia Professor of Law and Director of the Andrew & Renara Kaldor Centre for International Refugee Law at the University of New South Wales. She is also a non-resident Senior Fellow at the Brookings Institution in Washington, DC, and a Research Associate at the University of Oxford's Refugee Studies Centre. Professor McAdam publishes widely in the areas of international refugee law and human rights law, with a particular focus on climate change and mobility. She serves on committees of the International Law Association, the International Association of Refugee Law Judges, the Nansen Initiative on Disaster-Induced Cross-Border Displacement, and the International Bar Association. She is joint Editor-in-Chief of the *International Journal of Refugee Law*.

James Morrissey is a researcher who has been based at the University of Oxford for the past two years. His current research, on the nexus of climate change, human migration and human rights, funded by the MacArthur Foundation, is an extension of his PhD work, which explored the impacts of environmental stress on human mobility in northern Ethiopia. James has an established background exploring environment–society interactions, with an undergraduate degree in Ocean and Atmosphere Science from the University of Cape Town, and graduate degrees (Master's and PhD) in Development Studies, from the University of Oxford.

Anthony Oliver-Smith is Professor Emeritus of Anthropology at the University of Florida. He also held the Munich Re Foundation Chair on Social Vulnerability at the United Nations University Institute on Environment and Human Security in Bonn, Germany from 2005–09. His work on disasters- and development-driven displacement and resettlement focuses on post-disaster aid and reconstruction, vulnerability analysis, reconstruction, resettlement impacts, resistance and social consensus and conflict in Peru, Honduras, India, Brazil, Jamaica, Mexico, Japan and the United States. He has authored, edited or co-edited eight books and more than 70 journal articles and book chapters.

W. Courtland Robinson PhD is an Associate Professor in the Department of International Health at the Johns Hopkins Bloomberg School of Public Health, where he also serves as core faculty with the Center for Refugee and Disaster Response. He has been involved in refugee research and policy analysis since 1979 and is the author of numerous studies on refugee issues, particularly in Asia. His current research and teaching activities focus on populations in migration, whether displaced by conflict, natural disaster or widespread human rights abuse, or in the context of migrant labor and human trafficking.

Abbie Taylor, from Scotland, is a graduate of the Master of Arts in Arab Studies at Georgetown University and currently works as a Research Associate at the Institute for the Study of International Migration, based in Washington D.C. Abbie's research focus lies in the Levant, where she has worked with Iraqis, Palestinians and Syrians displaced by conflict and violence to neighboring countries. In addition to exploring histories of those displaced, their experiences in host countries and perceptions of the future and return, Abbie is examining the relationship between health and displacement in the context of humanitarian crises afflicting the region.

Alice Thomas joined Refugees International (RI) in 2010 to launch the Bacon Center for Climate Displacement. The Bacon Center seeks to increase understanding of the impacts of extreme weather and climate change on displacement, and advocates to the US government, other donors and the UN for improved policy, legal and institutional responses. Prior to joining RI, Ms. Thomas was a staff attorney in the international program at Earthjustice, a non-profit environmental law firm, and Deputy Director of the American Bar Association's Asia Law Initiative. She has also worked in the Balkans and Central Asia on programs to promote environmental good governance.

Koko Warner is the Head of the Environmental Migration, Social Vulnerability and Adaptation (EMSVA) Section at the United Nations University Institute for Environment and Human Security (UNU-EHS). She is a Lead Author for IPCC's 5th Assessment Report, Working Group 2 on Adaptation (Chapter 20). She received her PhD in Economics as a Fulbright Scholar and has been published in *Nature Talks Climate Change, Scientific American, Climate Policy, Global Environmental Change, Disasters, Environmental Hazards, Natural Hazards, Population and Environment, The Geneva Papers on Risk and Insurance – Issues and Practice*, and other journals. In addition, she serves on the editorial board of the *International Journal of Global Warming*.

Sanjula Weerasinghe (LL.B. (Hons) (Monash), LL.M. (Dist) (Georgetown)) is an Australian attorney based at the Institute for the Study of International Migration at Georgetown University in Washington D.C. Sanjula began her career in commercial law at Mallesons Stephen Jaques (now King & Wood Mallesons) before directing legal aid centers in Thailand and Hong Kong where she represented asylum claimants seeking refugee status recognition before UNHCR. Sanjula's research interests lie at the intersection of international law and migration with a current focus on individuals and groups falling outside, or unable to benefit from, existing frameworks for protection.

Roger Zetter is Emeritus Professor, University of Oxford, retiring as the Director of the Refugee Studies Centre (RSC) in 2011. He was Founding Editor of the *Journal of Refugee Studies* from 1998–2001. His research, teaching and consultancy over a 35-year-long career has focused on Sub-Saharan Africa, the EU and the Middle East, covering many aspects of forced displacement – exile, reception and settlement, protracted displacement, integration social cohesion and social capital, repatriation and post-conflict reconstruction. He was editor of the *IFRC World Disasters Report 2012*– themed on forced migration and displacement. Research on climate change and population displacement has been funded by the UNHCR, the Governments of Norway and Switzerland, and the John T. and Catherine D. MacArthur Foundation.

PREFACE AND
ACKNOWLEDGMENTS

Humanitarian crises have generated large-scale migration and displacement for all of human history. Yet, it was only in the twentieth century that international responses developed to respond to the assistance and protection needs of those forced to leave their home countries. The efforts of the League of Nations to address displacement from the Russian Revolution, and the Turkish-Greek population exchanges of the 1920s, led to the formation of the office of the High Commissioner for Refugees. This office failed to save the lives of countless refugees seeking to escape Nazi Germany in the 1930s, but efforts to protect refugees and displaced persons resumed in the aftermath of World War II, culminating in the establishment of the UN High Commissioner for Refugees (UNHCR) and adoption of the UN Convention relating to the Status of Refugees. The UNHCR continues to bear principal responsibility for protecting refugees, and its mission has been expanded to cover those displaced by conflict whether they move across international borders or internally.

While significant problems still face refugees and conflict-affected displaced persons, and protection is far from certain, as witnessed in places such as Syria, the international community is facing new challenges in the twenty-first century in responding to movements from a much wider set of crises: acute natural hazards, slow-onset environmental degradation, nuclear and industrial accidents, gang and communal violence, to name a few. This book examines three aspects of these crises: "voluntary" migration in anticipation of crises; emergency displacement and evacuation as a result of crises; and trapped populations unable to get out of harm's way and in need of relocation.

The book is the first stage of a multi-year project, which is undertaken by the Institute for the Study of International Migration (ISIM) at Georgetown University with generous support from the John D. and Catherine T. MacArthur Foundation. It presents a conceptual framework for understanding crisis migration and presents case studies that help elucidate the issues that need to be addressed. Subsequent stages of the project will focus on the identification of guiding principles and effective practices to respond to the complexity in movements related to humanitarian crises.

The chapters were presented and discussed at a workshop in September 2012 organized by ISIM. In addition to the authors represented in the volume, participants included the late Peter Benda, Roberta Cohen, Joel Charny, Vincent Cochetel, Bill Frelick, Elżbieta M. Goździak, Stefanie Haumer, Shea Houlihan, Walter Kälin, Michelle Leighton, Sue Le Mesurier, Sean Loughna, Silva Meybatyan, Tara Polzer Ngwato, Karoline Popp, Patrice Quesada, John Slocum and Alissa Walter. Some of their papers have been published as working papers, available at isim.georgetown.edu. We also presented the conceptual framework at the 2013 International Association for the Study of International Migration conference in Kolkata, a workshop on "The Ethics and Politics of the Global Refugee Regime" at Princeton University, and a meeting organized by the MacArthur Foundation on crisis migration. Our appreciation goes to participants at all of these conferences for their cogent comments and suggestions for improving the book.

Our colleagues Patricia Weiss Fagen and Natasha Hall were involved at the inception of the project and contributed to the conceptualization of the issues to be discussed. We especially want to thank Patricia for also reviewing and commenting on drafts of all of the papers. Thanks are also due to Margaret Okole for her expert copy-editing.

Several research assistants at Georgetown University contributed to the project, including Adina Appelbaum, Melanie Banesh, Grace Benton, Sarah Cechvala, Wendy Crompton, Olga Creamer, Sarah Drury, John Flanagan, Joanna Foote, Aaron Gregg, Shea Houlihan, Samantha Howland, Pitchaya Indravudh, Nina Laufbahn, Narintohn Luangrath, Justin Simeone and Alissa Walter. Throughout the project, Shirley Easton, ISIM's business manager, provided invaluable administrative support.

Finally, we thank the many migrants, refugees and displaced persons whom we have encountered in our own research. They have given freely of their time in answering our many questions and affording a human face to the phenomenon of crisis migration.

Authors' note

All URLS were accessed on 25 March 2014 unless otherwise stated.

ABBREVIATIONS

AADMER	ASEAN Agreement on Disaster Management and Emergency Response
ABG	Autonomous Bougainville Government (Papua New Guinea)
ADF	Australian Defence Force
AHTF	ASEAN Humanitarian Task Force
AMISOM	African Union Mission in Somalia
ARRA	Administration for Refugee and Returnee Affairs (Ethiopia)
ART	Antiretroviral Treatment
ASCG	Arctic Slope Consulting Group (Alaska, United States)
ASEAN	Association of Southeast Asian Nations
AU	African Union (formerly Organization of African Unity – OAU)
CAT	Convention Against Torture
CDC	Center for Disease Control (United States)
CERF	Central Emergency Response Fund
CIDEHUM	Centro Internacional para los Derechos Humanos de los Migrantes
CIRP	Carteret Integrated Relocation Project (Papua New Guinea)
CLOPADS	Comité Local para la Prevención y Atención de Emergencias y Desastres (Colombia)
CNDH	Comisión Nacional de los Derechos Humanos (Mexico)
CNIC	Computerized National Identity Card (Pakistan)
CREPADS	Comité Regional para la Prevención y Atención de Desastres (Colombia)
CSA	Country Social Analysis (World Bank)
DDMA	District Disaster Management Authority (Pakistan)
DFDR	Development-Forced Displacement and Resettlement
DGR	Risk Management Directorate (Colombia)

DIDR	Disaster-Induced Displacement and Resettlement
DISERO	Disembarkation Resettlement Offer
DRC	Democratic Republic of Congo
DRM	Disaster Risk Management
DRR	Disaster Risk Reduction
ECHR	European Convention on Human Rights and Fundamental Freedoms
ECOWAS	Economic Community of West African States
ECOSOC	United Nations Economic and Social Council
EPRDF	Ethiopian People's Revolutionary Democratic Front
ERWG	Early Recovery Working Group
EU	European Union
EXCOM	Executive Committee
FAO	Food and Agricultural Organization of the United Nations
FEMA	Federal Emergency Management Agency (United States)
FISDL	Fund for Social Investment for Local Development (El Salvador)
FMSP	Forced Migration Studies Programme (South Africa)
FNSAU	Food Security and Nutrition Analysis Unit (Somalia)
FY	Fiscal Year
GAO	Government Accountability Office (United States)
GDP	Gross Domestic Product
GFDRR	Global Facility for Disaster Reduction and Recovery
H1N1	Influenza A virus subtype H1N1 (Swine Flu)
HCT	Humanitarian Country Team
HIV/AIDS	Human Immunodeficiency Virus Infection / Acquired Immunodeficiency Syndrome
HRW	Human Rights Watch
IASC	Inter-Agency Standing Committee
ICC	International Criminal Court
ICCPR	The International Covenant on Civil and Political Rights
ICMC	International Catholic Migration Commission
ICMW	UN International Convention on the Protection of the Rights of All Migrant Workers and their Families
IDB	Inter-American Development Bank
IDMC	Internal Displacement Monitoring Center
IDP	Internally Displaced Person
ICRC	International Committee of the Red Cross
ICTY	International Criminal Tribunal for the former Yugoslavia
IFRC	International Federation of Red Cross and Red Crescent Societies
IHL	International Humanitarian Law
IHR	International Health Regulations (World Health Organization)
IHRL	International Human Rights Law

ILO	International Labor Organization
IML	International Migration Law
IMO	International Maritime Organization
INFON-AVIT	Mexican National Institute for Workers' Housing
INGO	International Non-governmental Organization
INM	Instituto Nacional de Migración (Mexico)
INS	Immigration and Naturalization Service (United States)
IO	International Organization
IOM	International Organization for Migration
IPCC	Intergovernmental Panel on Climate Change
IRIN	Integrated Regional Information Networks
IRL	International Refugee Law
ISR	International Sanitary Regulations
KP	Khyber Pakhtunkhwa
LCHR	Libyan Committee for Humanitarian Aid and Relief
LRCS	Libyan Red Crescent Society
MCOF	International Organization for Migration's Migration Crisis Operational Framework
MINUSTAH	United Nations Stabilization Mission in Haiti
MSF	Médecins Sans Frontières
NADMO	National Disaster Management Organization (Ghana)
NATO	North Atlantic Treaty Organization
NDMA	National Disaster Management Authority (Pakistan)
NGO	Non-governmental Organization
NLC	National League for Democracy (Burma)
NPG	Newtok Planning Group (Alaska, United States)
NRC	Norwegian Refugee Council
NS–NDPAM	National Strategy for Natural Disaster Prevention, Response and Mitigation (Vietnam)
NTC	Newtok Traditional Council (Alaska, United States)
OAS	Organization of American States
OAU	Organization of African Unity
OECD	Organization for Economic Cooperation and Development
OHCHR	Office of the High Commissioner for Human Rights
PDMA	Provincial Disaster Management Agency (Pakistan)
PHEIC	Public Health Event of International Concern
PLHIV	People Living with HIV/AIDS
R2P	Responsibility to Protect
RASRO	Rescue at Sea Resettlement Offer
RI	Refugees International
RMMS	Regional Mixed Migration Secretariat
SAR	Search and Rescue
SARC	Syrian Arab Red Crescent
SDC	Swiss Development Cooperation

SNPAD	National System for Disaster Management and Prevention (Colombia)
SOLAS	Safety of Life at Sea
SPDC	State Peace and Development Council (Burma)
TCG	Tripartite Core Group (ASEAN/Burma/United Nations)
TCLM	Temporary and Circular Labor Migration (Colombia and Spain)
TFG	Transitional Federal Government (TFG)
TPS	Temporary Protected Status (United States)
UK	United Kingdom
UN	United Nations
US	United States
UNAIDS	Joint United Nations Program on HIV/AIDS
UNCLOS	United Nations Convention on the Law of the Sea
UNDP	United Nations Development Program
UNEP	United Nations Environment Program
UNFCCC	United Nations Framework Convention on Climate Change
UN-HABITAT	United Nations Human Settlements Program
UNHCR	United Nations High Commissioner for Refugees
UNICEF	United Nations Children's Funds
UNOCHA	United Nations Office for the Coordination of Humanitarian Affairs
UNRWA	United Nations Relief and Works Agency for Palestine Refugees in the Near East
USACE	United States Army Corps of Engineers
USAID	United States Agency for International Development
WCD	World Commission on Dams
WFP	World Food Program
WHO	World Health Organization

PART I

Introduction and a
Theoretical Perspective

1

SETTING THE SCENE

Susan F. Martin, Sanjula Weerasinghe and Abbie Taylor

Introduction

Not a week passes in which humanitarian crises do not precipitate some form of movement. The movements that occur in the context of humanitarian crises are complex and diverse. People move within their countries and across international borders, on a temporary or permanent basis, in a legal or irregular manner. They move on their own or with assistance from external actors. Some benefit from evacuation mechanisms, voluntary migration programs, or social and diaspora networks. Others resort to clandestine networks, travelling by land or sea and taking enormous risks. Many move in direct response to acute events, while others migrate in anticipation of future harm. Yet others remain trapped in their home communities or in transit, their movement inhibited by the unfolding crisis, or by lack of resources or capacity to reach safety. Both citizens and non-citizens of countries experiencing crises are affected. These movements have implications that extend well beyond immigration control and national interests, touching upon human rights, humanitarian and development principles, and the frameworks for international protection, cooperation and burden sharing, among others. Existing legal and institutional frameworks manifest limited capacity to accommodate all those with protection needs, and yet a coherent set of principles for addressing such movements and protection is still to be devised. This volume is a first step toward this endeavor.

At the time of writing in 2013, the ongoing *conflict* in Syria has cost the lives of more than 100,000 people. There are more than six million displaced persons inside Syria and more than two million refugees who have fled to neighboring countries, with a few thousand people leaving Syria every day according to the United Nations (OCHA 2013; UNHCR 2013). Among the millions of people affected by the crisis in Syria, thousands remain trapped inside houses and makeshift shelters without basic supplies, as their neighborhoods have become battlegrounds between government and opposition forces. In Yemen, severe food insecurity and malnutrition overshadowed by violence and *political instability* is affecting more than ten million people (IRIN 2013a), including refugees and internally displaced persons (IDPs) residing inside and outside of camps, Yemeni migrant workers deported from the Gulf countries, and transit migrants arriving from the Horn of Africa. The intersection of *drought and conflict* continues to play out in the Sahel region, too, where more than one million remain displaced and 11 million are food insecure (WFP 2013). In the villages of northern India and Nepal, where flooding displaced some five million people in 2012, heavy seasonal rains have prompted *flooding* and *landslides* that have overwhelmed communities and submerged dozens of villages once again. More than 6,000 are presumed to have died during the flooding and evacuation efforts, and more than 250,000 local residents, pilgrims and tourists were forced to flee to higher ground (BBC 2013; Reliefweb 2013). Meanwhile, many of those living in coastal regions, such as the shore towns of Pakistan's Sindh province, are being relocated or are moving inland by their own means, confronted with *rising sea levels, cyclones* and *storm surges* that have destroyed homes and depleted their livelihoods as fishermen and farmers (IRIN 2013b).

In the background, protection gaps remain for those who moved in the context of crises that peaked not long ago—the returned and stranded migrant workers who fled in the wake of protests in Libya, conflict-affected and malnourished individuals within and outside famine-hit Somalia, the hapless communities devastated by back-to-back floods in Pakistan and Colombia, the evacuees of Japan's *tsunami* and *nuclear accident*, and the IDPs existing in Haiti's camps, many of whom suffered a cholera *epidemic* following the *earthquake* that ravaged the country in 2010. Equally, ongoing and intense *violence* in Mexico and chronic *governance failures* in countries such as North Korea create silent and often hidden needs for citizens and non-citizens alike, compounding the complexity of movements and necessary responses.

These recent examples, many of which are elaborated in the contributions to this volume, highlight the perpetual reality and significance of movement as a crucial and time-honored response to humanitarian crises. Indeed, the number and frequency of these crises may well increase substantially in the years ahead. *Climate change* is expected to generate internal and international movements due to increases in the intensity and frequency of natural hazards, rising sea levels, persistent drought and desertification, and, potentially, new conflicts over scarce resources. At the same time, recent and historical lessons demonstrate that processes of political change have a destabilizing effect, provoking violence and influencing new movements and protracted needs. Such circumstances underline the need to respond to crisis-related movements through effective, humane and implementable principles and practices.

Among them, the chapters in this volume illustrate: (1) the disparate ways in which humanitarian crises manifest; (2) the diverse forms of movements (and non-movements) that occur in such contexts—described for the purposes of this book as "crisis migration," with the descriptive term "crisis migrants" encompassing all those who move and those who remain trapped but in need of relocation in the context of crises; (3) protection and assistance needs of crisis migrants; and (4) factors to consider when formulating and analyzing responses targeted to address crisis migration. In doing so, the contributors shed light on the commonalities and differences in movements across diverse crisis situations, the commonalities and differences in associated protection needs of crisis migrants, and the potential and shortcomings of existing legal, policy and institutional frameworks for affording protection and assistance to crisis migrants.

The volume is divided into four sections: the first, including this chapter, introduces and conceptualizes the analytical premise of crisis migration upon which this book is based, and reflects on key considerations and preliminary recommendations for protecting crisis migrants; the second highlights the types of movement, protection needs and responses in a collection of case studies that feature humanitarian crises; the third focuses on populations "at risk" due to their status, the nature of their movement (or non-movement) and/or the context in which they move; and the fourth maps the global governance of crisis migration, examining the interplay between the local, national and international levels. The remainder of this chapter guides the reader through the components of a humanitarian crisis, the concept of crisis migration, and types of crisis migrants. In outlining the existing protection framework, identifying important parallels between diverse crises and guiding future action, this framing chapter is enriched by observations and recommendations drawn from the findings of the contributing authors.

What is a humanitarian crisis?

For the purposes of this volume, a "humanitarian crisis" is *any situation in which there is a widespread threat to life, physical safety, health or basic subsistence that is beyond the coping capacity of individuals and the communities in which they reside.* Humanitarian crises may be triggered by events or processes, and can unfold naturally, in combination with anthropogenic factors and/or through human accident or ill will. Hurricanes, cyclones, tsunamis, earthquakes, epidemics and pandemics, nuclear and industrial accidents, "acts of terrorism," armed conflict, environmental degradation, drought, famine, other climate change impacts, and situations of generalized violence and political instability are all potential triggers.

While events and processes may be the trigger or immediate "cause" of a humanitarian crisis, in most cases, underlying structural factors or contemporaneous stressors provide the context in which they occur. Lack of, or poor, national and local governance and emergency preparedness, high levels of poverty and inequality, human rights violations, insufficient access to basic services, and weaknesses in local and national capacity combine to precipitate (and at times perpetuate) humanitarian crises. Wealthy countries are not immune, although stable and more economically advanced countries generally have greater capacity to assist affected populations. Even the most precipitous events, such as earthquakes, do not necessarily result in humanitarian crises, if individuals, households and communities have the capacity to mitigate their most destructive impacts (for example, through earthquake-resistant buildings, disaster preparedness frameworks and established early warning systems), supported by good governance.

Each of the case study contributions to this volume illustrates the threats to life, physical safety, health and/or subsistence encountered during the crisis in question. The impact of structural and contemporaneous stressors on the threats and the coping capacities of the individuals and communities are elaborated in many cases. With more than half of the world's population now living in urban areas, and towns and cities increasingly overcrowded sites of human insecurity and depleted infrastructure, urban landscapes feature in numerous chapters.

Examples of events that have the potential to trigger acute crises and influence movements include extreme natural hazards. Cyclone Nargis, which struck Burma in 2008, inflicted massive loss of life and caused significant numbers to be internally displaced or trapped in the worst affected areas. The damage was due, in large part, to a delayed and inadequate response from the government, which was also loath to cooperate with the international community in providing assistance. In her case study of Haiti, Elizabeth Ferris (Chapter 4) situates both the 2008 hurricanes and the 2010 earthquake in Haiti within a history of crisis migration amid chronic political and economic instability, state fragility and endemic poverty. Ferris argues that while the type of movement may vary depending on the trigger, Haiti has endured decades of an intractable humanitarian crisis. Accordingly, an analysis reveals the merits of considering the interconnections between underlying stressors, triggers and subsequent patterns of movement, in crafting appropriate and durable responses.

Hazards can also be human-made, such as nuclear, chemical and biological accidents and attacks, accidental or deliberate setting of fires, and similar situations that make large areas uninhabitable. The accident at the Chernobyl nuclear plant in 1986, for example, resulted in the evacuation of more than 100,000 people within days (World Nuclear Association 2009). Twenty-five years later, the earthquake and tsunami in Japan led to nuclear power plants losing their capacity to cool reactors, forcing the evacuation of thousands. In both cases, to varying degrees, a lack of disaster preparedness and insufficient information and consideration for the extent of the impact on affected populations, including effects on at-risk demographics such as the elderly, contributed to the crisis. Epidemics and pandemics, namely the cholera outbreak in Zimbabwe and the SARS, H1N1 and HIV/AIDS viruses, also carry movement implications. Based on their case studies, Michael Edelstein, Khalid Koser and David Heymann (Chapter 5) find little evidence of cross-border movement in the context of health emergencies. They note, however, that short-term internal migration, and instances of people becoming trapped or contained, are common features that require further consideration and action from governments and international health bodies to coordinate appropriate preventive mechanisms and timely responses.

Conflict, political instability and generalized violence can also trigger acute crises, particularly if the state is party to the violence or unable and/or unwilling to respond. Recent events in North Africa and the Middle East fit into this category. Violence following contested elections in Kenya (2007), Zimbabwe (2008) and Côte d'Ivoire (2011) are other examples. Each of these situations generated large-scale internal and cross-border movements, the latter involving neighboring host countries. Communal violence that does not rise to the level of armed conflict has displaced large numbers in and from the Karamoja region of Uganda, Bangladesh, Ethiopia and elsewhere. The violence can be between clans, ethnic groups, economic competitors, religious groups or pastoralists claiming the same land. Gang and cartel violence is increasingly a factor in movements from and within Central America and Mexico. In his discussion of this particular issue, Sebastián Albuja (Chapter 6) observes the complexities involved in identifying those directly and indirectly affected, including both Mexicans and Central Americans caught up in pervasive violence in Mexico. Between 2007 and 2012, an estimated 70,000 people are thought to have died in Mexico, with hundreds of thousands moving throughout the country and across borders to escape the effects of a complex and omnipresent "war" involving gangs and drug cartels, and federal, state and local authorities. In one city in the north of Mexico, some 230,000 people have moved, as violence has threatened lives, physical safety and the ability to sustain a livelihood. Due to the extensive span of violence and subsequent movements, identifying and meeting the protection needs of those moving in the context of violence has proved challenging. These difficulties are exacerbated by a lack of willingness and capacity on the part of the Mexican government to acknowledge and respond adequately to the situation.

Slower-onset crises arise in a number of different contexts. Jane McAdam prefaces the selection of case studies in this volume by unsettling traditional notions of a "crisis" as a finite event, particularly in the context of slow-onset emergencies and protracted situations (Chapter 2). Such gradual processes make it imperative to understand the so-called "tipping point" of any given crisis, and to observe the individual agency and coping capacities of households and communities as they make the decisions when to move, how to move and for how long, if at all. Prolonged drought is a principal trigger of slow-onset crises and a major reason why millions who are reliant on subsistence agriculture and pastoralist activities decide to move. Recurrent droughts undermine livelihoods when crops fail and livestock are sold or die because of inadequate rain and depletion of other water sources. When markets do not function in a manner that allows a redistribution of food to drought-affected populations, migration within and across borders becomes one of the main ways to cope with losses caused by the environmental change. In worst-case examples, when drought intersects with conflict or other political factors to preclude migration and/or food distribution in communities of origin, famine may result, as has occurred repeatedly in the Horn of Africa. Manifestations of climate change may also precipitate a crisis, particularly if people do not have the resources to adapt—whether through moving or staying put. In Guatemala, for example, isolated and food-insecure rural communities are highly sensitive to increases in rainfall, and are faced with few options to diversify their livelihoods and limited finances to fund migration.

What is "crisis migration"?

Categorizing movements related to humanitarian crises presents many dilemmas for scholars and policy makers alike. In the early 1990s, Richmond put forward the continuum of "proactive" (voluntary) and "reactive" (involuntary) migration (Richmond 1993). Since then, numerous scholars have referenced this continuum, arguing that the categories of "voluntary" and "forced" are unsatisfactory and may be misleading. According to Van Hear, it is increasingly recognized that few migrants are "wholly voluntary" or "wholly forced" (2009). Almost all migration involves a degree of compulsion, just as almost all migration involves choices. Those who move in anticipation of threats make choices, but they do so within constraints and may have few alternatives to migration. Equally, forced migrants are compelled to move, but they make choices, albeit within a limited range of possibilities, particularly as to where they will move. And, while initial movement may be forced, precipitated by persecution, armed conflict or some other imminent threat, any secondary movement, including the choice of destination, may also be shaped by economic livelihood, betterment or other life-chance considerations (Van Hear 2009, 4). Even in the direst humanitarian crises, there is still an element of choice, since some may choose to stay, risking their lives rather than leave their homes. Unfortunately, too often, those who remain behind have no choice in the matter, as poverty, disability or other factors impede their ability to get out of harm's way.

In charting the evolution of normative and operational responses to movement in the context of humanitarian crises, causality has been of paramount concern in framing responses since the world wars of the twentieth century. The forced–voluntary dichotomy has influenced conceptions of causality and shaped classification systems that place those who migrate into specific categories. For example, the 1951 UN Convention relating to the Status of Refugees (1951 Refugee Convention) gives specific recognition, as refugees, to those who have crossed an international border and cannot return due to a well-founded fear of persecution for reasons of race, religion, nationality, membership of a particular social group or political opinion, and who are unable or unwilling to avail themselves of the protection of their country of nationality or habitual residence (Article 1A(2)). The 1969 Organization of African Unity (OAU) Convention Governing the Specific Aspects of Refugee Problems in Africa (1969 OAU Convention) extends this recognition to those who are compelled to leave their place of habitual residence in order to seek refuge in another place outside their country of origin or nationality owing to external aggression, occupation, foreign domination or events seriously disturbing public order in either part or the whole of their country of origin or nationality (Article 1(2)). Yet this dimension alone has become increasingly porous, with the convergence of factors such as drought and conflict or the interplay of drivers and motivations that hinder a straightforward assessment of causation in many cases.

"Mixed migration" and the "migration–displacement nexus" are terms that have come to represent this growing realization of the difficulties inherent in clearly differentiating between forced and voluntary movement and, further still, theorizing and classifying causes of movement (and associated responses) across a linear continuum, with "forced" at one end and "voluntary" at the other. In deconstructing the so-called nexus, Khalid Koser and Susan Martin put forward various manifestations. One example is the "intersection of categories", when migrants simultaneously fit two or more pre-existing categories, such as internally displaced non-citizens in Libya and Iraqi refugee returnees who have subsequently become IDPs after enduring additional displacement upon their return to Iraq. Another is "mixed flows," or migrants with different motivations utilizing the same routes to enter transit and destination countries, such as those people who board boats and risk their lives at sea in an attempt to flee from harm or to seek economic opportunity (Koser and Martin 2011, 5–6).

Irregular migrants, as Judith Kumin discusses in Chapter 15 on the challenges of mixed migration by sea, have resorted to crossing oceans since the Cold War. In recent years, parts of Africa and Asia in particular have seen increased numbers of people risking their lives in a bid to reach Europe and Australia. While the UN High Commissioner for Regfugees (UNHCR) has made some progress in its calls for the protection of refugees and asylum-seekers traveling by boat, governments remain more focused than ever on interdiction, criminal enforcement and de facto control over what they consider "extended borders," at the expense of rights protection and practical actions to ensure the safety of those traveling by sea.

"Mixed strategies," in which different types of migrants adopt similar coping mechanisms, should also be taken into consideration. A few examples from those elaborated upon by Patricia Fagen in her chapter on urban adaptation to crisis migration (Chapter 16) include rural–urban migrants, refugees, returnees, IDPs, former combatants and gang members who face similar obstacles in making a life for themselves among slum dwellers in the urban centers of Monrovia, San Salvador and Kabul. In observing the commonalities and differences in the experiences and vulnerabilities of these migrants and the urban poor, Fagen underlines the need for humanitarian and development projects to target the broader urban population defined by need, rather than by individuals' origin and the reasons for their presence in cities.

Many argue that the experiences and strategies of those who fall outside existing categories that address forced migration tend to disappear from sight, which leads to neglect or worse, underestimating widespread problems and needs during and following movement (Bakewell 2011, 26). Existing and emerging research substantiates these views. Alexander Betts has broadened the lens of inquiry by developing the concept of "survival migration" and proposing wider consideration for the protection needs of "persons outside of their country of origin because of an existential threat to which they have no access to a domestic remedy or resolution." In this sense, Betts goes some way in encapsulating the complex interaction of the environment with other social, economic and political factors—in particular, loss of livelihood and state fragility (Betts 2010, 362; Betts 2013). In this volume alone, throughout his comparative case study of North Koreans in China and Burmese in Thailand, W. Courtland Robinson (Chapter 7) explores the unrealistic and potentially harmful implications of the dichotomous question, "Are they or are they not refugees?" In China, North Korean asylum-seekers are treated as irregular migrants who have entered the country for economic reasons, and are consequently bereft of protection through a state-sponsored framework. The implications span generations. In Thailand, while the Burmese in border camps are considered "temporarily displaced" and receive some assistance from UNHCR, Burmese with similar protection needs who bypassed the camps in favor of work, and asylum-seekers from other countries residing outside camps, suffer from a lack of protection.

Similarly, Anna Lindley's analysis (Chapter 8) of combined factors focuses on the intersection of intractable conflict, poor governance and drought that contributed to the 2011 famine in Somalia, killing 258,000 people and displacing a quarter of the population. While four million people who remained inside Somalia were trapped or displaced in the face of starvation, state failure, violence, limited access to aid and a scarcity of resources, others who crossed borders were confronted with limited rights protection and assistance due to host government concern over granting prima facie recognition to those fleeing the effects of the drought.

Dominating efforts to address those falling outside existing categories is the recognition by governments, academics, institutional and civil society actors of protection gaps for those who move across international borders because of environmental and climate change. Commendable as these efforts are, questions remain as to the benefits of isolating and privileging these factors as a "cause" of movement, particularly in light of the abundance of literature explaining the diversity of factors influencing individual and household movement-related decisions. Empirical evidence does not show climate or environmental change as the sole "cause" of movement. Rather, most research suggests that climate- and environmental change-related impacts have a multiplier effect on other drivers that influence movement-related decisions (Government Office of Science, London 2011). In some cases, environmental change-related impacts may be the trigger for movement, but not necessarily the cause, and to view them in such a manner risks oversimplifying the context in which they are embedded.

This narrower focus on environmental and climate change also assumes frameworks for addressing movements in other contexts are adequate, or ignores this dimension. In essence, one may ask "Why hurricanes and not earthquakes?" or "Why so-called natural disasters and not human-made ones?" In other words, should those displaced by disasters that are more numerous or intense because of climate change deserve greater international attention than those stemming from other forces? Even in slower-onset scenarios, where loss of livelihood from intensified drought or rising sea levels occurs, is the connection to climate change relevant to the formulation of solutions? Should those who are forced to move from these situations be treated any differently or more generously than those who move because they fear for their lives, safety or health because of a nuclear accident or persistent gang violence? Should responses privilege certain "causes"? These are not easy questions to answer.

The crisis migration umbrella, which provides the analytical framework for this edited volume, is a deliberately broad lens. Rather than organize categories around the specific causes of movement, the commonalities and differences in *all* movements across various crisis situations and the associated protection needs of those who move (and those who remain trapped and in need of relocation) in times of humanitarian crisis are considered. The contributions to this volume indicate that, across crisis situations, some movements occur due to the real or perceived imminence of real or perceived threats to life, physical safety, health or basic subsistence, exhibiting a coercive element, while others take place in anticipation of such harm. Still other movements that should take place do not, leaving individuals and communities at considerable risk.

Based on this understanding of the complexity of crisis migration, we posit three principal ways in which humanitarian crises affect movement. These categories are not intended as legal definitions of various "types" of migrant, but rather reflect our attempt to describe the phenomenon of crisis migration with a view to informing future discussions around protection. These categories are not mutually exclusive, as people may move from one category to another or fall into more than one.

1. **Displacement:** This category encompasses those who are directly affected or directly threatened by a humanitarian crisis. For those who move in the aftermath of a humanitarian crisis, their life, physical safety, health or subsistence must be directly affected. For those who are evacuated or move just prior to the humanitarian crisis, their life, physical safety, health or subsistence must be directly threatened. Those who are evacuated or move following the issuing of an early warning would fall into this category. This category is intended to encompass those who are compelled to move by events beyond their direct control. The displacement may be temporary or it may become protracted.

2. **Anticipatory movement:** This category encompasses those who move because they anticipate future threats to their lives, physical safety, health or subsistence, for example because of slow-onset processes that are beyond their control. This category includes (but is not limited to) those who live in areas that are predicted to experience intensified and recurrent climatic hazards, increased drought and desertification, rising sea levels, and other results of climate change. In some cases the movements may involve entire communities, while at other times individuals and households migrate.

3. **Relocation for trapped populations:** This category encompasses those who are directly affected or threatened by a humanitarian crisis (i.e. those that are in the same situation as persons in the first category above) but who do not/cannot move due to physical, financial, security, logistical, health and/or other reasons and are in need of relocation.

Rethinking categories based on forms of movement, rather than causes, does not mean that causation is unimportant. On the contrary, ascertaining the reasons people move may be critical at the assessment stage, particularly in trying to understand needs, in evaluating the potential and shortcomings of existing legal, policy and institutional frameworks, and in determining the type and content of necessary solutions—for example, whether people can return home safely or will be in need of more permanent relocation. When considering solutions, causal considerations may also prove especially salient in ascribing roles and responsibilities to certain actors, particularly in situations where such actors have contributed to the crisis migration and protection vacuum in question.

In this context, it is worth noting that the notion of mixed migration and its various manifestations, as discussed earlier, often pervades movements that occur in the context of humanitarian crises. For example, the cases discussed in the contributing chapters highlight many situations of "mixed flows" in which those who are displaced (category 1 above) or anticipate future harm (category 2) use the same modes of movement (for example, smuggling operations) as those who are migrating for other purposes. The cases also highlight situations in which migrants have "mixed motives"—at the same time, migrants may be fleeing insecurity at home and seeking better economic opportunities elsewhere. These situations of mixed migration may become humanitarian crises in themselves when, for example, the modes of movement endanger people's lives, physical safety, health or subsistence, as detailed in Kumin's chapter.

Who are the so-called "crisis migrants"?

A "crisis migrant" is a descriptive term used to encompass all those who move and those who become trapped and are in need of relocation in the context of humanitarian crises. Men, women and children who are citizens of the country undergoing a crisis may all fall within the description, provided they move or require relocation. Equally, non-citizens—refugees, asylum-seekers, stateless persons, tourists, students, business travellers, diplomats and migrant workers, residing temporarily or habitually in legal or irregular status—may become crisis migrants.

While the term "crisis migrant" is intended to capture the breadth of those who move or become trapped and require relocation in the context of crises, specific types of crisis migrants may experience varying levels of vulnerabilities, even if their movements and non-movements are similar. At the height of certain crises, everyone may be vulnerable because of the chaos that results. Yet, to a large degree, even when crises strike communities indiscriminately, vulnerability is a measure of demographic and socio-economic factors. Some groups are inherently vulnerable during humanitarian crises, in spite of any ability to call upon reserves of resilience—young children, the elderly, disabled and those previously displaced are a few examples. Others may become vulnerable because they lose their social or economic support systems in periods of crisis, and during and following movement (e.g. women heads of household, the poor, unaccompanied minors, persons who are trafficked). Yet others may be vulnerable because of their status—non-citizens and those without legal status, such as Palestinians in Syria, female domestic workers caught up in the conflict in Lebanon in 2006 and Libya in 2011, or stateless

Rohingya in Burma exposed to recurring disasters. In his chapter on the protection of non-citizens in situations of conflict, violence and disaster (Chapter 13), Khalid Koser outlines the particular risks to non-citizens, and the gaps in protection and barriers to assisting them, through a series of case studies. These include the political instability and civil war that afflicted Libya in 2011, during which an estimated 2.5 million migrant workers from around the world were affected, in addition to more than 10,000 refugees and asylum-seekers. While tens of thousands were evacuated by their countries of origin together with the cooperation of other governments and international agencies, the majority were forced to cross borders without assistance or became internally displaced. Still more remained trapped in place, unable to get to places of greater safety.

Coping capacities may also evolve or erode, according to the life cycle of a humanitarian crisis and the stage at which people move. In the case of slow-onset crises associated with climate change and environmental degradation, people's resilience may erode gradually over time, as described by Robin Bronen in her chapter on coastal communities in the Arctic and the South Pacific (Chapter 11). Those who move sooner rather than later, before the "tipping point" of a crisis, may be less at risk (particularly if they are consulted from the outset and treated as active participants in assisted community relocation programs), than those whose coping capacities are diminished and who may become trapped. In their chapter on human mobility in the context of climate change, Koko Warner and Tamer Afifi (Chapter 10) further explore the relationship between movement, vulnerabilities and resiliencies. Their research on rainfall variability, food and livelihood security and migration reveals four broad household profiles: households that use migration to improve resilience; households that use migration to survive, but not flourish; households that use migration as a last resort and as an erosive coping strategy; and households that cannot migrate and are struggling to survive in areas of origin. In so-called "resilient" households, migration is one of a variety of adaptation mechanisms to diminish the impact of climate change, whereas for "vulnerable households" lacking resources, migration may serve only to worsen their plight.

Relatedly, not all are able to benefit equally when solutions are promoted. For example, those rendered landless or who have disabilities resulting from a crisis may continue to have pressing needs even after return is feasible and desirable for others. During episodes of widespread flooding in Pakistan and Colombia, the majority of people returned to their homes within one year. However, as noted by Alice Thomas (Chapter 3) in her comparative case study analysis of those floods, quick return had little to do with a cessation of needs and improved conditions in an abating crisis. Rather, ill consideration for the specific needs of vulnerable populations, neglect for early recovery programming and rigidity among governments and the international community in responding to the evolution of the humanitarian crisis played a role in prompting premature return and provoking further displacement. Anthony Oliver-Smith and Alex de Sherbinin in their examination of development-forced displacement and resettlement (DFDR) and disaster-induced displacement and resettlement (DIDR) (Chapter 12), focus on community-oriented initiatives, and underscore the tension between resettlement as both a solution and a threat to communities displaced. More often than not, protection of affected communities is compromised due to a lack of agency afforded to those resettled, as well as planning and design deficiencies that ignore historical, geographical, political and socio-economic considerations. The authors argue that, while existing and emerging guidelines can (in theory) protect human lives and rights, these frameworks are not yet grounded in the reality that governments of many developing countries in the Global South face, where the most significant climate impacts are predicted to happen (and are already happening).

For those people who do not move in times of crisis, immobility may pose an immediate threat to their lives—as was the case for those left behind during Hurricane Katrina—or erode resiliency to respond to slower-onset crises. Richard Black and Michael Collyer, in their chapter on trapped populations (Chapter 14), distinguish between the *desire* to move and the *need* to move, going some way toward clarifying the blurred lines between aspiration and ability. By reference to case studies from both wealthier and poorer countries and a survey of existing literature on the subject, the authors argue for, and lay out, a research agenda on this at-risk and largely invisible population of those "involuntarily immobile." Distinctions are also made between those who become trapped on the move in what is described as "fragmented migration" and those who are never able to commence their flight, due to a lack of various types of capital. In looking ahead to the crafting of policy responses, Black and Collyer posit that workable solutions (some of which are already in existence) must be informed by an understanding of the nature of the problem, including the particular reasons underlying involuntary immobility in any given case.

Expanding protection for crisis migrants

When identifying the need for responses, all crisis migrants and their unique vulnerabilities, coping capacities and protection needs must be considered, although some form of prioritization may need to be undertaken to determine who gets protection and what status and content that protection should entail. In formulating responses, a number of factors need to be taken into account, including the roles and responsibilities of different actors such as countries of origin, transit and destination, and the protection afforded under existing mandates and frameworks.

Protection is a concept that enjoys a long history and manifests today in a myriad of forms: diplomatic protection, consular protection, surrogate protection, complementary protection, temporary protection and humanitarian protection, to name a few. As its use has grown, no single definition has developed within international law; the term itself is not defined in any international or regional refugee or human rights instrument. Indeed, the absence of a clear definition and the proliferation of terms arguably compound the uncertainty associated with the content and boundaries of protection. This is further aggravated by the diverse range of actors involved in the delivery of protection, and the confusion that can arise from the intersection of a wide array of overlapping frameworks, including international refugee, migration, human rights and humanitarian law, as discussed in the chapter on the global governance of crisis migration (see Betts, Chapter 17 in this volume).

Historically, protection has been the responsibility of sovereign states, such that a state has, at a minimum, the duty to protect its citizens. Evolving from this historical context, and apposite to humanitarian crises, the concept was linked in particular to refugees, being akin to redressing the situation of those who had crossed an international border and were unable to benefit from the diplomatic and consular protection of the state to which the person had previously belonged (having not acquired another nationality). The protection to be provided to refugees was "'legal', in that it related to their status, rights and interests in other countries [and] it was 'political', in that the situation of refugees and the solutions to their problems required presentation to governments at the political level, if the requisite measures of assistance and necessary solutions were to be found" (Goodwin-Gill 1989, 7). And, at least initially, this is how protection was envisioned under the 1951 Refugee Convention.

The prevailing system of protection for individuals satisfying the legal definition of a refugee continues to accept implicitly the importance of the state apparatus in respecting and protecting the rights of its nationals and habitual residents, and embodies a surrogate state accepting responsibilities toward persons who have the "status" of refugees. Under this form of "international protection," the principal duty relates to *non-refoulement*—an obligation, subject to limited exceptions, against the forcible return of a refugee to the frontiers of territories where his or her life or freedom would be threatened on account of those traits outlined in the refugee definition discussed earlier (1951 Refugee Convention, Article 33(1)). While some rights are automatic and apply to asylum-seekers not yet recognized as refugees, according to Hathaway's cogent treatise, "additional entitlements accrue as a function of the nature and duration of the attachment to the asylum state" (Hathaway 2005, 154).

Developments in human rights law build on the parameters of international protection. The interplay, and indeed the interpretation, of the 1951 Refugee Convention and definition in light of evolving human rights law and practice has expanded the breadth of persons eligible to benefit from recognition as a refugee. Moreover, human rights law continues to clarify and extend states' obligations to refrain from returning individuals to serious harm. For example, the 1984 Convention Against Torture and Other Cruel, Inhuman or Degrading Treatment or Punishment contains an express absolute prohibition against *refoulement* or extradition of a person to a state where there are substantial grounds for believing the person would be subjected to torture (Article 3(1)). The International Covenant on Civil and Political Rights contains an implied prohibition against *refoulement* to serious

harm (Article 7). This type of international protection is better known as "complementary protection," a term that encompasses "protection granted to individuals
on the basis of a legal obligation other than the principal refugee treaty" (McAdam
2007, 2).[1] At the same time, regional expansions of the refugee definition have seen
the scope of international protection extend to those who have fled across borders
due to conflicts and other events seriously disturbing public order.

Regional and national temporary protection mechanisms and other ad hoc
national mechanisms—some based largely on humanitarian and/or other discretionary considerations rather than legal obligations—further expand avenues of
protection for those who flee across borders. These temporary mechanisms may
have the capacity to provide access to more permanent solutions, particularly as
those granted temporary protection establish ties that make it difficult for them to
return to their home countries. In general, temporary protection mechanisms fall
into two categories: (1) those that respond to new movements of people leaving
countries experiencing humanitarian crises and (2) those that permit individuals
already on the territory of the destination country to remain for at least a temporary
period. Falling within the former, the 2001 European Union (EU) Temporary
Protection Directive establishes a mechanism for providing temporary protection
during mass influx situations with, at a minimum, those who have fled areas of
armed conflict or endemic violence, and persons at serious risk of, or who have
been the victims of, systematic or generalized violations of human rights potentially
able to access protection (Article 2). To impose obligations on Member States in
any given case, the European Council, upon a proposal from the European Commission, must adopt a decision establishing the scope of the Directive's operations
(Articles 4–6). Despite the potential benefits of this mechanism, since its adoption
in 2001 the Directive is yet to be invoked, even in the case of Libya or Syria. An
example of a national mechanism falling into the latter category is the United
States' Temporary Protection Status (TPS). Persons who are unable to return safely
to their home country because of ongoing armed conflict, an earthquake, flood,
drought, epidemic, or other environmental disaster, may remain temporarily in
the United States provided a discretionary designation is made to that effect by the
relevant US authority (Immigration and Nationality Act, INA § 244(b), 8 USC
§1254a(b)). Brazil's provision of humanitarian visas to Haitians following the 2010
earthquake represents a mechanism that falls into both categories—Brazil provided
humanitarian visas to Haitians who were already in the country and those who
arrived on its territory without documentation, and also issued humanitarian visas
to Haitians at the Brazilian embassy in Port-au-Prince (see Ferris, Chapter 4 in this
volume).

While the dimension of geography—or more specifically, the fact of being outside one's country of nationality or habitual residence—has been crucial to circumscribing those able to benefit from international protection, the end of the
Cold War and subsequent interventions increased the visibility of persons forcibly
uprooted within country borders because of armed conflict, internal strife and systematic violations of human rights. Even so, only since 1998 have international

norms been clarified regarding the rights of those who are forcibly displaced within their own countries. The Guiding Principles on Internal Displacement (Guiding Principles) are a non-binding soft law framework based on existing international human rights law and international humanitarian law and refugee law, by analogy. IDPs are described as "persons or groups of persons who have been forced or obliged to flee or to leave their homes or places of habitual residence, in particular as a result of or in order to avoid the effects of armed conflict, situations of generalized violence, violations of human rights, or natural or human-made disasters, and who have not crossed an internationally recognized state border" (Guiding Principles 1998, Introduction, para. 2). The Guiding Principles cover a broad range of IDP needs during all phases of displacement: prior to displacement, during displacement, and during return, resettlement or reintegration. Importantly, they highlight that protection is the primary responsibility and duty of the national authorities within whose jurisdiction the displacement takes place, but welcome international assistance in upholding the rights and needs of IDPs (Guiding Principles 1998, Principle 3 and 25, respectively).

In 2009, the African Union (AU) adopted the Convention for the Protection and Assistance of Internally Displaced Persons (Kampala Convention), which entered into force in 2012. More generally, however, even among countries that have ratified the convention, adherence to the norms articulated therein continues to pose a problem. Some governments have been slow to implement the policies and laws they have adopted into their national legal frameworks, while most have failed to do so at all (see Brookings Institution—London School of Economics 2011), particularly in countries where there is a lack of convergence between domestic political interests and IDP legislation. Moreover, some countries, such as Colombia, which has well-established domestic laws regarding conflict displacement, do not have similar provisions regarding natural disasters (see Thomas, Chapter 3 in this volume). Roger Zetter and James Morrissey offer an explanation for these gaps in protection at the national level in their chapter on environmental displacement and the challenges of rights protection (Chapter 9). Through case studies of five countries affected by crisis migration, they posit that the way in which categories of crisis migrants and subsequent rights are conceived and articulated, and the formulation of national discourses and legislation dealing with crisis migration must be attributed to a web of underlying historical, political and socio-economic intricacies.

The chasm between international or regional legal obligations and domestic implementation is not unique to the domain of internal displacement. As Kumin, Betts and other contributors make clear, it is evident in the context of international protection, whether based in international or regional refugee law or human rights law.

At the practical level, the most widely accepted definition of protection used by humanitarian actors was developed during a series of workshops and consultations sponsored by the International Committee of the Red Cross (ICRC): "The concept of protection encompasses all activities aimed at obtaining full respect for the rights of the individual in accordance with the letter and the spirit of the relevant bodies of law (i.e. HR law, IHL, refugee law)" (ICRC 1999). This definition, which was reached by some fifty humanitarian, human rights and academic organizations and institutions was subsequently adopted by the Inter-Agency Standing Committee (IASC), a unique forum involving key UN and non-UN humanitarian partners and the primary inter-agency mechanism for inter-agency cooperation of humanitarian assistance (IASC 1999).

Some argue that the breadth of this maximalist, non-specific catalogue definition, covering *all* activities aimed at respecting *all* rights, stretches the concept of protection so far that it encompasses all humanitarian and development work, and therefore risks becoming meaningless (see Ferris 2011, 275). There is merit to this argument, and the idea that a more limited and precise definition of protection can assist in finessing concepts and acquire greater cogency in advocacy with states such that in the long term it has the power to be regarded as an obligation by states (see Goodwin-Gill 2013).

Nonetheless, in the context of humanitarian crises, the value of this definition arguably lies in its ability to accommodate the plethora and diversity of needs exhibited by those who move, and those who are trapped and require relocation. The contributing chapters show that crisis migrants' needs are manifold. Some crisis migrants may need immediate protection, be it evacuation from areas facing imminent threats, protection of physical safety and security, or access to life-saving services and basic subsistence. Some of these people will be internally displaced, in need of the protection of their own country, while others may need the protection of another country. In the latter case, protection may not be forthcoming unless the state is willing to refrain from forcibly returning crisis migrants to countries where their lives, safety, health or subsistence might be threatened. For some, the need for protection is short term, ending when they can return safely to their homes, while others will require more sustained interventions. Even in cases of return, however, there may be need for compensation, restitution or remedial mechanisms to protect fundamental human rights. Finally, all of those potentially affected by crises require preventative and development action that addresses structural and systemic concerns, builds in mechanisms for livelihood diversification, and enhances resilience to allow for individuals and communities to manage their choices in a way that respects their inherent dignity and integrity.

In Chapter 17, Alexander Betts posits that, although crisis migration is a relatively new term, it is not to say there is no global governance in the area. In revealing the potential utility of what he describes as the "complex tapestry of multiple and fragmented institutions," he uses three timely case studies to illustrate four types of movement: cross-border displacement that falls outside the refugee framework in the context of drought and displacement in the Horn of Africa; trapped populations in Libya; and anticipatory movement and mixed migration in Zimbabwe. Based on this analysis, Betts argues that, rather than new institution building, addressing protection gaps for crisis migration requires finding creative and gradualist ways to make existing frameworks and institutions function better across three levels: implementation, institutionalization and international agreements.

A framework for enhancing protection

Ultimately, perhaps the most pressing challenge in providing protection to crisis migrants is to determine who is in need of international protection. As noted earlier, the current system of protection for individuals satisfying the legal definition of a "refugee" implicitly accepts the importance of the state apparatus in respecting and protecting the rights of its nationals. It substitutes surrogate international protection for those who are unwilling or unable to accept the protection of their own country, in the refugee case, because of their well-founded fear of persecution. Following this line of reasoning, one can divide those who move in the context of humanitarian crises into three categories according to their relationship to their own governments, in order to determine if international protection is needed because of an absence of state protection.

In the first category are individuals whose governments are willing and able to provide protection. Those affected by acute and slower-onset humanitarian crises in wealthy, democratic countries usually, though not always, fall into this category. There are also examples of poorer and more authoritarian governments that have good records in protecting and assisting those affected by acute events and slow-onset processes.[2] Generally, movements in these contexts are internal, not international, since the crisis migrants are able to find assistance from their own governments and have few reasons to cross an international border. There is a limited role for the international community, although other governments and international organizations may offer assistance, as was evident during the 2011 Fukushima disaster, when countries around the world provided support to Japan. Other forms of assistance might include: the deployment of search and rescue teams, financial aid for rebuilding homes, health professionals and other experts in disaster relief, and long-term development-related programs and interventions. In these situations, there is generally no need for surrogate protection from the international community.

The second category includes individuals in situations where governments are willing but unable to provide adequate protection. Certainly, poor countries that do not have the financial capacity to provide assistance may fall into this category. They would like to protect their citizens from harm but do not have the capacity or resources to do so. If those affected by the crisis move within the country of origin to find safety, a government may well attempt to fulfill its protection responsibilities by calling upon the international community to assist. In some cases, those who are already out of the country may need a temporary status that permits them to remain abroad until the capacity of the home country enables safe return. Ultimately, in these situations, the international community has an important role to play in the short, medium and long term, by ensuring that it buttresses the willing states' ability to provide protection through financial and other aid, helping this group of countries to recover from current crises, and preparing for and reducing risk from future emergencies.

The third category includes situations in which governments are unwilling to provide protection to their citizens or non-nationals on their territory. In some cases, the government has the capacity to provide protection, but is unwilling to offer it to some or all of its residents. For example, the government may not spend its resources on political opponents or ethnic or religious minority groups. Alternatively, it may limit assistance and protection to citizens and not address the needs of non-nationals (some of whom may be illegally in the country). In the other situations, the government is both unwilling and unable to protect its citizens. Failed states would fit into this category because they have neither the willingness nor the ability to protect those living on their territory. These situations produce extremely high levels of vulnerability for those who are not afforded the protection of the state. In these situations, international protection may well be essential, regardless of the cause.

The third category presents significant challenges for the international community. In cases of internal displacement, if state sovereignty or security conditions preclude direct access, the international community could still play an important role as an advocate for unprotected persons, up to and including encouraging the Security Council to intervene. In the past, international organizations have had considerable success in gaining access to conflict-induced displaced persons. The humanitarian diplomacy that has enabled such access is a model that should be applied or extrapolated more generally to non-conflict-induced crises.

In considering the interplay among each of these categories and crisis migration, it should be borne in mind that in contrast to cross-border movements (discussed below), there is substantial consensus on a normative framework for addressing all phases of internal displacement resulting from a broad range of crises. As noted earlier, the definition of IDPs in the non-binding Guiding Principles on Internal Displacement encompasses people who are forced to leave their homes because of natural or human-made disasters. The Kampala Convention adopts the same definition (2009, Article 1(k)). In this sense, these instruments apply broadly to crisis migrants that move internally, including, arguably, those who move in anticipation of harm associated with slow-onset situations related to environmental change, provided that a notion of coercion impelled their movement (Cohen 2013, 8–9). In fact, with respect to Africa, Article 5(4) of the Kampala Convention explicitly recognizes that governments must take action in these situations: "States Parties shall take measures to protect and assist persons who have been internally displaced due to natural or human made disasters, including climate change" (2009).

Roberta Cohen, in her working paper on Lessons Learned from the Development of the Guiding Principles on Internal Displacement, notes that the definition of an IDP is not limited to those who are citizens of their country and that the notion of one's own country implicit in the definition should be interpreted in a flexible manner (Cohen 2013, 9). "It could mean 'the country of nationality or, if nationality is uncertain the country of usual [or habitual] residence'" (Cohen 2013, 9; internal citation omitted). She goes on to indicate that the term IDP "thus goes beyond citizens to include: stateless persons who were born in the country or took more than temporary residence there; and long-term residents, i.e. persons with a foreign nationality who without having become citizens were born in the country concerned or settled there permanently and legally" (Cohen 2103, 9; internal citation omitted).

Certain categories of non-citizen crisis migrants who move within a country experiencing a humanitarian crisis, however, may not be covered under the Guiding Principles or the Kampala Convention. Again, according to Cohen, "[t]ourists and other visitors, and also migrant workers were not intended to be included since they come to a country 'temporarily' and can return home" (2013, 9; internal citation omitted). Experts have since begun to question whether these types of non-citizen should also come under the IDP definition (see Koser 2012).

Equally, those who are trapped in the context of the types of humanitarian crises envisaged under the Guiding Principles or the Kampala Convention, may not, by virtue of their failure to flee or leave their homes or places of habitual residence, benefit from the protections articulated under these frameworks. Whether the IDP umbrella has the elasticity and tenacity to encompass trapped crisis migrants requiring relocation is also a question that remains to be answered.

A core precept of the Guiding Principles and the Kampala Convention applicable to relevant crisis migrants is the right to be protected against arbitrary displacement (1998, Principle 6; 2009, Article 4(4)(a)–(h), respectively). In the case of humanitarian crises triggered by natural or human-made disasters, any displacement is regarded as arbitrary, unless the safety and health of those affected requires their evacuation (Guiding Principles 1998, Principle 6(d); Kampala Convention 2009, Article 4(4)(f)). With respect to situations of armed conflict, displacement is considered arbitrary in circumstances where it is not compelled for imperative military reasons or to ensure the security of civilians (Guiding Principles 1998, Principle 6(b); Kampala Convention 2009, Article 4(4)(b)). Notably, and apposite to some of the other types of crises discussed in this volume, in its non-exhaustive listing of circumstances that amount to arbitrary displacement, the Kampala Convention explicitly references displacement caused by generalized violence and violations of human rights (2009, Article 4(4)(d)).

In such situations, pursuant to the Guiding Principles, concerned authorities are required to ensure that all feasible alternatives are explored in order to avoid displacement altogether and where no alternatives exist, to minimize displacement and its adverse effects (1998, Principle 7(1)). Authorities must also ensure that proper accommodation is provided to the displaced persons, that members of the same family are not separated, and displacements are effected in satisfactory conditions of safety, nutrition, health and hygiene (1998, Principle 7(2)).

When contemplating conditions for return, guidance provided to state authorities by the UN Emergency Relief Coordinator and the Secretary General's Special Representative on Internally Displaced Persons with respect to natural disasters is relevant to considering solutions for crisis migrants, including those affected by other trigger events and processes. These experts note that the right to freedom of movement of affected persons should be understood as "including the right to freely decide whether to remain in or to leave an endangered zone. It should not be subject to restrictions except those which are: (1) provided for by the law; (2) serve exclusively the purpose of protecting the safety of the persons concerned, and (3) are used only when there are no less intrusive measures. In the case of evacuation, temporary relocation should not last longer than absolutely necessary" (IASC 2011, 46: internal reference omitted). "Permanent prohibitions of return without the consent of affected persons and communities should only be considered and implemented if the area where people live or want to return to is indeed an area with high and persistent risk for life and security that cannot be mitigated by available adaptation and other protective measures" and, in such contexts, any such prohibitions must respect a range of conditions (IASC 2011, 48). Conversely, people should not be required to return to areas in which their safety may be compromised: "Persons affected by the natural disaster should not, under any circumstances, be forced to return to or resettle in any place where their life, safety, liberty and/or health would be at further risk" (IASC 2006b, 19).

Despite the existence of these normative frameworks and guidance, and notwithstanding the gaps in coverage for some crisis migrants, as noted earlier, implementation gaps abound; be it as a result of the failure to ratify conventions or adopt national legislation and policies or the lack of capacity, resources or political will to carry out the requisite action. In this context, norms and guidance alone will not protect crisis migrants without significant improvements in international responses as well. International involvement is crucial where governments are unable and/or unwilling to protect those within their jurisdiction—that is, situations comprising the third (and perhaps to a lesser extent the second) category discussed earlier in this section.

A first step toward improving international responses would be to strengthen the UN cluster for protection of those who are internally displaced. In 2005, the United Nations established the cluster coordination system, in which a single UN agency is responsible for coordinating activities in a particular sector. The IASC "Guidance Note on Using the Cluster Approach" explains that the role of cluster leads "is to facilitate a process aimed at ensuring well-coordinated and effective humanitarian responses in the sector or area of activity concerned" (IASC 2006a, 7). While sector leads aren't expected to carry out all the necessary activities themselves, they are required to commit to being the "provider of last resort" within the sector or area of activity in question when necessary and when access, security and availability of resources make this possible (IASC 2006a, 7). The Note recognizes that "the 'provider of last resort' concept is critical to the cluster approach, and without it the element of predictability is lost" (IASC 2006a, 10).

In cases of conflict, UNHCR has taken on responsibility for the protection cluster but responsibility for protection of those uprooted by other crises is more diffuse, with the designation made on a case-by-case basis. As such, the absence of clear responsibility for protection often leaves gaps, putting the internally displaced persons at risk when no agency is prepared to intercede on their behalf. The UN Office for the Coordination of Humanitarian Affairs has pointed out that "major protection concerns in recent large-scale natural disasters such as Haiti" has raised questions about the current process (OCHA n.d.).

In cases where crisis migrants move across borders because of the unwilling-ness of their own government to provide protection—the third category discussed above—the host country, whether transit or final destination, has principal respon-sibility for determining what, if any, form of protection to offer. If the host country is willing and able to protect the cross-border population, there is little reason for the international community to become involved. On the other hand, if the host country is unwilling or unable to protect these crisis migrants, or attempts to return them to the home country without adequate guarantees of their safety, the inter-national community may well have a reason to offer its protection. As in refugee situations today, the international community could offer its assistance as a way to encourage the host country to permit the cross-border crisis migrants to remain until it is safe to return or other solutions are found for them. This aid could take the form of financial and technical assistance and temporary protection, as well as offers of longer-term resettlement in situations where there is a real and lasting danger to the life, security or physical integrity of affected persons.

More effective legal and policy frameworks that articulate the rights of cross-border crisis migrants are also imperative. In some cases, crisis migrants may fall within the purview of the Refugee Convention. For example, those fleeing because of competition for resources arising from acute natural hazards or other crises may qualify if they are unable to access lifesaving resources because of a pro-tected characteristic (that is, race, religion, nationality, membership of a particular social group, or political opinion). In Africa, the scope of coverage might be greater because of the 1969 OAU Convention, which, as discussed earlier, includes those who are compelled to flee owing to events seriously disturbing public order in either part or the whole of his country of origin or nationality. To the extent that humanitarian crises seriously disturb public order, this normative framework may cover crisis migrants who flee across borders. As with internal movements, how-ever, this broader normative coverage does not address or resolve implementation vacuums on the part of host governments, due to inability and/or unwillingness on their part.

Most crisis migrants, however, are unlikely to meet the strict standard as applied by refugee law. In some cases, national laws may offer some measure of protection, stepping in to fill the void in international and regional frameworks. Temporary protection policies, for example, are a means through which some states grant some security to persons whose lives and safety would be at risk if they were returned to their home countries. Many governments already apply this principle, albeit in an ad hoc fashion. After the 2004 tsunami, for example, Switzerland, the United Kingdom and Canada suspended deportations of those from such countries as Sri Lanka, India, Somalia, Maldives, Seychelles, Indonesia and Thailand. A number of governments announced similar plans after the 2010 earthquake in Haiti. The US TPS affords a model for other states to consider. TPS offers work permits and residency to persons already in the United States at the time of a crisis in their home country. It is lacking, however, in two respects: it does not address the situation of those who flee after the relevant authority makes a discretionary determination to grant TPS to persons in the US from the crisis-affected country, and it does not provide for a permanent status for those who cannot return after years of exile. Both of these issues would need to be addressed in a more universal application of the concept.

To date, there are no examples of legislation or policies that expressly facilitate the migration of persons from slow-onset crises that may destroy habitats or livelihoods in the future. For the most part, movements from situations that limit economic opportunities or affect longer-term survival are treated in the same manner as other economically motivated migration. Persons moving outside of existing labor and family migration categories are considered to be irregular migrants. In the absence of a strong humanitarian basis for exempting them from removal proceedings (which is unlikely in the slow-onset situation), these migrants would be subject to the regular systems in place for mandatory return to their home countries. As their immediate reasons for migrating would be similar to that of other irregular migrants—that is, lack of economic opportunities at home and better economic opportunities abroad—there would be little reason for destination countries to manage these movements outside of their existing immigration rules. Yet, returning people to situations that will only worsen over time also presents challenges, not only for the migrants themselves, but also for the communities to which they return. Eventually, and perhaps inevitably, slow-onset scenarios are likely to precipitate and transform into acute humanitarian crises, with the potential to amplify the impact on the international community.[3]

Planned relocation is one mechanism through which crisis migration from slow-onset scenarios could be addressed. Nonetheless, as discussed more fully in Oliver-Smith and de Sherbinin's chapter (Chapter 12), suitable policies and practices regarding planned relocation are also lacking. Experience with resettlement in the context of development projects raises questions about the efficacy of these programs from both an economic and a human rights perspective. Cernea cites eight processes affecting those resettled that often leave them more vulnerable after relocation: landlessness, joblessness, homelessness, marginalization, food insecurity, increased morbidity and mortality, loss of access to communal property and services, and social disintegration (Cernea 1999). Among the issues requiring attention are definitions of when land is to be considered uninhabitable, identification of available land for resettlement, procedures for establishing ownership and clear legal title, and compensation for lost property.

Conclusion

As highlighted throughout this chapter, a number of challenges and uncertainties surround the protection of crisis migrants, leaving people at serious risk of harm and upheaval as crises continue to permeate daily life. Solutions, as the authors in this volume reveal, are not beyond reach, but require a concerted effort from legal experts, policy makers, practitioners and scholars. In this respect, when analyzing existing frameworks, evaluating existing responses and developing new frameworks and responses to protect crisis migrants, a range of factors must necessarily inform such an inquiry.

In reviewing the contributing chapters and contemplating possible solutions, it is important to bear in mind that the sovereign prerogative of states to regulate the movement of foreign nationals across their borders is not absolute. Rather, it is tempered by human rights law and refugee law, which accords rights to individuals even when they are outside their country of origin or habitual residence. Most importantly, individuals have a right to leave their country (although there is no direct corresponding obligation on states to admit non-nationals) and as previously discussed, certain classes of non-citizens such as refugees and those able to benefit from complementary protection cannot be *refouled*. With these framing thoughts in mind, we leave the reader to delve into the contributing chapters of this volume for a rich and varied discussion of humanitarian crises and associated movements, protection implications and responses.

Notes

1   McAdam notes that "[i]n contemporary practice, [complementary protection] describes the engagement of States' legal protection obligations that are *complementary* to those assumed under the 1951 Refugee Convention (as supplemented by its 1967 Protocol), whether derived from treaty or customary international law. Importantly, it stems from legal obligations preventing return to serious harm, rather than from compassionate reasons or practical obstacles to removal" (2007, 2–3). In continuing, she asserts "the 'complementary' aspect of 'complementary protection' is not the form of protection or resultant status accorded to an individual, but rather the *source* of the additional protection. Its chief function is to provide an alternative basis for eligibility for protection. Understood in this way, it does not mandate a lesser duration or quality of status, but simply assesses international protection needs on a wider basis than the dominant legal instrument, presently the 1951 Convention" (2007, 23).
2   China's response to the earthquake in Sichuan province is an example.
3   The Nansen Initiative (http://www.nanseninitiative.org/), which is a state-led process, seeks to develop a protection agenda to address the needs of people displaced across international borders in the context of natural disasters including the effects of climate change.

References

Bakewell, O. (2011) "Conceptualising Displacement and Migration: Processes, Conditions, and Categories," in K. Koser and S. Martin (eds.), *The Migration–Displacement Nexus: Patterns, Processes, and Policies*, New York: Berghahn Books.

BBC (2013) "India Floods: Thousands Flee Homes in Assam," available online at: http://www.bbc.co.uk/news/world-asia-india-23237130.

Betts, A. (2010) "Survival Migration: A New Protection Framework," *Global Governance*, 16(3): 361–382.

Betts, A. (2013) *Survival Migration: Failed Governance and the Crisis of Displacement*. Ithaca: Cornell University Press.

Brookings Institution—London School of Economics (2011) *From Responsibility to Response: Assessing National Approaches to Internal Displacement*, Washington, DC: Brookings Institution.

Cernea, M. (1999) *The Economics of Involuntary Resettlement: Questions and Challenges*, Washington, DC: World Bank.

Cohen, R. (2013) "Lessons Learned from the Development of the Guiding Principles on Internal Displacement," available online at: http://isim.georgetown.edu/work/crisis/products/workingpapers/.

Ferris, E. (2011) *Politics of Protection: The Limits of Humanitarian Action*, Washington, DC: Brookings Institution.

Goodwin-Gill, G.S. (1989) "The Language of Protection," *International Journal of Refugee Law*, 1(1): 6–19.

Goodwin-Gill, G.S. (2013) "Protection," available online at: http://www.rsc.ox.ac.uk/events/protection.

Government Office of Science, London (2011) "Foresight: Migration and Global Environmental Change," Final Project Report, available online at: http://www.bis.gov.uk/assets/foresight/docs/migration/11-1116-migration-and-global-environmental-change.pdf.

Hathaway, J. (2005) *The Rights of Refugees in International Law*, Cambridge: Cambridge University Press.

Inter-Agency Standing Committee (IASC) (1999) "Protection of Internally Displaced Persons: Interagency Standing Committee Policy Paper," available online at: http://www.humanitarianinfo.org/iasc/pageloader.aspx?page=content-products-products&productcatid=10.

IASC (2006a) "Guidance Note on Using the Cluster Approach to Strengthen Humanitarian Response," available online at: http://www.refworld.org/docid/460a8ccc2.html.

IASC (2006b) "Protecting Persons Affected by Natural Disasters: IASC Operational Guidelines on Human Rights and Natural Disasters," available online at: http://www.brookings.edu/~/media/research/files/reports/2006/11/natural%20disasters/11_natural_disasters.pdf.

IASC (2011) "IASC Operational Guidelines on the Protection of Persons in Situations of Natural Disaster," available online at: http://www.brookings.edu/~/media/research/files/reports/2011/1/06%20operational%20guidelines%20nd/0106_operational_guidelines_nd.pdf.

International Committee of the Red Cross (ICRC) (1999) *Third Workshop on Protection: Background Paper*, 7 January.

International Regional Information Networks (IRIN) (2013a) *Forgotten Hunger in Yemen*, available online at: http://www.irinnews.org/report/98376/slideshow-forgotten-hunger-in-yemen.

IRIN (2013b) "Pakistan's Coast: Where the Sea is an Enemy, not a Friend," available online at: http://www.irinnews.org/report/98399/pakistan-s-coast-where-the-sea-is-an-enemy-not-a-friend.

Koser, K. (2012) "Responding to New Internal Displacement Challenges: The Displacement of Non-Citizens," Brookings Institution, available online at: http://www.brookings.edu/research/opinions/2012/12/20-displacement-noncitizens-koser.

Koser, K. and Martin, S. (eds.) (2011) "The Migration–Displacement Nexus: Patterns, Processes, and Policies," *Studies in Forced Migration*, 20, New York: Berghahn Books.

McAdam, J. (2007) *Complementary Protection in International Refugee Law*, New York: Oxford University Press.

Office for the Coordination of Humanitarian Affairs (OCHA) (n.d.) "Thematic Areas: Displacement," available online at: http://www.unocha.org/what-we-do/policy/thematic-areas/displacement.

OCHA (2013) "Syria Crisis Overview," available online at: http://syria.unocha.org/.

Reliefweb (2013) "Almost 6,000 People Missing in India Floods Presumed Dead," available online at: http://reliefweb.int/report/india/almost-6000-people-missing-india-floods-presumed-dead.

Richmond, A. (1993) "Reactive Migration: Sociological Perspectives on Refugee Movements," *Journal of Refugee Studies*, 6(1): 7–24.

United Nations High Commissioner for Refugees (UNHCR) (2013) "Syria Regional Response Portal," available online at: http://data.unhcr.org/syrianrefugees/regional.php.

Van Hear, N. (2009) "Managing Mobility for Human Development: The Growing Salience of Mixed Migration," New York: United Nations Development Program (UNDP), Human Development Research Paper 2009/20.

World Food Program (WFP) (2013) "Sahel Crisis Overview," available online at: http://www.wfp.org/crisis/sahel.

World Nuclear Association (2009) "Chernobyl Accident Appendix 2," available online at: http://www.world-nuclear.org/info/Safety-and-Security/Safety-of-Plants/Appendices/Chernobyl-Accident---Appendix-2--Health-Impacts/#.UjoCS2TXiR8.

2

CONCEPTUALIZING "CRISIS MIGRATION"

A theoretical perspective

Jane McAdam

Introduction

According to the United Nations (UN)'s Emergency Relief Coordinator, more frequent and severe disasters may be "the new normal" (Holmes 2008). On top of this, the slower-onset impacts of climate change, such as temperature rises, glacial melt, drought and sea-level rise, may ultimately force people away from their homes. Nuclear and industrial accidents, such as that witnessed in Japan in 2011, especially when combined with natural hazards, pose further risks to people's lives and livelihoods. Communal violence and civil strife continue to be major drivers of displacement, often on a large scale.

Protection and assistance issues may be as acute in the aftermath of a natural disaster as in conflict (Strohmeyer 2011). Those displaced may suffer from the same lack of access to basic rights and resources, and experience psychological distress. Until recently, however, the international community's focus has been on protecting those displaced by conflict, despite the growing (and larger) number of people being displaced by natural hazards. The UN High Commissioner for Refugees has stated that, "while the nature of forced displacement is rapidly evolving, the responses available to the international community have not kept pace" (Guterres 2011, 3).

This edited volume seeks to illuminate those protection gaps by examining different experiences of forced movement, which are conceptualized as "crisis migration." This is the first time that such a study—or indeed such a conceptualization—has been attempted. Each of the case studies in this book examines a particular type of hazard that may lead to displacement when certain variables are present. Hazards do not become disasters (or crises) for individuals or communities unless they face a certain degree of socio-economic vulnerability, which "conditions the behavior of individuals and organizations throughout the full unfolding of a disaster far more profoundly than will the physical force of the destructive agent" (Oliver-Smith and Hoffmann 2002, 3). As the UN High Commissioner for Refugees (UNHCR) observes, "[n]atural hazards do not in themselves constitute disasters; rather human actions exacerbate the effects of natural phenomena to create disasters. The impact of natural disasters is a function of both the severity of the natural hazard and the capacity of a population to deal with it" (UNHCR 2012a, 27).

The premise of this book is that "crisis migrants" fall outside the protective coverage of existing legal instruments and institutional mandates, and as such are especially vulnerable. How does framing each of these case studies as "crisis migration," as opposed to the more conventional and generic label of "forced migration,"[1] help to illuminate the nature of such movement and the kinds of policy responses required to address it? What commonalities can be extracted for the clarification of universal standards, and where might the idiosyncrasies of context demand highly tailored responses?

In this volume, a "crisis" is defined as "any situation in which there is a widespread threat to life, physical safety, health or subsistence that is beyond the coping capacity of individuals and the communities in which they reside" (see Martin *et al.*, Chapter 1 in this volume).[2] Importantly, this definition refers to *situations*, which can encompass sudden events as well as slower processes of change or deterioration. As this chapter explores, understanding the notion of "crisis" in this broad way is significant when it comes to finding appropriate legal and policy responses. When a "crisis" is understood as something more than a single, sudden event, we can start to contemplate interventions over longer timeframes, different combinations of institutional actors, new partnerships and more sustainable funding models.

"Crisis migration" refers to mobility in situations of crisis. What constitutes a "crisis" and spurs migration will depend upon resources and capacity—both of those who move, as well as the ability of the state into/within which they move to respond to their plight.[3] A helpful way to understand this is in terms of "tipping points": when does the cumulative impact of stressors such as those described above tip people over the edge, such that moving away seems preferable to staying put? Irrespective of whether a crisis is triggered by acute or chronic conditions, there will be tipping points involved, and these will vary from individual to individual. However, it is important not to assume that all crisis migration manifests as acute displacement. Rather, like the notion of "crisis" itself, "crisis migration" encompasses movement from both acute and chronic conditions. In this way, it captures anticipatory movement as well as "classic" displacement. In addition, some people may be trapped by their circumstances and unable to move at all. While their lack of mobility may render them less visible, it does not mean that they are less vulnerable. Indeed, an inability to move may be a sign of heightened vulnerability.

The present chapter examines the notion of "crisis migration" from a theoretical perspective to determine the extent to which it offers a useful conceptualization for advancing legal and policy responses to forced migration.[4] In particular, it seeks to show why it is important to interrogate and challenge the ways in which the concept of "crisis" is used, since its meaning is culturally constructed and behind it lie "specific ways of thinking about how the world works and specific, if often implicit moral orientations" (Calhoun 2010, 29). The chapter argues that a clear understanding of who and what is covered by the concept of "crisis migration" is necessary to develop well-attuned responses.

In doing so, the chapter critiques assumptions that tend to underlie the concept of "crisis": does it imply a single "event" or can it also encompass processes of destabilization? This is important because how the concept is conceived will necessarily affect the nature and location of policy interventions—for example, whether the focus is on responding to the physical impacts of sudden-onset events or is instead on longer-term structural changes. As Lakoff has observed: "The need to be prepared is not in question; what can be a source of dispute is, rather, how to prepare and what we need to prepare for" (Lakoff 2007, 249). "Crises" are posited as exceptions, requiring special solutions. From a policy perspective, this risks sidelining "everyday," systemic issues such as poverty, vulnerability and environmental fragility, which are central to how people experience hazards, and overlooking relevant legal frameworks, such as human rights law (Charlesworth 2002). Thus, integrated into the chapter is analysis of the extent to which *normative* legal and institutional gaps exist, as opposed to gaps in implementation and enforcement.

The narrative of crisis

> The production of meaning . . . can be regarded as a performance, because any given set of real events can be emplotted in a number of ways, can bear the weight of being told as any number of different kinds of stories. Since no given set or sequence of real events is intrinsically "tragic," "comic," or "farcical," but can be constructed as such only by the imposition of the structure of a given story-type on the events, it is the choice of the story-type and its imposition upon the events which endow them with meaning.
>
> *(White 1984, 20; see also Roe 1991)*

The narrative of crisis is one we all recognize. It is the tale of disaster, of impending destruction, which warrants immediate action (Bravo 2009).[5] It is a call to arms which has an intense, even if only brief, hold on public attention (Bravo 2009, 257). A "crisis" denotes a "vitally important or decisive stage"; "a state of affairs in which a decisive change for better or worse is imminent," often in times of difficulty or insecurity.[6] The notion of "crisis migration" therefore implies an extreme or exceptional situation where the movement of communities is compelled and passivity is not an option: a decision will be forced.[7] Indeed, forced migration may itself be an indicator of crisis, with dislocation signifying a disruption to "normality" (Calhoun 2010, 32). One scholar has even described refugees as the "prototypical face of the emergency" (Calhoun 2010, 33).

Yet, the notion of "crisis" is a constructed concept. Like Calhoun's "emergency imaginary," a crisis tends to be imagined (by the media, policy makers and others) as "a sudden, unpredictable event emerging against a background of osten-sible normalcy, causing suffering or danger and demanding urgent response" (Cal-houn 2010, 30). The potency of this *idea* helps both to constitute the emergency and drive responses to it. As Hewitt explains, the "geography of disaster is an *archi-pelago* of isolated misfortunes," with natural disasters set aside from the everyday, "quarantined in thought as well as practice" (Hewitt 1983, 12). But emergencies are not freestanding events; they are moments in time that form part of longer processes.

Roe suggests that crisis narratives are in fact a form of appropriation by policy makers—a means by which experts and institutions can "claim rights to steward-ship over land and resources they do not own. By generating and appealing to crisis narratives, technical experts and managers assert rights as 'stakeholders' in the land and resources they say are under crisis" (Roe 1995, 1066) and "sustain the narrative that defines communities 'at risk' in order to justify expert interventions" (Bravo 2009, 262). At times, this appropriation may imply that those directly affected are unable or perhaps even unwilling to take charge of their own lives. Nevertheless, it may be tolerated by the governments of some affected states which perceive it as their best chance of benefitting from aid money that may flow from this kind of characterization (Ferguson 1990; see also the discussion of Tuvalu's ambivalent position in McAdam 2012, 31–36).

Thus, there is a question about *whom* crisis migration narratives serve. Affected communities may have little voice in the institutional fora that develop and sustain these discourses, and accordingly their interests may be misrepresented or silenced (Bravo 2009, 268). Top-down "macro-scale narratives" (Bravo 2009, 257) may dilute and abstract very important local concerns, such that there is a dissonance between what is perceived to be needed on the ground, and what is perceived as necessary by international decision makers.

Crisis narratives may also be used strategically to advocate for particular forms of political action. For example, environmental groups have used the idea of the "cli-mate change refugees" and "disappearing states" to highlight the destructive force of climate change and lobby for the mitigation of greenhouse gas emissions. By contrast, migration scholars argue that this characterization misrepresents the likely nature of human movement related to climate-change-related impacts and may result in maladaptive policies (see generally McAdam 2012, 26–30). Others suggest that this perspective is imposed by outsiders and serves to marginalize the voices of affected communities by framing movement as a "pathological condition" rather than a sign of resilience (Farbotko and Lazrus 2012, 384; see also Tacoli 2009). As has been argued in relation to the small island state of Tuvalu:

> Migration and cultural change are not necessarily crises, as they are currently ordinary practices of everyday life . . . It is not migration in and of itself that involves significant threat to the way Tuvaluan people imagine their future, but how sea level rise is framed and governed.
>
> *(Farbotko and Lazrus 2012, 388)*

This is why context matters. If policy makers do not appreciate the way in which mobility has (or has not) featured historically within a particular community, for example as a long-standing adaptation strategy, then interventions may be misplaced. Importantly, we need to distinguish between "crisis migration" as a descriptive category and analytical tool, and as a category with any special normative import. Whereas the concept of "crisis migration" might be helpful to identify tipping points and to draw out commonalities between situations not ordinarily examined together, it is arguably too blunt a concept to drive a single legal or normative response.

Finally, the narrative of crisis may frighten and paralyze with its seemingly insurmountable challenges for policy makers. It may cloud judgement and lead to short-term snap decisions or overly emotive scholarship. For this reason, it is essential that the alarmism it may engender does not trump considered, empirically grounded research, from which constructive solutions can be crafted. As Bravo notes, it is important to understand "how objective information should be framed by narratives . . . for sustaining public dialogues" about the issue (Bravo 2009, 257).

What is "crisis migration"?

> Whenever we want to research or discuss the consequences of any phenomenon, we need to have a clear idea of what that phenomenon is.
>
> *(Quarantelli 1985, 41)*

The term "crisis migration" has been used to describe a variety of different concepts, but is not a term of art (see e.g. Prabhakara 1986; Pooley and Turnbull 1998; King and Mai 2008; Bijak 2011). It is often used in a very descriptive—and unselfconscious—sense to refer circularly to movement away from various "crises." For instance, "the removal of crofters from the Scottish Highlands following the clearances, Irish famine migration in the mid-nineteenth century, the mass movement of Jews from Central and Eastern Europe following persecution and, more recently, the exodus of refugees from the former Yugoslavia" have all produced "highly emotive pictures of migration as a traumatic and disruptive event stimulated by crisis conditions beyond the control of individual migrants" (Pooley and Turnbull 1998, 241).[8]

However, natural disasters and other hazards will not manifest as "crises" unless certain variables are present. Although the concept of "crisis" may connote a pivotal *moment* that is dramatic, sudden or acute, and outside the realm of the normal or "everyday," hazards are "*normal* features of specific environments" and "environmental processes are not novelties . . . but periodic regularities" (Oliver-Smith and Hoffman 2002, 8, emphasis added; see also Hewitt 1983). They will transform into "crises" only when individual and/or community coping capacity is overwhelmed.

It should also be recalled that migration is a normal, rational response to natural disasters and the more gradual impacts of environmental change. In some contexts, such as the Pacific islands, mobility is a core part of historical (and present) experience. Movement therefore needs to be understood as an adaptive strategy that is part of a historical continuum. This is not to say that it should always be assumed to be voluntary, but rather that it should not automatically be pathologized.

Thus, while we might instinctively think that "crisis migration" entails movement in response to an objectively perceptible hazard, such as a flood or earthquake—and, indeed, while a "crisis" *may* be conceptualized as an external event or pressure—it is the underlying social dimension that will transform it from a merely hazardous encounter into a situation of stress that tests the resilience of the individual and the community. As Bankoff observes,

> disasters are considered to be primarily about processes in which hazardous events represent moments of catharsis along a continuum whose origins lie buried in the past and whose outcomes extend into the future. It is the pre-disaster conditions that mainly affect a society's ability to cope with hazard; it is its reconstruction operations that largely determine the frequency and magnitude of subsequent events.
>
> *(Bankoff 2001, 30)*

Even if a discrete event results in temporary displacement, robust disaster risk reduction and management policies and sustainable development practices may facilitate fast relief and rehabilitation, such that "crisis migration" is avoided. As Warner notes, policy interventions—and the timing of them—will play a major role in shaping outcomes and determine whether migration is a form of adaptation, or a sign of a failure to adapt (Warner 2010).

A sudden event may therefore act as a "tipping point," interacting with pre-existing stressors such as poverty, overcrowding, environmental fragility, development practices and weak political institutions. Isolating a single cause of movement is not possible, as this Bangladeshi government official explained in an interview with the author through the following analogy:

> Let's say for example, one person is able to carry only 40kg on his shoulders. That's his limit, and he's a poor man. Now on the top of that, I come, and I give him one kilogram. So now the question is: who is responsible for killing him? Is this the 40 kilograms he was already carrying on his head, or the one kilogram I have now put on the top of that?
>
> *(Uddin 2010)*

Since resilience to disasters depends upon underlying vulnerabilities, it is not surprising that reactions are highly personalized (see e.g. Richmond 1994). Particular groups may be more vulnerable and liable to displacement than others (for example, on account of factors such as age, health or gender).[9] Even in the absence of a trigger "event," there may come a point when individual or household coping capacity is simply overwhelmed by the cumulative impact of stressors. However, as with all forms of migration, individuals will "try to calculate the *relative advantages* of moving against the relative advantages of remaining behind" (Morrissey 2008, 29, emphasis added), which will depend on such things as their kinship and social ties, their financial resources, and the extent to which their skills and livelihoods will be recognized or valued elsewhere. What constitutes a crisis for one community or individual may not for another.

Pooley and Turnbull make the interesting observation that while major events, like war, "are likely to be widely viewed as producing crisis conditions, their very universality may make the impact of the event less severe" (Pooley and Turnbull 1998, 241). They suggest that during the world wars, for example, many people continued relatively normal lives, even though one way of framing their situation was as a crisis. This observation should not be read as dismissing the very extreme circumstances faced by some individuals and groups during wartime, but rather that even though war may be "exceptional," it is erroneous to assume that the coping capacity of everyone who experiences it is automatically overwhelmed. Hence they advocate for a more nuanced understanding of "crisis" as being "a personal and subjective process" that varies between individuals and may change over time (Pooley and Turnbull 1998, 242).

Lubkemann's analysis of migration from civil war in Mozambique (1977–92) shows why this matters in terms of understanding and responding to "crisis migration." He argues that "predominant demographic theories of forced migration rest on a highly reductionist model of migratory decision-making in acute crisis contexts that fails to adequately examine actor agency and the social and cultural factors that inform agency in producing demographic outcomes" (Lubkemann 2004, 371). He suggests that part of the flaw is located in the assumption that a "crisis" is constituted by an acute event, rather than appreciating decisions about mobility in their broader context. For example, he notes that migration was already a strategic option in Mozambique prior to the war, and as such it "significantly shaped wartime migration" (Lubkemann 2004, 371). This accords with migration dynamics generally, which show that past migration patterns are the best indicators of future movement (see Barnett and Webber 2009; Hugo 2010).

"Crisis migration" is therefore best understood as a response to a complex combination of social, political, economic and environmental factors, which may be *triggered* by an extreme event, but not caused by it. In that sense, and perhaps counter-intuitively to the notion of "crisis" as event, it is a response to a series of cumulative pressures that make life at home intolerable and unsustainable for the particular individual or household.[10] When conceptualized in this way, "crisis migration" references a situation of acute pressure on the person or group that moves, rather than necessarily indicating the presence of an extreme or sudden event. It therefore encompasses a considerable degree of subjectivity since it depends on personal or household coping thresholds. While these may be aided or undermined by external support frameworks (e.g. disaster risk reduction strategies, disaster relief, institutional assistance, social security, livelihood diversification and so on), decisions about whether to stay or leave are typically highly individualized.

Thus understood, the notion of "crisis" implies a set of circumstances in a particular environment, which at some point in time leads people to believe that leaving is a better option than staying.

This may say as much about *perception* of risk as it does about physical risk. A perception of crisis—and action in response to it—may be triggered even where an emergency has not yet materialized. For example, behavioral responses to false rumors that a dam had broken were indistinguishable from the responses of groups that fled when a dam actually broke (Quarantelli 1985, 47). This is particularly apposite in the context of climate change. Some scholars argue that in small Pacific island states, "the result of *lost confidence* in atoll-futures may be the end of the habitability of atolls," rather than the physical impacts of climate change per se (Barnett and Adger 2003, 330, emphasis added). Fear and uncertainty about the future may engender a sense of crisis, which impacts on people's decision-making processes, including about migration. If a discourse of crisis comes to dominate the sphere of law and policy making, and overshadows interventions that acknowledge people's resilience, adaptation and the positive dimensions of migration, then the negative assumptions on which that discourse is built may well become self-fulfilling prophecies. They may contribute to "a common *expectation* of serious climate impacts leading to changes in domestic resource use and decreased assistance from abroad" (Barnett and Adger 2003, 330, emphasis added). Similarly, "large-scale migration may be an impact of climate change affected by policy responses *in anticipation of* climate impacts rather than by material changes in the environment per se" (Mortreux and Barnett 2009, 111, emphasis added).

This is *not* to say that the impacts of climate change on small Pacific states are innocuous, but rather that embracing a vulnerability discourse may be counterproductive.[11] It can entrench inequitable power relations between affected communities and sites of "knowledge" production (Farbotko and Lazrus 2012, 382)—such as international institutions, academia, non-governmental organizations (NGOs) and the media (see further Bravo 2009). It therefore "needs to be used with caution and with a sensibility to its negative connotations" (Barnett and Campbell 2010, 165). As Oliver-Smith has observed:

> The physical existence of disasters establishes an agency of nature that exists independently of human perception. However, human beings are deeply implicated in the construction of the forms and scale in which that agency expresses itself. The impact of these hazards, when they are perceived and cognized, rapidly confirms this agency, but just as rapidly constructs a social text around it that may either *reduce or accentuate* the impact.
>
> *(Oliver-Smith 2002, 39, emphasis added)*

Interviews conducted by myself and other researchers in the Pacific island states of Kiribati and Tuvalu reveal that people's *perceptions* about the risk of climate change there seem to be a far more significant factor in their decision-making processes about future migration than actual biophysical change (Grothmann and

Patt 2005). This is why Barnett and Webber argue that "social processes that create poverty and marginality are more important determinants of migration outcomes than environmental changes per se" (2010, 38).

Why do we need to theorize "crisis migration"?

With states' attention increasingly drawn to the notion of crisis migration as an area for developing migration policy, it is essential that it is adequately theorized. Its meaning is not self-evident, since it is largely a discursive construct as much as a material phenomenon (Farbotko and Lazrus 2012, 382). A "crisis" is not simply a matter of empirics; its meaning rests "fundamentally on the conceptual definitions and theoretical approaches used" and the "goals underlying whatever work is undertaken" (Quarantelli 1985, 42). Yet it is frequently used unselfconsciously, and sometimes synonymously, with concepts such as "migration crisis" and "managing migration in crisis situations"—the last of which was selected by states as the theme of the International Organization for Migration's (IOM) 2012 International Dialogue on Migration.

The lack of clarity about precisely what states and other actors understood this to mean was apparent at IOM's first workshop on the topic (IOM 2012a). The workshop's scene-setting presentation defined a "migration crisis" as having the following common features:

- large scale, sudden or slow complex migration flows due to a crisis (natural and/or human caused);
- affected individuals become vulnerable (e.g. human rights violations);
- impacts on host and transit communities;
- protection and assistance challenges;
- manifestation of migration flows in many forms/patterns (mixed flows);
- migration flow is fluid and evolving (not static).

(Haque and Abdiker 2012, 3)

Yet, despite acknowledging the relevance of "slow complex migration flows," all the examples cited were specific "crises," such as Afghanistan, Burma, Haiti, Libya, the Philippines and Yemen. The Deputy Director General of IOM similarly noted that IOM had witnessed many "migration crises"—in Haiti, the Horn of Africa, North Africa, Pakistan and so on (Thompson 2012, 1). In explaining the situations where the organization had implemented a migration crisis management approach, she referenced only responses to emergencies.[12] In articulating how a crisis management approach complements the existing humanitarian system, she revealed its "event"-oriented rather than "process"-oriented approach: "We ask ourselves questions such as: How do people move in the event of a disaster? What is the relationship between migration patterns before and after a crisis? What happens to those who do not or cannot move in crisis situations?" (Thompson 2012, 2). In this and other presentations, a "crisis" was conceived of as having a clear beginning and an end (Roe 1991, 288), but without any explanation of how these were to be identified or understood. Quarantelli has posited that a "crisis" may start and stop at different chronological points for different organizations and affected groups—it is "always a relative matter" (Quarantelli 1977, 102). Without an agreed meaning, it is impossible to formulate operational responses since misperceptions of others' positions will lead to confusion, uncoordinated approaches and miscommunication.

Crisis as event

Conceptualizing "crisis migration" as a response to an external event shifts the focus from a social situational construct to one that concentrates on a physical occurrence (Quarantelli 1985, 44ff.). It may also shift the focus from individual to group movement, which has considerable ramifications for policy development. If we concentrate on the "mobility behaviours of *populations* affected by crisis" ("Chair's Summary" in IOM 2012b, 12, emphasis added) and "large-scale" movement ("Chair's Summary" in IOM 2012b, 9), then the obvious operational focal points are those dealing with emergency responses and post-crisis assistance, rather than institutional actors that may effect longer-term structural changes to enhance resilience and/or facilitate migration as an adaptation strategy. Emergency interventions are typically reactive and ad hoc (Charlesworth 2002). When does a crisis begin and end? Who decides? Does this have to be determined at the level of a geo-political community, or can a crisis be understood in a more nuanced way, as having different temporalities for different individuals? Operational interventions tend to be fairly blunt in this regard.

Creating "strong, new and innovative partnerships," especially between governments and international and local agencies and NGOs, is crucial ("Chair's Summary" in IOM 2012b, 14). However, to effect meaningful change, these need to move beyond "effective coordination amongst the primary actors in crisis response" ("Chair's Summary" in IOM 2012b, 14) to a far more holistic approach, both within organizations themselves and across different sectors. This will take time and require sustained dialogue and political will, and in some cases a serious rethinking and reorienting of approach. This is encapsulated by Calhoun's observation of the divide between the humanitarian and development communities: "Advocates for development . . . often cite the adage that if you give someone a fish, you feed them for a day, but if you teach them to fish, you enable them to feed themselves for life. Humanitarians are unabashedly in the fish-for-a-day business. They stress that someone who dies today will not learn to fish tomorrow" (Calhoun 2010, 52).

Furthermore, if "crisis migration" is to be understood and addressed more holistically, then the *internal* organizational divisions of certain institutions will also need to be restructured (Payton 2011). For example, IOM's Emergency and Post Crisis Division is not really focused on longer-term development and livelihood diversification, which may help to make a community more secure over time (IOM 2013). If development agencies such as the United Nations Development Program

(UNDP) are not alert to the stresses to which that displacement/migration may point, then their own strategies may be maladaptive.

Organizational mandates can also hinder or limit interventions. UNHCR, for instance, is limited by its statute and subsequent UN General Assembly resolutions to assisting particular classes of forced migrants. This shapes its capacity to respond to complex situations and institutionally privileges some groups over others (refugees, stateless persons and some internally displaced persons)—groups that already have clearly articulated rights in international legal instruments. While there are historical (and legal) reasons for this, it necessarily affects the kind of role that UNHCR can play in responding to crises.

According to Hewitt, the "placement of the problem is a necessary founding act" (Hewitt 1983, 13), and institutional interest in an issue is rarely driven by epistemological inquisitiveness but rather a concern to stake a territorial claim (Hewitt 1983, 14). Furthermore, institutional actors may have an interest in prolonging a "crisis" (if funding flows from such a designation), or in curtailing it (if personnel and other resources are tied up). For this reason, different actors may adopt different measures to determine when a crisis exists or ceases, and, depending on their influence, may succeed in mobilizing or withdrawing support from the wider humanitarian community.

To have any utility as a policy-making tool, the concept of "crisis migration" needs to be deliberately confined. It needs to be differentiated from the broader concept of "forced migration," and the newer concept of "survival migration" (which, in turn, may benefit from, and be distinguished by, a tighter focus on movement spurred by chronic, less acute, impacts. This seems to be encapsulated in the idea that "survival migration" relates to "serious deprivations of socioeconomic rights related to the underlying political situation" in a country) (Betts 2010a, 362; also see Betts 2010b, 211). Given that the IOM discussions suggest that in policy-making circles, "crisis" is intuitively understood as a pivotal moment or turning point—an emergency situation—how might this element be harnessed to shift thinking about planning for and responding to displacement? Might such "migration" be best understood as *evacuation* (for example, in the aftermath of a nuclear disaster, or the uprising in Libya)? Conceptualized in this way, we get a clearer sense of the kinds of planning and response mechanisms that may be needed.

That said, there is something inherently conservative about diagnosing "crisis migration" as sudden-onset in nature, when it is arguably deeply structural. To draw on the ideas of Amartya Sen, writing on the causes of famine (Sen 1981), any concept of "crisis migration" as event-oriented is deceptive because it is rooted in systemic inequities that render particular groups more vulnerable to displacement. This chapter argues that distress migration can be understood as a coherent phenomenon only if these trigger points are recognized as just one aspect of the *process* of a crisis: "the social and technological construction of conditions of vulnerability" (Oliver-Smith 2002, 23). This is because the kinds of "crises" examined here are multidimensional, occurring "at the intersection of nature and culture" (Oliver-Smith 2002, 24).

Slow-onset crises

For the reasons set out above, the present project defines a "crisis" as something "beyond the coping capacity of individuals and the communities in which they reside." This recognizes both the relevance of individual tipping points, and that the kinds of situations that will require national, regional or international intervention are those where the community itself is unable to adequately function or respond. In other words, policy responses are necessary when ordinary social functions are seriously disrupted.[13]

For such responses to be timely and relevant, the notion of "crisis" must also be forward looking. Only if *anticipated* crises are encompassed within the concept—as crises in slow motion—do we have any hope of addressing migration arising from the impacts of slow-onset processes, such as climate change. Current legal frameworks for assessing and responding to protection needs do not adequately address the time dimension of anticipatory, or pre-emptive, movement. When considering whether or not a person will be at risk of harm if removed, decision makers insist that the threat is sufficiently direct and imminent.[14] This is a hallmark of the jurisprudence of the European Court of Human Rights on the risk of inadequate medical treatment in the country of origin. Thus, in *N v. United Kingdom* the court stated:

> The rapidity of the deterioration which she would suffer and the extent to which she would be able to obtain access to medical treatment, support, and care, including help from relatives, *must involve a certain degree of speculation*, particularly in view of the *constantly evolving situation* as regards the treatment of HIV and AIDS worldwide.[15]

Since the impacts of slow-onset climate change (and other) processes may take some time before they amount to "inhuman or degrading treatment" or a threat to life, which are the grounds on which protection may be forthcoming, the timing of a protection claim will be crucial.[16] Based on the law as it currently stands, the availability of protection will depend on the point in time at which the claim is made and the severity of the immediate impacts on return.

This is why recourse to traditional protection mechanisms, either based on complementary protection (human rights-based *non-refoulement*) or temporary protection mechanisms for sudden disasters, are an inadequate and incomplete response to movement from slow-onset processes. Rational, planned migration options in anticipation of the impacts of such processes need to be created. With inbuilt human rights safeguards, these can provide a relatively safe mechanism for enabling people to move away from such impacts without artificially treating them as being in need of international "protection" (from a persecutory or abusive state) (McAdam 2012, 201ff.).

Theorizing "crisis migration"

It is impossible to have vision from everywhere and yet from nowhere.
(Farbotko 2010, 55)

In order to understand a complex phenomenon, it is necessary to break it down into small parts before reassembling them to see what, if any, common themes emerge. Only then is it possible to identify accurately whether conceptualizing the phenomenon in that way is useful—both from a theoretical and a practical perspective.

The case studies examined in this book are not homogenous, and while some common guiding principles may be extracted from them, each raises unique concerns that may not be amenable to a universal response. There is a risk that drawing together different forms of movement within a single analytical framework can obscure the idiosyncrasies of each type, and generalize or oversimplify matters that need to be understood within their particular context in order for policy responses to be sufficiently well targeted and effective.[17] For example, the notion of "climate change-related movement" itself embodies a wide range of different scenarios, and it is unhelpful—indeed, counterproductive—to attempt to address them in a uniform way (McAdam 2012, 17–20). For this reason, one of the central aims in examining "crisis migration" is to determine the extent to which identifying the *cause* of movement is key to crafting solutions, as opposed to the *nature* of movement. For example, is there utility in examining forms of displacement by theme, such as development-forced displacement, conservation displacement, conflict displacement, nuclear displacement, environmental displacement, climate change-related displacement and so on? Or are effective outcomes better framed around how movement occurs—how rapidly, whether it is internal or cross-border, whether it is likely to be temporary or require more permanent solutions, and so on? To what extent does focusing on "tipping points" provide a sound conceptual basis for drawing together crises triggered by acute events and slow-onset processes? In each scenario, some flexibility will be required to ensure that legal and operational responses are appropriately targeted.

In most cases, movement is likely to be internal rather than across an international border, at least in the emergency phase. The Guiding Principles on Internal Displacement are a well-established soft law instrument that already applies to those who have been obliged to leave their homes "as a result of or in order to avoid the effects of armed conflict, situations of generalized violence, violations of human rights or human-made disasters" (1998, Introduction, para. 2). These are precisely the kinds of circumstances envisaged as constituting "crisis migration" scenarios. If there is a protection gap in responding to such scenarios, one must query whether it is a legal gap, or a political gap—that is, an absence of political will or capacity to implement existing legal norms. The latter is not solved by creating more law, but rather by strengthening implementation and enforcement of existing law. And the call on *international* law is less relevant: what is crucial is getting states to tailor effective national responses (which may be grounded in international human rights norms, but do not require the creation of additional norms).

There are other unresolved questions. For instance, how far does a person need to move for it to qualify as "migration"? In a study of people displaced by riverbank erosion in Bangladesh, on average households only moved one kilometer from their homes, with ten kilometers the furthest distance moved. Some said that they would prefer not to move far "because of attachment to the land where their forefathers had lived" and their deep respect for their ancestral homes (Abrar and Azad 2004, 46). However, these are some of the most vulnerable people in Bangladesh: landless, impoverished and without the social networks to facilitate movement elsewhere. Are they to be counted as "crisis migrants"? Or are they counted among those who are too vulnerable to move at all—a group generally invisible from the analysis of forced *migration*? Or do they fall into a gap in between? If migration is to be understood as one option among others, then "the study of migration in crisis contexts must include the investigation of non-migrants and thus of a larger population than that which have been typically studied in the analysis of so-called forced migration" (Lubkemann 2004, 391). For this reason, the present volume uses "crisis migration" to encompass both movement and non-movement in the context of crises.

Perhaps the concept of crisis "movement" is a more apt term to use. It sidesteps technical problems that the use of the term "migration" may create—for example, whether "migration" excludes refugee flows, or (short) internal movement, or repatriation. If one takes the example of Libya, the majority of those who were displaced were migrant workers stranded when the Gaddafi regime fell (UNHCR 2012b). What they required was evacuation, then repatriation. They were not really crisis *migrants*. Furthermore, any protection "gap" arose not as a matter of law, but rather as a result of logistical difficulties and/or an absence of political will. The legal obligations here were clear—as citizens, their national governments were obliged to provide them with diplomatic protection, including consular assistance.

How crisis migration is framed will elicit different kinds of policy outcome. If disasters are seen as purely physical events, then it is likely that funding will be channeled into technical assistance (mitigation) rather than addressing systemic socio-economic vulnerabilities (adaptation). An insurance-based approach will foster different interventions than a preparedness approach, for instance, which seeks to "reduce current vulnerabilities and put in place response measures that will keep a disastrous event from veering into unmitigated catastrophe" (Lakoff 2007, 254).

Similarly, until we have a fuller understanding of what "crisis migration" means, we cannot identify whether—and to what extent—there are gaps in the existing international law regime. International human rights law, for example, applies to all people, irrespective of their nationality. In addition, specialist treaties, such as those for women, children and the disabled, identify the rights of groups that may be more vulnerable. While it is certainly the case that there are barriers to accessing the full range of rights to which one is entitled as a matter of international law, that is a problem of political will in implementing and enforcing such rights, rather than a problem with the normative framework itself.

There may be utility in developing a compilation of rights applicable in particular crisis situations, similar to the Guiding Principles on Internal Displacement. The Guiding Principles have provided states with a useful tool in identifying which rights are at risk at particular stages of displacement, thereby assisting them in fulfilling their obligations and more effectively targeting assistance and relief. However, as noted above, since the Guiding Principles already apply to natural or human-made disasters, it will be important to articulate whether a normative protection gap actually exists and why. If it relates to the fact that the Guiding Principles only apply within states and not to cross-border movement, then it will be important to consider the extent to which refugee law may or may not apply and where human rights-based *non-refoulement* may fill the gap. Existing analysis of the scope of these forms of protection in the context of climate change-related movement is relevant, including in relation to movement in response to slow-onset changes (see McAdam 2011).

Conclusion

Notwithstanding the concerns above, this book has an important role to play. In framing its case studies as examples of "crisis migration," a shared vocabulary may be developed to link "diverse policy discussions and initiatives across different scales and contexts" (Bravo 2009, 261). As Bravo has noted, "the semantic malleability of concepts like resilience and adaptation gives them considerable metaphorical purchase, which is precisely what allows linkages between strategies in disparate contexts" (2009, 261).

In articulating and deconstructing the phenomenon of "crisis migration," this book therefore has the potential to make four significant contributions. First, a study of "crisis migration" can conceptually connect events that might otherwise be regarded as isolated, enabling common themes to be identified and policy "silos" to be connected. There may be utility in viewing disasters as "routine and normal, connected to one another along various social fault lines and a direct product of our culture, not something to be imagined as simply an exceptional event" (Button 2010, 17). Lakoff explains that, following Hurricane Katrina in the US, "it was common to see comparisons made between the failed governmental response to the hurricane and the more successful response to the attacks of September 11. To an observer a decade before, it might have been surprising that a natural disaster and a terrorist attack would be considered part of the same problematic" (Lakoff 2007, 249). Connecting different events may link different skill-sets and approaches that can be carried across from one policy area to another, simultaneously enriching yet streamlining expertise. Of course, on the flipside, it may also lead to undesirable connections. For example, in the climate change context, migration and security have sometimes been conflated, such that migration itself comes to be perceived as a security threat (rather than as evidence of a lack of human security for those who move).[18] This can serve to justify increasingly restrictive and/or militarized responses, which are clearly at odds with the protection frameworks advocated in the present project.

Much can be learned from a generic approach to crises as opposed to an agent-specific approach.[19] Looking at the commonalities of different kinds of events enables patterns to be discerned, and may facilitate more systematic planning in terms of assessing hazards, preparing response mechanisms and establishing clear post-impact procedures (Tierney 1980, 18–19). This is likely to be of assistance to local authorities in particular, which lack the capacity to develop parallel strategies.

That said, it is important not to over-generalize so that responses are maladaptive. Meaningful distinctions need to be developed, which may relate less to the *type* of disaster (e.g. flooding) than to features such as how rapidly or slowly it occurs (Quarantelli 1985, 58).

Second, a focus on "crisis migration" can illuminate the role of other processes in contributing to migration, such as population growth, rapid urbanization, unsustainable development and so on. The broader relevance of these may be obscured (or assumed to be idiosyncratic) if left to discrete case studies. For this reason, the examples studied in this book are perhaps best understood as "freeze frames of crisis, momentarily captured in time and space simply for the ease of analysis" (Button 2010, 17). "Crisis migration" may therefore provide a lens for surveying broader societal dynamics that contribute to mobility.

Third, the study can reconceive "crisis" as a process of deprivation, vulnerability and disempowerment, rather than a single event (echoing the reconception of "disaster" by Hewitt 1983). Social conditions prefigure disaster (Hewitt 1983, 24–27; see also Wisner *et al.* 2004), although the disaster "event" may become the metonym for the underlying vulnerability that transforms a hazard into a crisis (Oliver-Smith 2002, 28). Often, though, the event *obscures* these pre-existing fragilities, such that it comes to be understood as emblematic of environmental hazard or climate change impacts rather than more holistically. The danger of this is that sustained areas for intervention, such as improved development practices, poverty reduction schemes and so on, which already have strong institutional frameworks, may be overlooked in favor of emergency responses that address "symptoms but not causes" (Oliver-Smith 2002, 32). This risks solutions being only partially attuned to the needs of affected communities, and reactive rather than proactive. The challenge is to ensure that responses are locally relevant and meaningful.

Fourth, naming particular movements as "crisis migration" may focus attention and mobilize action. In this way, the concept may have an important political function to play, generating a set of policy (and funding) responses.

Ultimately, however, we need to be cautious and precise about how we use the concept of "crisis migration." First, there is a risk that the language of "crisis" may serve to pathologize all movement, instead of recognizing the positive, adaptive (and traditional) role that migration plays in some contexts—that is, migration as a form of resilience. Second, although the purpose of this book is to illuminate the vulnerability of those who are forced to move in circumstances of great stress, there is a risk that states may reflect the "crisis" back onto migration itself. "Crisis migration" may mean something very different if approached from the perspective of the state, rather than the migrant. As has been seen in the context of climate change-related movement and human security, some developed states have shifted concerns away from the vulnerability of those forced to move, to argue instead that such migration poses a security threat to their own citizens (McAdam 2012, 223–225; see also Hartmann 2010, 233). This sort of characterization is counterproductive, especially in a climate of general hostility toward "outsiders." Rather than sending a signal about the acute distress in which people find themselves (or in which people *perceive* themselves to be), the notion of "crisis migration" may instead put governments into a defensive posture, fearful of managing perceived "floods" of people on the move. In other words, the conflation of "crisis" with "migration" may be used to imply that movement itself is the crisis, rather than the crisis being embodied in the circumstances from which people are moving. Third, the notion of urgency may result in poorly thought-through interventions responding to the perceived "event"-hazard, rather than to the underlying processes of vulnerability which render that hazard a particular danger to a particular community.

We may achieve greater buy-in from states and more constructive policy interventions if we focus on migration as a normal form of adaptation to change. While this still requires creative and well-informed policy development, its starting point is more positive, positing movement as a productive force to be harnessed and developed, rather than as an overwhelming humanitarian calamity to be solved. In this way, the displaced are viewed as capable agents instead of helpless victims, which resonates with their own desires and capabilities as well (McAdam and Loughry 2009).[20]

Acknowledgments

I offer my sincere thanks to the participants in the ISIM Crisis Migration Project; the members of the Forced Migration Reading and Writing Group, Faculty of Law, University of New South Wales; Dr Matthew Gibney, Refugee Studies Centre, University of Oxford; and Professor Fleur Johns, University of New South Wales for their very helpful comments. Any errors or omissions are, of course, my own.

Notes

1 In the literature, the terms are often used interchangeably (see e.g. Lubkemann 2004, 374). See also the vast literature on the role of agency in "forced" migration (e.g. Richmond 1994; Van Hear 1998; Turton 2003).

2 Note the definition of "disaster" by the International Law Commission (ILC): "a calamitous event or series of events resulting in widespread loss of life, great human suffering and distress, or large-scale material or environmental damage, thereby seriously disrupting the functioning of society." The qualifier "calamitous" stresses the extreme nature of the events covered by the draft articles (see ILC 2010, 325). It is important to note that the definition of "disaster" used by the ILC involves a higher threshold than the present volume's conception of "crisis." This emphasizes the extreme importance of carefully articulating the intended meaning and scope of terms such as "crisis migration," especially when they may be used to agitate for legal or policy responses.

3 For a discussion of the notion of the "emergency" in the refugee context, and states' capacity to respond, see e.g. Durieux and McAdam 2004.

4 When this chapter was originally conceived, the editors of the volume had not yet formulated a definition of "crisis" or "crisis migration." Instead, these were used as shorthand terms to describe mobility in humanitarian crises.

5 As Bravo notes, crisis narratives typically describe large numbers of people—often entire regions— "being threatened by some large phenomenon such as famine or global warming" (2009, 262).

6 *Oxford English Dictionary Online* (accessed 8 May 2012), meaning 3.

7 Interestingly, the Greek origin of the noun (κρίσις) encompasses three possible meanings: discrimination, decision, crisis: *Oxford English Dictionary Online* (accessed 8 May 2012).

8 In a chapter entitled "Migration as a Response to Crisis and Disruption," Pooley and Turnbull (1998) critique this "public" understanding of crisis to offer a more nuanced approach that recognizes the role of individual circumstances in mobility decisions.

9 "Vulnerable populations are those most at risk, not simply because they are exposed to hazard, but as a result of a marginality that makes of their life a 'permanent emergency'" (Bankoff 2001, 25).

10 As has been noted in the refugee context, "[p]ersecution cannot and should not be defined *solely* on the basis of serious human rights violations. Severe discrimination or the cumulative effect of various measures not in themselves alone amounting to persecution, *as well as their combination with other adverse factors*, can give rise to a well-founded fear of persecution, or, otherwise said: make life in the country of origin so insecure from many perspectives for the individual concerned, that the only way out of this predicament is to leave the country of origin" (Feller 2002, 3, emphasis added).

11 See also Bankoff's critique of the vulnerability discourse generally: "The discourse of vulnerability, no less and no more than that of tropicality or development, belongs to a knowledge system formed from within a dominant Western liberal consciousness and so inevitably reflects the values and principles of that culture" (2001, 29). See also Farbotko (2010).

12 "[P]roviding emergency transport and conducting evacuations, for instance, can save lives or reduce tensions in conflict zones. IOM has helped identify assistance and protection gaps that exist, for example, for potential victims of trafficking or unaccompanied minors in large scale forced movements of disaster-affected populations. Or, by offering livelihood options during the transition and recovery phase following a disaster, IOM has helped minimize further forced displacement and promoted durable solutions" (Thompson 2012, 1–2).

13 The ILC's definition of a "disaster" refers inter alia to circumstances "seriously disrupting the functioning of society" (ILC 2010, 325).

14 See e.g. *Gounaridis and others v. Greece* App no 41207/98 (European Commission of Human Rights, 21 October 1998); *Tauira and others v. France* App no 28204/95 (European Commission of Human Rights, 4 December 1995).

15 *N v. United Kingdom* [2008] ECHR 453, para. 50 (Lord Hope), emphasis added.

16 This is apparent in some of the cases that have already been brought before the Refugee Review Tribunal in Australia and the Refugee Status Appeals Authority in New Zealand (see e.g. cases cited in McAdam 2012, 44–48).

17 For example, Lewis argues that coping mechanisms in particular societies are marked by singular "interpretations of hazardous uncertainty" and their "own context of geographic, topographic and cultural variety" (1990, 247).

18 See further McAdam (2012, 224). Elsewhere, Vásquez Lezama (2010) has warned of the dangers of "compassionate militarization"—the use of armed forces to respond to natural disasters where civilian authorities are overwhelmed.

19 This is borrowed from Quarantelli's approach to disasters (see Quarantelli 1985, 52).

20 The Chair of IOM's April 2012 inter-sessional meeting stated: "One strong message which emerged from the debates concerned the agency, capacity and resilience of affected communities, including strengths and skills acquired through the crisis itself. Participants strongly cautioned against perpetuating the victimization of populations while delivering needed assistance" (IOM 2012b, 3). See generally Bankoff (2001).

References

Abrar, C.R. and Azad, S.N. (2004) *Coping with Displacement: Riverbank Erosion in North-West Bangladesh*, RDRS Bangladesh, North Bengal Institute, and Refugee and Migratory Movements Research Unit, Dhaka.

Bankoff, G. (2001) "Rendering the World Unsafe: 'Vulnerability' as Western Discourse," *Disasters*, 25(1): 19–35.

Barnett, J. and Adger, N.W. (2003) "Climate Dangers and Atoll Countries," *Climatic Change*, 61(3): 321–327.

Barnett, J. and Campbell, J. (2010) *Climate Change and Small Island States: Power, Knowledge and the South Pacific*, London: Earthscan.

Barnett, J. and Webber, M. (2009) *Accommodating Migration to Promote Adaptation to Climate Change*, Stockholm: Commission on Climate Change and Development.

Barnett, J. and Webber, M. (2010) "Migration as Adaptation: Opportunities and Limits," Jane McAdam (ed.), *Climate Change and Displacement: Multidisciplinary Perspectives*, Oxford: Hart Publishing.

Betts, A. (2010a) "Survival Migration: A New Protection Framework," *Global Governance*, 16(3): 316–392.

Betts, A. (2010b) "Towards a 'Soft Law' Framework for the Protection of Vulnerable Irregular Migrants," *International Journal of Refugee Law*, 22(2): 209–236.

Bijak, J. (2011) *Forecasting International Migration in Europe: A Bayesian View*, Dordrecht: Springer.

Bravo, M.T. (2009) "Voices from the Sea Ice: The Reception of Climate Impact Narratives," *Journal of Historical Geography*, 35(2): 256–278.

Button, G. (2010) *Disaster Culture: Knowledge and Uncertainty in the Wake of Human and Environmental Catastrophe*, Walnut Creek: Left Coast Press.

Calhoun, C. (2010) "The Idea of Emergency: Humanitarian Action and Global (Dis)Order," Didier Fassin and Mariella Pandolfi (eds.), *Contemporary States of Emergency: The Politics of Military and Humanitarian Interventions*, New York: Zone Books.

Charlesworth, H. (2002) "International Law: A Discipline of Crisis," *Modern Law Review*, 65(3): 377–392.

Durieux, J.-F. and McAdam, J. (2004) "*Non-Refoulement* through Time: The Case for a Derogation Clause to the Refugee Convention in Mass Influx Emergencies," *International Journal of Refugee Law*, 16(1): 4–24.

Farbotko, C. (2010) "Wishful Sinking: Disappearing Islands, Climate Refugees and Cosmopolitan Experimentation," *Asia Pacific Viewpoint*, 51(1): 47–60.

Farbotko, C. and Lazrus, H. (2012) "The First Climate Refugees? Contesting Global Narratives of Climate Change in Tuvalu," *Global Environmental Change*, 22(2): 382–390.

Feller, E. (2002) "Statement by Ms Erika Feller, Director, Department of International Protection, UNHCR," Strategic Committee on Immigration, Frontiers and Asylum, Brussels, November 6, available online at: http://www.refworld.org/publisher,UNHCR,,,3dee02944,0.html.

Ferguson, J. (1990) *The Anti-Politics Machine: "Development," Depoliticization, and Bureaucratic Power in Lesotho*, Cambridge: Cambridge University Press.

Grothmann, T. and Patt, A. (2005) "Adaptive Capacity and Human Cognition: The Process of Individual Adaptation to Climate Change," *Global Environmental Change*, 15(3): 199–213.

Guiding Principles on Internal Displacement (1998) UN Doc E/CN.4/1998/53/Add.2 (February 11, 1998).

Guterres, A. (2011) "Statement by Mr. António Guterres, United Nations High Commissioner for Refugees, Intergovernmental Meeting at Ministerial Level to Mark the 60th Anniversary of the 1951 Convention relating to the Status of Refugees and the 50th Anniversary of the 1961 Convention on the Reduction of Statelessness," Geneva, 7 December, available online at: http://www.unhcr.org/4ecd0cde9.html.

Haque, M.S. and Abdiker, M. (2012) "Moving to Safety: Migration Consequences of Complex Crises," *Intersessional Workshop on Moving to Safety: Migration Consequences of Complex*

Crises, IOM International Dialogue on Migration 2012 on Managing Migration in Crisis Situations, April 24, available online at: http://www.iom.int/jahia/webdav/shared/shared/mainsite/microsites/IDM/workshops/moving-to-safety-complex-crises-2012/speeches-presentations/ICP-DOE-Setting-the-Scene-24-April-2012.pdf.

Hartmann, B. (2010) "Rethinking Climate Refugees and Climate Conflict: Rhetoric, Reality and the Politics of Policy Discourse," *Journal of International Development*, 22(2): 233–246.

Hewitt, K. (1983) "The Idea of Calamity in a Technocratic Age," Kenneth Hewitt (ed.), *Interpretations of Calamity from the Viewpoint of Human Ecology*, London: Allen & Unwin.

Holmes, J. (2008) "Opening Remarks at the Dubai International Humanitarian Aid and Development Conference and Exhibition," DIHAD 2008 Conference, Dubai, UAE, April 8, available online at: http://www.reliefweb.int/rw/rwb.nsf/db900sid/YSAR-7DHL88?OpenDocument.

Hugo, G. (2010) "Climate Change-Induced Mobility and the Existing Migration Regime in Asia and the Pacific," Jane McAdam (ed.), *Climate Change and Displacement: Multidisciplinary Perspectives*, Oxford: Hart Publishing.

International Law Commission (ILC) (2010) Commentary: Report of the ILC, 62nd session (May 3–June 4, and July 5–August 6, 2010), UN Doc A/65/10.

International Organization for Migration (IOM) (2012a) "Overview," available online at: http://www.iom.int/cms/idmcomplexcrises.

IOM (2012b) *Moving to Safety: Migration Consequences of Complex Crises*, Geneva: IOM (International Dialogue on Migration, No. 20), available online at: http://publications.iom.int/bookstore/free/RB20_ENG_web.pdf.

IOM (2013) "IOM in Emergency and Post Crisis Scenarios," available online at: http://publications.iom.int/bookstore/free/iom_in_emergency_and_post_crisis.pdf.

King, R. and Mai, N. (2008) *Out of Albania: From Crisis Migration to Social Inclusion in Italy*, Oxford: Berghahn Books.

Lakoff, A. (2007) "Preparing for the Next Emergency," *Public Culture*, 19(2): 247–271.

Lewis, J. (1990) "The Vulnerability of Small Island States to Sea Level Rise: The Need for Holistic Strategies," *Disasters*, 14(3): 241–249.

Lubkemann, S.C. (2004) "Situating Migration in Wartime and Post-War Mozambique: A Critique of 'Forced Migration' Research," Simon Szreter, Hania Sholkamy and A. Dharmalingam (eds.), *Categories and Contexts: Anthropological and Historical Studies in Critical Demography*, Oxford: Oxford University Press.

McAdam, J. (2011) *Climate Change Displacement and International Law: Complementary Protection Standards*, UNHCR Legal and Protection Policy Research Series, PPLA/2011/03, May.

McAdam, J. (2012) *Climate Change, Forced Migration, and International Law*, Oxford: Oxford University Press.

McAdam, J. and Loughry, M. (2009) "We Aren't Refugees," *Inside Story*, June 30, available online at: http://inside.org.au/we-arent-refugees/.

Morrissey, J. (2008) "Rural–Urban Migration in Ethiopia," *Forced Migration Review*, 31: 28–29.

Mortreux, C. and Barnett, J. (2009) "Climate Change, Migration and Adaptation in Funafuti, Tuvalu," *Global Environmental Change*, 19(1): 105–110.

Oliver-Smith, A. (2002) "Theorizing Disasters: Nature, Power, and Culture," Susanna M. Hoffmann and Anthony Oliver-Smith (eds.), *Catastrophe and Culture: The Anthropology of Disaster*, Santa Fe: School of American Research Press.

Oliver-Smith, A. and Hoffmann, S.M. (2002) "Why Anthropologists Should Study

Disasters," Susanna M. Hoffmann and Anthony Oliver-Smith (eds.), *Catastrophe and Culture: The Anthropology of Disaster*, Santa Fe: School of American Research Press.

Payton, D. (2011) Remarks by David Payton of UNDP in a roundtable discussion at the Brookings Institution, April 3, available online at: http://www.brookings.edu/~/media/events/2012/4/03%20climate%20roundtable/roundtable%20on%20climate%20change%20and%20human%20mobility%20report.pdf.

Pooley, C. and Turnbull, J. (1998) *Migration and Mobility in Britain since the 18th Century*, London: University College London Press.

Prabhakara, N.R. (1986) *Internal Migration and Population Distribution in India: Some Reflections*, New Delhi: Concept Publication Co.

Quarantelli, E.L. (1977) "Social Aspects of Disasters and Their Relevance to Pre-Disaster Planning," *Disasters*, 1: 98–107.

Quarantelli, E.L. (1985) "What is Disaster? The Need for Clarification in Definition and Conceptualization in Research," in Barbara J. Sowder (ed.), *Disasters and Mental Health: Selected Contemporary Perspectives*, Rockville: US Department of Health and Human Services.

Richmond, A.H. (1994) *Global Apartheid: Refugees, Racism and the New World Order*, Oxford: Oxford University Press.

Roe, E.M. (1991) "Development Narratives, or Making the Best of Blueprint Development," *World Development*, 19(4): 287–300.

Roe, E.M. (1995) "Except-Africa: Postscript to a Special Section on Development Narratives," *World Development*, 23(6): 1065–1069.

Sen, A. (1981) *Poverty and Famines: An Essay on Entitlement and Deprivation*, Oxford: Clarendon Press.

Strohmeyer, H. (2011) "Informal Summary Note: ECOSOC Humanitarian Affairs Segment 2011: Side Event on Protection and Displacement in Natural Disaster Situations," Policy Development and Studies Branch, Office for the Coordination of Humanitarian Affairs (OCHA), available online at: http://www.un.org/ecosoc/julyhls/pdf11/has_protection_and_displacement.pdf.

Tacoli, C. (2009) "Crisis or Adaptation? Migration and Climate Change in a Context of High Mobility," *Environment and Urbanization*, 21(2): 513–525.

Thompson, L. (2012) "Introductory Remarks," Inter-sessional Workshop on Moving to Safety: Migration Consequences of Complex Crises, IOM International Dialogue on Migration 2012 on Managing Migration in Crisis Situations, April 24, available online at: http://www.iom.int/jahia/webdav/shared/shared/mainsite/microsites/IDM/workshops/moving-to-safety-complex-crises-2012/speeches-presentations/IDM-Introductory-remarks-DDG-April-2012-2.pdf.

Tierney, K.J. (1980) *A Primer for Preparedness for Acute Chemical Emergencies*, Newark: Disaster Research Centre.

Turton, D. (2003) "Conceptualising Forced Migration," Lecture, Refugee Studies Centre, University of Oxford, International Summer School in Forced Migration, available online at: http://repository.forcedmigration.org/show_metadata.jsp?pid=fmo: 2531.

Uddin, A.M.K. (2010) Comprehensive Disaster Management Program, Dhaka, Bangladesh. Interview with the author, June 16.

UNHCR (2012a) *The State of the World's Refugees: In Search of Solidarity: A Synthesis*, Geneva: UNHCR.

UNHCR (2012b) "2012 UNHCR Country Operations Profile: Libya," available online at: http://www.unhcr.org/pages/49e485f36.html.

Van Hear, N. (1998) *New Diasporas: The Mass Exodus, Dispersal and Regrouping of Migrant Communities*, London: University College London Press.

Vásquez Lezama, P. (2010) "Compassionate Militarization: The Management of a Natural Disaster in Venezuela," Didier Fassin and Mariella Pandolfi (eds.), *Contemporary States of Emergency: The Politics of Military and Humanitarian Interventions*, New York: Zone Books.

Warner, K. (2010) "Assessing Institutional and Governance Needs Related to Environmental Change and Human Migration," Study Team on Climate-Induced Migration, German Marshall Fund of the United States.

White, H. (1984) "The Question of Narrative in Contemporary Historical Theory," *History and Theory*, 23(1): 1–33.

Wisner, B., Blaikie, P., Cannon, T. and Davis, I. (2004) *At Risk: Natural Hazards, People's Vulnerability and Disasters* (2nd edn.), London: Routledge.

PART II

Case Studies of Humanitarian Crises: Movements, protection implications and responses

3

RISING WATERS, BROKEN LIVES

Experience from Pakistan and
Colombia floods suggests
new approaches are needed

Alice Thomas

Introduction and overview

Natural hazards, such as earthquakes, hurricanes, floods and droughts, can have
severe impacts on human displacement. When acute, these events have the potential not only to drive large numbers of people from their homes to avoid harm, but
also to prevent them from returning by damaging or destroying houses, cutting
off access to basic necessities like food and clean water, impairing livelihoods and
creating unsafe conditions. Acute hazards become disasters when they overwhelm
the ability of national authorities and affected populations to respond. The Internal
Displacement Monitoring Centre (IDMC) estimates that, in 2010 and 2012 (two
recent years that saw a high number of disasters), 42 million and 32.4 million people, respectively, were newly displaced by sudden-onset natural disasters (IDMC
and OCHA 2009; Yonetani 2011; IDMC and NRC 2013). Notably, these estimates do not include millions more people displaced by slow-onset disasters such as
droughts. As IDMC points out, "[t]he sheer scale of displacement should leave no
doubt as to the seriousness and immediacy of the challenge this poses for affected
populations, governments, and the international community" (Yonetani 2011, 4).

This case study focuses on displacement from flooding, drawing upon Refugees
International's research and observations regarding severe floods in Pakistan and
Colombia in 2010 and 2011.[1] In both Pakistan and Colombia, the floods affected
millions of people across a wide geographical area, rendered huge numbers homeless, resulted in significant damage to agriculture and public infrastructure, and presented massive operational challenges for governments and humanitarian agencies
responding to the emergencies. Moreover, the floods exposed serious shortcomings
in the ability of both national governments and the international community to
protect people displaced by natural disasters—particularly the most vulnerable sectors of society—both during their displacement and upon return or relocation. Significantly, the flooding occurred in two countries that had extensive displacement

from protracted and ongoing conflict that increased vulnerabilities and challenges, and highlighted disparities between the systems for addressing conflict and natural disaster-induced displacement.

Although often considered "sudden-onset" in nature, floods such as those that occurred in Pakistan and Colombia unfolded over several months and therefore, could be described as both "sudden-onset" and "slow-onset." For example, in Pakistan, the majority of deaths resulted from massive flash floods in the northern areas that occurred very early on. However, it took months for the deluge to move from the north to the south. In Colombia, flooding that started in August reached its peak in December 2010; however, the rain continued thereafter into the country's second rainy season, meaning that floods persisted or recurred in many areas over a period of eighteen months.

In this context, important implications regarding human movement can be drawn from the Pakistan and Colombia floods. The nature of the floods meant that different types of movement, including forced displacement (both emergency flight and evacuation), return and, to some extent, resettlement, occurred simultaneously in different parts of the country. In addition, the period of displacement proved to be relatively short term. Those displaced by the floods returned home as soon as possible, particularly as compared to other natural disasters like the Haiti earthquake, or to conflict environments wherein populations are often prevented from returning by ongoing insecurity and violence.

This had significant implications for the response that were often overlooked. Returning populations faced many of the same needs and vulnerabilities as displaced populations yet received limited priority and assistance. Early recovery programs that would have helped them to get back on their feet were slow and underfunded. Moreover, opportunities were missed to address factors that rendered certain groups more vulnerable to begin with, which would have helped to build resiliency to future shocks.

There is anecdotal evidence that some of the displaced chose not to return, primarily because they encountered better opportunities in receiving areas (author interviews, August 2011 and March 2012). However, those who migrated to cities were likely to face discrimination, and lack access to secure shelter and basic services. There is no available data on whether the floods resulted in out migration to foreign countries.

As described in greater detail below, at present even countries like Pakistan and Colombia, with significant experience in responding to natural disasters and relatively advanced disaster management frameworks, proved unprepared and ill equipped to assist and protect people displaced during the 2010–11 flood emergencies. These shortcomings occurred regardless of whether the national government (Colombia) or the international community in cooperation with the national government (Pakistan) took the primary role in the response. In order to effectively meet the growing threat that extreme weather presents, national governments and the international community must take proactive steps to address these shortcomings and institute measures to protect affected populations more effectively during natural disasters, and to facilitate temporary and permanent migration as a coping strategy in their aftermath.

Background

Floods, climate change and displacement

There are two primary reasons why a focus on flooding as a cause of current and future displacement is of particular relevance. First, recent data on the number of recorded natural disasters indicate that floods occur more frequently than any other type of natural hazard, and that they routinely displace millions of people. In both 2010 and 2011, for example, floods accounted for close to 50 percent of recorded natural disasters (CRED n.d.), and 85 percent of disaster-related displacement (Yonetani 2011, 11).[2] Not captured by these figures are small-scale, recurrent floods, which often go unreported. Thus, it is hard to know to what extent recurrent, smaller-scale floods result in temporary displacement, or influence the decisions of those affected regarding whether to continue to return and rebuild/recover or to move away.

Second, while no individual flood event can be solely attributed to climate change, climate scientists predict that one of the most significant impacts of climate change will be an increase in severe weather events, including flooding (IPCC 2012, 6). Available data indicate that this may already be occurring. Over the past fifty years, the number of recorded floods has increased more than any other type of natural disaster (CRED n.d.). This may be due, in part, to human factors, including deforestation, population expansion and lack of flood control infrastructure. Nonetheless, in the coming decades, climate change—together with human factors—will likely result in more frequent and severe flood events, which, in turn, will displace growing numbers of people.

The 2010–2011 floods in Pakistan and Colombia

In late July 2010, two weather systems collided over northwest Pakistan, dumping a massive and unprecedented amount of rain. Flash floods tore away roads, bridges and entire villages. As the rain continued, the deluge moved south along the Indus River, eventually submerging one-fifth of the country's landmass and damaging or destroying vast expanses of water, power and transportation infrastructure. Close to five million acres of agricultural land were destroyed. Ultimately, more than twenty million were affected—more than the 2004 Indian Ocean tsunami, the 2005 Kashmir earthquake, and the 2010 Haiti earthquake combined. The floods destroyed 1.8 million homes, leaving an estimated nine million people without secure shelter (OCHA 2010).

Overwhelmed by the magnitude of the disaster, the Pakistani government appealed to the international community for assistance. In early August, a "flash appeal" was launched seeking $460 million in emergency funds to meet immediate needs, followed by a more comprehensive call to donor governments in September to provide more than $2 billion in assistance (United Nations News 2010). In addition, the humanitarian "cluster system" was activated, wherein the Emergency

Relief Coordinator (through the UN Office for the Coordination of Humanitarian Affairs, or OCHA), set up working groups composed of humanitarian agencies to coordinate their activities around specific humanitarian services (e.g. protection, shelter, food and nutrition, water, sanitation and hygiene).

Meanwhile, halfway across the globe, Colombia was also experiencing record-breaking flooding. Repeated and heavy rains brought on by La Niña, which started in April 2010, persisted through the normally dry months, depriving soil and communities of the opportunity to recover. By December 2010, 93 percent of the country's municipalities were experiencing floods and landslides, forcing President Juan Manuel Santos to declare a state of emergency. By the following June more than four million people had been affected (OCHA 2011a). More than 3.3 million acres of agricultural land were flooded, more than 370,000 head of livestock perished and more than 800 roads were damaged or washed away (OCHA 2011b). The rains also overwhelmed the existing water management infrastructure—most notably the Dique Canal, a major waterway that connects the Magdalena River to the port city of Cartagena. A major breach in the canal submerged half of the northern state of Atlántico, affecting 225,000 people (Thomas 2012). While there are no exact figures regarding the number of people displaced, more than 440,000 homes were reportedly damaged or destroyed by floods and landslides (RI 2012, 1–2).

The flood conditions brought on by the 2010–11 La Niña were anticipated to subside in the summer of 2011. However, in late 2011, La Niña conditions reappeared, causing more flooding and landslides. Between September and December 2011, more than 900,000 people throughout the country were affected. An additional 66,000 people were reportedly affected during the first six weeks of 2012 (OCHA 2011c; Government of Colombia 2012). In short, the back-to-back La Niñas meant that there was insufficient time for existing floods to dissipate before heavy rains started again.

In response to the initial emergency, the Colombian government launched *Colombia Humanitaria*, a campaign aimed at mobilizing resources from government, and private and public sources. The international community also provided direct assistance at the height of the crisis, mobilizing $14.7 million in emergency aid from the UN Central Emergency Response Fund (CERF), the UN Emergency Response Fund for Colombia (ERF) and donor governments. However, given the large amount of money raised by the *Colombia Humanitaria* campaign, the Colombian government turned down a request by the humanitarian community to launch an appeal, limiting somewhat the international community's role in the response.

Challenges in responding to acute flooding and protection gaps

Colombia and Pakistan's national disaster frameworks

Both Colombia and Pakistan are exposed to a wide range of natural hazards, and experience frequent flooding. Colombia is ranked tenth on the list of countries with the highest level of risk to natural hazards including floods and landslides, and has the highest recurrence of extreme events in South America (World Bank and GFDRR 2013a). Pakistan is also exposed to a number of natural hazards and has the highest risk of floods in South Asia. The monsoon rains, which occur from July through September, result in frequent and severe flooding in the Indus River Basin where millions of people live in low-lying areas (World Bank and GFDRR 2013b).

Given their exposure, both Colombia and Pakistan have developed natural disaster risk management (DRM) laws and frameworks aimed at enhancing disaster preparedness and response. In Colombia, at the time the flooding occurred in 2010–11, the country's DRM system was based upon the National System for Disaster Management and Prevention (SNPAD), which included both a central response authority (the Risk Management Directorate, or DGR), as well as regional committees presided over by the provincial governors (known as CREPADS) and local committees (known as CLOPADs) presided over by mayors.[3]

At the time of the Pakistan floods, Pakistan's DRM framework was less well developed as recent legal and institutional changes adopted in the wake of the 2005 earthquake in Kashmir had yet to be fully implemented. The National Disaster Management Authority (NDMA), established in early 2007, was made responsible for spearheading a network of national, provincial and local disaster management activities (Cochrane 2008, 7). However, making the provincial and local DRM bodies operational took time, and at the time of the 2010 floods, not all Provincial Disaster Management Authorities (PDMAs) or local-level District Disaster Management Authorities (DDMAs) had been established, although plans were in place to do so (White 2011, 5–6).

Challenges and shortcomings of the response

The vast scale of the Pakistan and Colombia floods would have presented serious operational challenges under any circumstances. Nonetheless, the severity of the disasters in both instances exposed serious flaws in the existing DRM systems. Highlighted below are specific shortcomings that appear to have further prolonged displacement and exposed affected communities to unnecessary harms and risks.

Lack of capacity and failure to implement DRM laws, especially at the local level

Despite its well-established DRM framework, and leadership in disaster preparedness in Latin America, Colombia proved ill equipped to deal with the 2010–11 floods. The central DRM response authority (the DGR) was significantly under-staffed and under-resourced to address the widespread nature of the disaster, which at its height affected 93 percent of municipalities. Meanwhile, at the local level, there was a serious lack of capacity and inconsistent implementation of the disaster response system. While local disaster response authorities in some areas proved successful in preparing for the floods, in many municipalities, citizens complained that their CLOPADs either did not exist or did not know what they were doing. Even where CLOPADs were operational, municipalities had extremely limited funds to prepare for and respond to the floods. As a result, they were largely dependent on outside assistance, which was insufficient and often extremely slow in arriving.

The newly elected Santos Administration, recognizing that the existing system did not have sufficient human and technical capacity—and skeptical that local governments would distribute millions of dollars in flood aid fairly—decided to adopt a new structure, *Colombia Humanitaria*, to implement the response and distribute funds. *Colombia Humanitaria* developed its own procedures, instituted through a series of Presidential decrees, for distributing aid directly to affected populations using designated "operators," which included a variety of public and private organizations such as the Colombian Red Cross, local chambers of commerce, and private charitable foundations.

Despite strong political will at the national level and significant financial resources, *Colombia Humanitaria* encountered serious challenges in distributing emergency assistance in a timely and effective manner. In brief, the system did not aim to bolster existing government capacity but to bypass it, and consequently proved woefully slow and fraught with problems. Contractual delays, bureaucratic obstacles, corruption, and incomplete information, meant that thousands received little to no assistance (RI 2011b, 11). Moreover, lack of government access to many flood-affected areas of the country due to the presence of illegal armed actors meant that conflict-affected populations often received limited flood assistance (e.g. the Pacific Coast) (OCHA 2010).

Pakistan encountered similar problems owing to its own lack of capacity and resources, and to incomplete implementation of the DRM system at the provincial and local level. At the time the floods hit, the NDMA had only twenty-one staff, and an extremely limited annual budget (US$740,000). Moreover, recent constitutional changes aimed at decentralizing power meant that the central government had no authority over the PDMAs (NDMA 2011, 2). The PDMA in the hardest hit province, Sindh, was also under-resourced and unprepared, while the PDMAs in Punjab and Baluchistan, two other heavily affected provinces, did not yet exist (Oxfam 2011, 18). The exception was in Khyber Pakhtunkhwa (KP), where the response proved relatively coordinated and effective given the province's experience in responding to the 2005 earthquake and to ongoing, conflict-related humanitarian emergencies.

Lack of coordination and cooperation

This leads to a second, and related, problem: lack of coordination between both national government and provincial/local governments, and government officials at all levels and international aid agencies. As mentioned above, in Pakistan, the relief effort in KP was relatively quick and effective due in part to cooperation and coordination between the government and international humanitarian agencies and previous experience with the cluster system. In the south, however, where the international humanitarian community did not have a significant presence prior to the floods, the response was far slower and less coordinated. Newly established clusters set up to respond to the floods faced greater challenges in coordinating with local authorities, with whom they lacked pre-existing relationships. Communication among the clusters was also weak. Overall, these shortcomings made for an incomplete understanding of the numbers, needs, and locations of affected people, and the activities being executed throughout the region (RI 2011a, 5–6).

In Colombia, a lack of coordination between the central government and the provinces, and between CREPADs and CLOPADs, caused significant delays in the response. The new system for distributing flood relief under *Colombia Humanitaria* instituted a lengthy process for verifying data before releasing funds. At the provincial level, many governors' offices lacked the capacity to collect data accurately, while at the municipal level there was confusion regarding the overall process. As a result, significant time was lost in an effort to collect and verify data, leaving thousands of people with little to no government assistance for months (RI 2011a, 10–13).

More broadly, the response to both the Pakistan and Colombia floods would have benefitted from a clearer understanding of the role of the international community, and from better communication between government authorities and international agencies. In Pakistan, tensions that resulted from the international community's lack of communication with the government (including cases in which international agencies were seen to have overstepped their role) may partly explain the government's decision to increasingly restrict international agencies' access. In the immediate aftermath of the 2010 floods, Pakistani authorities allowed international humanitarian agencies wide access to provide relief in flood-affected areas, including previously restricted parts of KP affected by conflict and insecurity. Over time, however, it became harder for foreign humanitarian workers to obtain official permission from provincial authorities to allow them to operate in affected areas. Moreover, when severe floods occurred again the following year, the NDMA waited more than a month before calling for international assistance, insisting it could handle the response alone. As a result, relief to flood victims was delayed. While the less open approach by Pakistani authorities toward the international community may have been due to political issues unrelated to the floods (e.g. escalating security-related tensions between the US and Pakistan), many felt that it reflected a reluctance and lack of trust stemming from poor coordination and communication during the initial flood response (e.g. the UN decision to move ahead and launch the humanitarian appeal to donors in New York without the consent of the Pakistan government).

In Colombia as well, the lack of clarity regarding the role of the UN and other international agencies appears to have hindered a more effective response. The refusal of the Colombian government to allow international humanitarian agencies to launch an appeal for response funds no doubt inhibited their ability to mount a more comprehensive flood relief effort. Nonetheless, many members of the Humanitarian Country Team (HCT) felt that the HCT was caught off guard by the severe and protracted nature of the floods, and that better flood contingency plans should have been developed and quickly implemented (author interviews, May 2011 and March 2012). Noting the technical support role the HCT plays in response to conflict displacement, some wondered why similar arrangements could not be implemented in the response to Colombia's frequent floods and other natural disasters. For example, international agencies could have contributed greater technical support in the areas of information sharing, needs assessments, emergency response standards, and humanitarian response capacity building for local officials (RI 2011b, 4–5).

Challenges to early recovery programs

In both Pakistan and Colombia, early recovery programs provided an important opportunity for helping displaced populations to recover more quickly and increase resilience to future shocks. In Pakistan, for example, where four out of five people affected by the floods were reliant on agriculture for their livelihoods, the floods not only wiped away thousands of acres of crops but also put at risk the upcoming harvest, since most farmers lacked seeds for planting (OCHA 2011d, 43–45). Recognizing the potential for long-term food shortages, a seeds distribution program led by the Food and Agricultural Organization (FAO) and funded by the US and other donors was implemented early on in the response, allowing millions of farmers in flood affected areas to plant in time for the winter harvest, thereby avoiding a food crisis (FAO 2011). Early interventions in the health and water and sanitation sectors were lauded for averting the spread of water-borne illnesses. In addition, both government and humanitarian agency cash compensation and cash for work programs provided a much-needed injection of resources that allowed affected families to meet early recovery needs (RI 2011a).

Yet, in both countries, funding for and implementation of early recovery programs proved challenging, in some instances exacerbating vulnerabilities (Polastro *et al.* 2011, 42) and undermining resilience. For example, in Colombia, the slow pace of construction of transitional shelters meant that many families who lost homes were displaced three or four times while they awaited completion of transitional housing by the government and its partners. Many spent months living in shacks along the road, or in overcrowded and unsanitary conditions in schools or churches, under the constant threat of eviction (RI 2011b, 4–5). This appears to have stemmed in part from the tendency of governments, international humanitarian agencies and donors to split the disaster response into "emergency" and "recovery" phases, rather than to start early recovery programs simultaneously with the emergency response.

In Pakistan, for example, the government announced that the emergency response phase, including OCHA leadership under the cluster system, would end on January 31, 2011, other than in the worst affected areas, and would be followed by a recovery phase (OCHA 2011f). Thus, up to that time, limited emphasis was placed on assisting the large number of people who already had returned home to recover. In addition, although the scheduled "end" of the emergency response phase was announced well in advance, it took months for the UN Development Program (UNDP), the agency responsible for the early recovery phase of the flood response, to set up the early recovery working groups (ERWGs) and complete the early recovery gap analysis, thereby further delaying early recovery activities (RI 2011a). Moreover, the dismantling of the clusters and establishment of the ERWGs created confusion on changing roles and responsibilities, and created coordination challenges ultimately slowing the response as aid agencies struggled to understand the new systems (Oxfam 2011, 14). A year after the disaster hit, there were still reports that the ERWGs were largely ineffective, disorganized or not completely functional. In addition, the Strategic Early Recovery Action Plan, launched in April 2011, failed to garner strong support from either donor governments or the Pakistani government, which was emphasizing developmental interventions instead. By August, with only four months left to complete the government-designated early recovery phase, a $413 million funding gap for early recovery activities remained (RI 2011a, 2).

In Colombia, the lack of livelihood programs made it difficult for poor households to recover from the floods. Many of the hardest-hit communities (for example, La Mojana region in southern Córdoba, and Afro-Colombian and indigenous groups) lived below the poverty line prior to the disaster. Many were farmers who lost crops and animals during the floods and who, due to persistently wet conditions, were unable to replant. The lack of robust livelihood programs was particularly concerning because the government stopped delivering food aid to most areas at the end of 2011 as it wound down the "humanitarian response" phase of the emergency. Several national and local government officials expressed concerns that continued food deliveries might create dependencies. Yet it was unclear how affected populations—especially in areas that were still flooded—could be expected to become more self-sufficient when their livelihoods had been significantly diminished and alternative sources of income remained scarce (RI 2011b, 2).

Protection gaps and specific risks to vulnerable populations

As is the case with most natural disasters, the floods in Pakistan and Colombia had a disproportionate impact on the most vulnerable sectors of society, including women, children, the elderly, the disabled, ethnic and religious minorities, the poor, and those who had been previously displaced or affected by conflict. What made these groups particularly vulnerable to the floods in Colombia and Pakistan was not only their exposure and susceptibility to harm as a result of the floods (as compared to the population at large), but also limitations on their ability to recover and/or seek redress due to factors such as poverty, discrimination/disenfranchisement, lack of physical and/or legal access to services and inability to exercise rights.

In both Pakistan and Colombia, initial flood relief activities did not fully account for the vulnerabilities of these groups, and there were numerous reports documenting these oversights (Engeler 2010; Islamic Relief 2011, 10–15, 22; UNHCR 2011a, 3–5; Oxfam 2011, 11; Moloney 2012). In Pakistan, for example, there were complaints that the layout of camps set up to accommodate flood-displaced populations made it difficult for Muslim women to observe *purdah*, a practice that requires the physical separation of men and women. Muslim women were often confined to closed tents despite overcrowded, uncomfortable conditions and the 100 degree-plus heat outside (IDMC and NRC 2011, 5, 9; UNHCR 2011a, 16). In Colombia, unsanitary and unsafe conditions were reported in many of the local government-run shelters, and measures to ensure child safety and welfare were lacking (RI 2012, 5).

In both countries, at-risk populations also encountered difficulties and discrimination in accessing assistance. For example, the March 2011 Inter-Agency Real Time Evaluation of the Pakistan floods response noted the following:

> Targeting was particularly weak as there was no systematic registration or verification process—often there were no beneficiary lists or selection criteria established. When lists were prepared, these were not always drawn up on the basis of vulnerability. As a result, unknown quantities of assistance have reportedly reached those that were the least vulnerable, close to feudal landlords or connected through certain political affiliations. Many people from ethnic and tribal minorities and most vulnerable individuals and groups, such as widows or other female-headed households, were not prioritized and therefore deprived from any assistance at all. People that went into organized camps were better assisted than those in spontaneous camps; while those in host families received limited assistance.
>
> *(Polastro* et al. *2011, 36)*

Pakistani women in particular struggled to access assistance. The government cash compensation program for flood victims (WATAN cash cards) were issued based on possession of a Computerized National Identity Card (CNIC). However, 92 percent of women were registered in the name of a male relative, making it difficult for women who lacked their own CNIC cards, such as widows or female-headed households, to access flood relief. Minority groups in Pakistan's Sindh Province reported that they did not have equal access to flood assistance due to discrimination on religious and ethnic grounds (UNHCR 2011a, 16). Poor communities living in remote areas also found it particularly hard to access assistance or seek redress (Polastro *et al.* 2011, 37). In certain hard-to-reach districts in Pakistan's KP Province, for example, the UN reported that only about 25 percent of the affected population received humanitarian assistance (Oxfam 2011, 12). Accessing registration and aid delivery points proved particularly challenging for such groups, especially because so much of the local transportation infrastructure had been wiped away.

Those who lacked secure land tenure—many of whom lived on flood-prone, marginal lands—also proved particularly vulnerable to discrimination, including those who had been affected by armed conflict. Consider the case of Afghan refugees who were displaced from Azakhel, a settlement outside of Peshawar, which was 90 percent destroyed by the floods. While many of the residents had lived there for more than thirty years and considered it home, they were prevented from returning and rebuilding once the flood waters subsided, ostensibly to permit the government to conduct a safety assessment. A subsequent UN assessment deemed many of the same areas safe enough for return. While it may be that the selective nature of Pakistan's zoning enforcement was unintentional, the situation nonetheless highlighted the risk of discrimination faced by vulnerable groups (here, refugees, and those who lacked secure land title) in the absence of protection mechanisms and advance planning (RI 2011a, 12).

Similarly, in Colombia, where millions of people have been internally displaced by decades of fighting between government forces and illegal armed groups, the floods disproportionately affected IDPs, who are far more disadvantaged than the rest of the population. Half of Colombia's IDPs live below the poverty line and only 15 percent have access to secure housing (IDMC and NRC 2010, 5). According to the national ombudsman's office, IDPs were more susceptible to flooding because conflict had driven them into otherwise undesirable land in high-risk areas. The Colombian human rights organization CODHES (*La Consultoría para los Derechos Humanos y el Desplazamiento*) estimated that 40 percent of people affected by the floods were conflict IDPs (OCHA 2011e, 4). Thus, IDPs' and refugees' diminished resilience meant that they were more susceptible to harm when the floods hit.

Also worth noting are differences in Colombia's and Pakistan's frameworks for responding to displacement from man-made versus natural disasters. In 1997, Colombia enacted Law 387 on internal displacement, which entitles those displaced by conflict to specific rights emanating from their situation of vulnerability. The law requires the Colombian government to take actions to prevent displacement, and assist and protect those who flee. Municipalities are required to provide emergency assistance to IDPs within the first 72 hours and, thereafter, the Presidential Agency for Social Action and International Cooperation (*Acción Social*), which is in charge of coordinating the government response to the humanitarian needs of IDPs, steps in. The system has proven relatively effective in providing emergency relief within a matter of days to conflict IDPs, including in areas under the control of illegal armed groups.

However, because Law 387 does not include within the definition of "internally displaced persons" people displaced by natural disasters (ECOSOC 1998), Colombians displaced by the floods were not entitled to the same rights and protections as those displaced by conflict. During the floods response, none of the existing institutions, protocols and procedures for providing emergency relief in the case of conflict-related "mass displacements" (incidents involving the displacement of more than fifty people) was triggered. Rather, an entirely different set of government institutions and procedures came into play, which unfortunately proved far less effective. As the Norwegian Refugee Council has pointed out, while the protection of people internally displaced by conflict has gradually been informed by international human rights standards, those affected by natural disasters continue to be viewed as "objects of care, rather than rights-holders" (Albuja and Cavelier-Adarve 2011).

In contrast, Pakistan's system relies on the same disaster management authorities to respond to both conflict-related and natural disaster emergencies. This legal arrangement proved effective in strengthening response capacity in KP during the 2010 floods, for example, as the provincial disaster response authorities had substantial prior experience working with humanitarian agencies (as well as the Pakistani army) in response to displacement caused by both the 2005 earthquake and the 2008–09 Taliban insurgency and government-led counter-insurgency. Pre-existing relationships between internal and international actors, and the established and practiced disaster response systems, helped to facilitate a timely and organized response by the provincial disaster response authorities.

Nonetheless, the overall response would have benefitted from a disaster response system that incorporated mechanisms and procedures to protect affected individuals and protect their human rights, such as those prescribed under the UN Guiding Principles on Internal Displacement (Guiding Principles). The key elements of a rights-based approach include an explicit connection to accountability, participation, empowerment, non-discrimination and attention to vulnerable groups, all of which are vital to recovery and increased resilience in the face of future disasters (Koskinen 2006). A rights-based approach to displacement—whether caused by conflict or natural disasters—would have meant that affected communities were not viewed as passive victims but as rights holders engaged in their recovery (Koskinen 2006, 5). However, Pakistan has not incorporated the Guiding Principles into national legislation, meaning that mechanisms and procedures that would have better protected the rights of flood-affected persons (e.g. an ombudsman's office; legal recourse) were lacking.

Understanding the relationship between flooding and displacement

Specific impacts of acute floods on displacement

Short-term nature of flood-induced displacement

As noted above, the 2010 Pakistan floods displaced approximately nine million people. While no firm figures are available for Colombia, the 2010–11 protracted and recurrent floods there are estimated to have displaced several million people. Yet despite the large numbers of people displaced by these disasters, the vast majority returned home within a year.

The quick rate of return was due to a number of factors. First and foremost was the desire of affected populations to return as soon as possible in order to salvage any remaining assets and begin rebuilding their lives. In Pakistan, where many lacked secure land tenure, and many land demarcations had been wiped away in the floods, the need to secure property was particularly acute. As a result, it was often women and children who remained in the camps or shelters, while men returned to flooded land.

In the heavily affected La Mojana river basin region of northwest Colombia (where seven out of ten families live below the poverty line), many poor fishing communities never left their homes throughout 18 months of floods, opting to live on *tambos*, or elevated wooden planks and platforms. Accustomed to living on or near the water, and partly dependent on fishing for their livelihoods, they appeared better able to adapt to the flood conditions although forced to endure water-borne illnesses and lack of food. Given the deplorable conditions in many shelters (see below), without further analysis it cannot be concluded that they fared significantly worse than those who were displaced to shelters, other than in terms of increased levels of food insecurity.

A second factor propelling the return of flood-displaced people was the poor conditions in shelters and camps. In Colombia, shelter management by many local governments was minimal, and numerous people described unsafe and unsanitary conditions including overcrowding, lack of proper sanitation, and insufficient supplies of food, water, mattresses, stoves and other provisions (RI 2011b). In Pakistan, where both the government and international agencies were responsible for camp management, conditions were also often overcrowded and uncomfortable (RI 2011a).

Third, many were forced to return home as a result of government policies requiring that camps and shelters be closed after a certain period of time following the disaster. In Pakistan, the government declared the end of the "response phase" in January 2011 (except in several of the worst-affected districts), ordered all but a few camps closed and prohibited aid delivery (IFRC 2012b). The displaced were given transportation home and some supplies. In Colombia, where schools were used to house flood-displaced populations, pressure from both the government and local communities to reopen schools at the end of the rainy season drove many people back home or resulted in secondary displacement (RI 2011b).

Returned but still displaced

The overall short-term nature of displacement had significant implications for the response that were often overlooked. As one UN official involved in the Pakistan floods noted, "By the time we finished setting up camps, they were empty" (author interview, October 2010). Returning populations, while no longer officially recognized as displaced, nonetheless faced many of the same needs and vulnerabilities as displaced persons. Most returned to houses and belongings that were damaged or destroyed, and were forced to live in unsafe, makeshift shelters next to their former houses. Many lacked access to clean water and sanitation, and children especially suffered from dehydration, diarrhea, rashes and other illnesses related to the digestion of or exposure to standing or contaminated water (Warraich, Zaidi and Patel et al. 2010). The floods also destroyed crops and livestock, resulting in increased food insecurity and loss of livelihoods. A year after the Pakistan floods, approximately 5.6 million people in flood-affected areas remained food insecure and alarmingly high numbers were malnourished (Oxfam 2011, 11; WFP 2011).

Compounding the situation, the fact that affected populations were no longer concentrated in camps or shelters made it far harder for local governments and humanitarian agencies to respond. Those living in remote areas, or further away from towns or main roads, often received little to no assistance (RI 2011a; see also Polastro *et al.* 2011, 38). Finally, as noted above, early recovery activities that would have expedited home rehabilitation and restored livelihoods were slow to get off the ground, poorly coordinated and underfunded.

Addressing the risk of recurrent displacement

In both Pakistan and Colombia, the governments' failure to adequately address the risk of recurrent displacement also left vulnerable populations without protection. Following the floods in Pakistan, the government did not require people living in at-risk areas (e.g. along levees, within the river flood plain) to relocate, meaning that many moved right back into harm's way. Consequently, large numbers of people in southern Sindh Province who were displaced by the 2010 floods were displaced a second time when severe floods occurred again in 2011. At the same time, many homes rebuilt in flood-prone areas were subsequently considered illegal. The exception was KP, where following the floods, the PDMA put in place a plan to address illegal encroachment. Unfortunately, the plan was subsequently put on hold after it met resistance from parties with political or other vested interests (DAWN 2012).

Moreover, in both Pakistan and Colombia, the governments were slow to repair damage to dams, dykes and other flood control infrastructure. In Colombia, the government acted quickly to close a major breach in one of the levees along the Dique Canal (Ejercito Nacional de Colombia 2010). However, there were reports that damage elsewhere was not addressed quickly, resulting in renewed flooding in some affected areas when the rains commenced the following year (Mannon 2011). In Pakistan, the failure of the government to properly repair dykes and levees following the 2010 floods allegedly made flooding the following year far worse (Oxfam 2011, 15–16). Many families in southern Sindh were reportedly displaced twice (UNHCR 2011b; Zafar 2011). As late as May 2012, almost two years after the floods hit, there were media reports that the PDMA in Sindh had not yet made critical structural alterations as recommended by experts (Arif Riar 2012).

Rebuilding houses destroyed by flooding to incorporate flood-proof elements can be a cost-effective measure to avoid future displacement, especially in areas that are exposed to annual flooding. However, in both Pakistan and Colombia, insufficient priority was given to rebuilding homes to incorporate flood resistant elements. In Pakistan, of the 1.6 million homes damaged or destroyed by the 2010 flooding, only a small fraction of new homes were retrofitted to incorporate flood-proofing elements (Polastro *et al.* 2011, 44). The government strategy for addressing the large number of destroyed homes was to provide cash compensation to flood displaced households through the WATAN card system. But this self-help approach did not take into account ways to encourage homeowners to incorporate disaster risk reduction (DRR) measures. Housing interventions by various international agencies were uncoordinated and did not always follow technical guidance aimed at "building back better" (Oxfam 2011, 8). Lessons learned coming out of the shelter response in Pakistan include requiring organizations and donors to adhere to a shelter strategy that takes into account the diverse housing needs across the country, ensuring that shelter programs include ways to transfer DRR technical know-how to local populations, and standardization of cost-effective, minimum DRR and construction standards (IASC 2012).

The shelter response also did not take into account the protection needs of women—many shelter projects were cash-transfer based, meaning that funds for building shelter would go to men, who may or may not choose to focus on the reconstruction needs of women. In addition, six months into the disaster, there appeared to be few, if any, shelter assistance projects targeted at female-headed households or that offered additional shelter assistance needed by women and other vulnerable groups including labor, materials and technical advice (IOM 2010, 19).

In Colombia, the government took a much more proactive approach and prohibited certain communities from rebuilding. For example, in Atlántico and Bolívar Departments, the provincial governments undertook ambitious plans to relocate displaced populations away from flood-prone areas. But more than a year later, the government and contracting agencies were still in the process of securing land and building permanent shelters. The dismally slow pace of the relocation process meant that thousands of people remained displaced, living in poor conditions in temporary shelters designed to last three to six months (RI 2011b, 2). In addition, many communities were divided about government relocation plans because the selected sites were too far away, the houses and/or plots of land were too small, or the housing designs or construction materials were inadequate (RI 2012).

Understanding and identifying less visible impacts of floods on displacement and migration

Estimates of permanent displacement from Pakistan and Colombia floods

When Refugees International visited KP and Sindh Provinces a year after Pakistan's floods, only a few hundred of the very poorest individuals remained squatting near closed camps, either because they were indebted to their landlords or had nothing to return to (author interviews, October 2011). There is anecdotal evidence, however, that in Sindh, where poverty and malnutrition rates were alarmingly high prior to the floods (Oxfam 2011, 10), as many as several hundred thousand people did not return because they had found better opportunities in the towns and cities to which they fled. For example, some families and single young men who had evacuated to Karachi during the height of the floods reportedly chose to stay there. It can be presumed that, for them, urban life offered more economic opportunities, and freedom from the oppressive life of a peasant farmer. However, because the government did not recognize their rights as displaced persons to relocate, most were forced to squat in vacant buildings where they faced the threat of eviction, discrimination and lack of access to basic public services (Aslam 2011; RI 2011a).

In Colombia, the situation was similar. Eighteen months after the floods began, only a handful of the very poorest were still occupying the few shelters that had not been closed by the government. There were also a number of Colombians who were evacuated during the floods but chose not to return because the receiving areas offered better opportunities. For example, in the heavily affected district of Ayapel, hundreds of families living in flood-prone, marshland areas (*cienagas*) were evacuated to a vacant plot on the outskirts of town. Each family was given a sizeable plot of land for growing vegetables and raising small animals. Not surprisingly, many of them decided to stay after the floods subsided, citing more economic opportunities, more educational opportunities for their children, and no risk of recurrent floods. Others, however, did return. When one older fisherman was asked why he had chosen to return to his flooded home along the *cienaga*, he replied, "[w]e know only how to fish. What would we do in the town?" Lack of skills and fear of discrimination were other reasons cited by families who decided to return to flood-prone areas (author interview, March 2012).

In both Pakistan and Colombia, there are no available data on the number of people who migrated internationally due to the floods. In Colombia, there was some anecdotal evidence that the floods resulted in an increase in seasonal migration to large cities both within Colombia and abroad. For example, in the town of Sucre, also in the La Mojana region, it is customary for young women to migrate to cities both in Colombia and abroad to work as domestic help. Following the floods, out-migration of young women, including mothers, reportedly increased thereby providing an additional source of income to flood-affected households struggling to recover (author interview, March 2012).

The lack of evidence of out-migration from Pakistan and Colombia in the wake of the floods is perhaps due to the fact that there were no migration arrangements available to flood-affected persons for doing so. Temporary and circular migration arrangements have the potential to facilitate migration as a coping strategy in the aftermath of natural disasters. For example, after the Galeras volcano eruption in 2006, a Temporary and Circular Labor Migration (TCLM) program between Colombia and Spain was employed to provide a migration opportunity for thousands of affected Colombians (de Moor 2010, 5). This TCLM program offered a livelihood alternative through temporary work abroad to families affected by natural disasters. Unfortunately, the program was suspended in 2009, and there is limited evidence of its effectiveness (McLoughlin and Münz 2011, 31–32). Nonetheless, it serves as a useful model to the extent that it seeks to provide an additional source of income to families affected by disasters by maximizing the impact of remittances on the recovery of the affected area, and prioritizes the needs of the most marginalized populations among the rural communities (IOM n.d.). If replicated elsewhere, TCLM could provide a coping strategy for vulnerable populations in post-disaster situations and offer an alternative to permanent migration to urban slums, where displaced populations often lack access to protection and basic services (de Moor 2010).

In sum, in most instances, the lack of sufficient assistance to and protection of flood-displaced populations, even for those for whom displacement was temporary, resulted in long-term, significant, adverse impacts that left people worse off and with fewer available coping mechanisms. Most chose to return either because they had no choice or because they wanted to return to their land, communities and way of life. Some displaced, however, chose not to return, primarily because they encountered better opportunities in receiving areas, although those who migrated to cities were likely to face discrimination and lack access to secure shelter, basic services and protection.

Lessons going forward

Preventing and mitigating displacement

Both Colombia and Pakistan have made substantial improvements to their disaster management frameworks based on lessons learned from the floods. Nonetheless, the fact that natural disasters are affecting more and more people, and the likelihood that climate change will cause an increase in extreme weather events, requires more emphasis on the positive obligations of states to anticipate and take measures to prevent or mitigate conditions likely to bring about displacement and threaten human rights.

Improved physical and technical interventions—such as flood-resistant housing, better flood control infrastructure and land use planning, and improved early warning systems—need to be prioritized.

Equally important is the need for states to improve their laws and institutions to enhance disaster management capacity. Of critical importance is the need for governments to ensure that disaster management systems are implemented at the local and community level. During the flood disasters in Pakistan and Colombia, local governments and communities were not only the first responders, they were often the *only* responders. New procedures and institutions must also be put in place to strengthen accountability and oversight mechanisms, and allow greater input by civil society organizations and affected communities. In countries affected by both conflict and natural disasters, a recommended approach to DRM is to place responsibility for responding to both man-made and natural disasters within the same ministry or institution, as is the case in Pakistan, thereby building capacity, promoting accountability and maximizing allocation of resources.

International humanitarian agencies will also need to improve risk management and contingency planning, especially in climate-vulnerable countries. In natural disaster-prone countries, the UN should ensure that a detailed contingency plan has been developed in consultation with the UN Country Team (UNCT) and national and local governments, and that the capacity to implement that plan can be mobilized if necessary. Humanitarian agencies should work with national governments to improve information sharing and coordination mechanisms, and to clearly articulate roles and responsibilities. As experience from Pakistan and Colombia demonstrates, the failure of international actors to coordinate fully with national and provincial governments can significantly undermine an effective response when disaster strikes.

Financial investments in disaster prevention, preparedness and response must be significantly increased.[4] States must be encouraged to adopt improved laws and systems that require a certain percentage of money to be set aside for disaster preparedness in their budgets, including at the provincial and local levels. In low-income countries that are susceptible to climate-related disasters, donors must prioritize and increase funding for DRM and look for opportunities to share expertise and provide technical support.[5]

Increasing disaster preparedness also obligates states to act proactively to implement land use planning regulations that prevent human settlements including return to at-risk areas, as well as programs that take people out of harm's way. This may entail relocating populations living in at-risk areas (e.g. along rivers, in floodplains, on steep hillsides, or living in poorly constructed housing or urban slums). However, given the abysmal track record of government-led relocation projects, new relocation models will be needed that better protect affected populations and allow communities to drive their own relocation process.

Enhance protection of those displaced

Central to the protection of people displaced by natural disasters is the need for states to develop and implement rights-based disaster management frameworks that treat those affected or displaced by natural disasters as rights holders, not as beneficiaries of disaster relief. Such frameworks must include both procedures to protect vulnerable populations from specific risks, and accountability mechanisms to guard against discrimination (e.g. complaints mechanisms, ombudsmen's offices and legal recourse). National governments should be encouraged to develop national IDP legislation and implement regulations that extend the protections provided under the Guiding Principles to persons displaced by natural disasters.

Donor governments can enhance protection by supporting OCHA to strengthen its work on IDP protection in natural disasters, as well as clarifying internal roles and responsibilities for IDP protection. For example, the US Department of State and US Agency for International Development (USAID) should review and update current US IDP policy, and encourage more IDP training for their staff working with displaced populations in post-disaster situations.

The UN must also get its house in order and ensure clarity of roles and responsibilities for protection in natural disasters. One obstacle in this regard is the current

UN arrangement for determining which agency will lead the protection cluster in the case of natural disasters. Unlike in the case of conflict emergencies, where UNHCR is the designated protection cluster lead agency, in the case of natural disasters, the Resident Coordinator/Humanitarian Coordinator must decide on a case-by-case basis whether UNHCR, the Office of the UN High Commissioner for Human Rights (OHCHR) or the UN Children's Fund (UNICEF) will lead the protection cluster (Global Protection Cluster 2013). This can lead to delays and leadership being designated to an agency lacking sufficient operational humanitarian protection experience. While there has been a call for UNHCR to act as de facto protection cluster lead, there is not yet consensus on the issue. Regardless of which agency heads up the protection cluster, strong leadership must be prioritized. For example, the selected agency must designate a person to head the cluster on a full-time basis, rather than creating a part-time position for a person who retains responsibilities within his or her own agency (i.e. double-hatting). At the same time, protection must be strengthened across *all* clusters. The terms of reference for all Humanitarian Coordinators should include clearly outlined responsibilities over IDPs, including those displaced by disasters. Finally, donors must increase funding for the protection cluster, which often lags far behind other clusters.

Ensuring that early recovery programs are robust and start early

The slow pace of early recovery programs in both Pakistan and Colombia following the floods proved to be a substantial reason for prolonging displacement and exposing displaced populations to risk of harm. The lag in early recovery (as well as the focus on camps) also meant that populations who decided to return quickly, or remain in their flooded homes during the crisis, received little to no assistance for months. National governments, international agencies and donors must do far more to ensure that early recovery programs are funded and implemented *as early on in the response as possible*. Coordination mechanisms for response activities must be more closely aligned and integrated with early recovery programs.

Improved shelter strategies

In both Pakistan and Colombia, one of the largest impacts of the floods—and the greatest cause of prolonged displacement—was the destruction of housing. In acute floods such as these, housing demands will be impossible to meet under any circumstances. Nonetheless, experience from Pakistan and Colombia (and elsewhere) starkly highlights the need for both national governments and international agencies to improve plans and strategies for providing shelter to populations displaced by natural disasters. Local governments charged with maintaining camps and temporary shelters (such as schools) must adopt improved shelter management measures and require staff training to ensure a safe and secure environment for displaced populations, and to take into account the unique protection needs of vulnerable populations. Laws must be put in place that require money to be set aside in local budgets for this purpose.

More broadly, both national governments and international agencies should develop more flexible, streamlined shelter strategies that take into consideration the quick rate of return in post-flood situations and that shift the focus away from camps to providing displaced populations with more permanent shelter assistance back in their home areas. In addition, national governments must coordinate closely with international agencies and other actors to ensure that the shelter response is coordinated, that the most vulnerable populations are prioritized, that limited resources are maximized by emphasizing cost-effective housing designs and that, where appropriate, cost-effective DRR elements are incorporated into technical standards. The rebuilding effort should not prioritize those with secure land rights but rather use the crisis as an opportunity to address underlying land and property inequities and protect against land dispossession.

Facilitating migration

In addition to increasing financial and technical assistance to climate vulnerable countries, the international community must seek to implement arrangements to promote temporary or permanent migration as a strategy to adapt to a changing environment, and to increase resilience of disaster-affected populations. Facilitating international migration for persons affected by natural disaster and climate change can prevent them from being forcibly displaced in already overpopulated and environmentally fragile places within their own region. Temporary international migration could also provide an alternative to permanent and urban migration, and may mitigate pressure on overcrowded urban areas. Through earning a livelihood abroad, migrants might also reduce the vulnerability of their communities of origin, so as to cope better with future environmental disruptions.

Notes

1   The information contained in this report is based on both desk research and field visits by Refugees International (RI) staff to Pakistan in September 2010 and July–August 2011, and to Colombia in March 2011 and February 2012. During field visits, RI visited affected areas and conducted confidential interviews with affected individuals, and local and international stakeholders For more information on RI's findings, please see Thomas and Rendon 2010; RI 2011a, 2011b, 2012.
2   Flooding also results in significant financial costs. Total losses in exceptional years such as 2008 and 2010 exceeded $40 billion (GFDRR 2012).
3   In April 2012, Colombia adopted a new disaster management law (Act 1523), which replaced SNPAD with the "National System for Disaster Management." For a discussion, see IFRC (2012a, 6).
4   For a comprehensive analysis of DRR as a percentage of humanitarian assistance, see Kellett and Sparks (2012).
5   Between 2000 and 2009, Pakistan received more than US$1 billion for DRR, the highest of any of the top 40 humanitarian recipient countries over the decade. However, only 2 percent of this (US$16 million) was directly related to flooding (Kellett and Sparks 2012).

References

Albuja, S. and Cavelier-Adarve, I. (2011) "Protecting People Displaced by Disasters in the Context of Climate Change: Challenges from a Mixed Conflict/Disaster Context," *Tulane Environmental Law Journal*, 24, available online at: http://www.law.tulane.edu/uploadedFiles/Tulane_Journal_Sites/Tulane_Environmental_Law_Journal/Vol.24_2.pdf.

Arif Riar, M. (2012) "Flood Warning Issued," DAWN, available online at: http://dawn.com/2012/05/17/flood-warning-issued/.

Aslam, S. (2011) "Pakistani Flood Victims Demand Relief," DEMOTIX, available online at: http://www.demotix.com/news/774520/pakistani-flood-victims-demand-relief#media-774508.

Cochrane, H. (2008) "The Role of the Affected State in Humanitarian Action: A Case Study of Pakistan," London, Working Paper, Overseas Development Institute, available online at: http://www.odi.org.uk/sites/odi.org.uk/files/odi-assets/publications-opinion-files/3417.pdf.

CRED (Centre for Research on the Epidemiology of Disasters) (n.d.) "Emergency Events Database, Natural Disasters Trend (EM-DAT)," *Numbers of Natural Disasters Reported 1900 to 2011*, available online at: http://www.emdat.be/natural-disasters-trends.

DAWN (2012) "Khyber Pakhtunkhwa Braces for Mild Flooding," DAWN, available online at: http://dawn.com/2012/06/19/khyber-pakhtunkhwa-braces-for-mild-flooding-2/.

de Moor, N. (2010) *Temporary Labour Migration for Victims of Natural Disasters: The Case of Colombia*, Hohenkammer: Universiteit Gent, available online at: http://www.ehs.unu.edu/file/get/5403.

ECOSOC (UN Economic and Social Council) (1998) "Guiding Principles on Internal Displacement," July 22, E/CN.4/1998/53/Add.2, available online at: http://www.refworld.org/docid/3c3da07f7.html.

Ejercito Nacional de Colombia (2010) *Army Does Repair Work at Canal del Dique*, Barranquilla: Ejercito Nacional de Colombia, available online at: http://www.ejercito.mil.co/?idcategoria=268698.

Engeler, E. (2010) *Mass Communications Programme Talks and Listens to Pakistan's Flood Victims in Migration*, Geneva: IOM, available online at: http://publications.iom.int/bookstore/free/Migration_Winter2010_EN.pdf.

FAO (Food and Agriculture Organization of the United Nations) (2011) "In Flood-stricken Pakistan, Good Wheat Harvest is Expected," available online at: http://www.fao.org/news/story/en/item/54043/icode/.

GFDRR (Global Facility for Disaster Reduction and Recovery) (2012) *Cities and Flooding: A Guide to Integrated Urban Flood Risk Management for the 21st Century*, Washington, DC: GFDRR, available online at: http://www.gfdrr.org/sites/gfdrr.org/files/urbanfloods/pdf/Cities%20and%20Flooding%20Guidebook.pdf.

Global Protection Cluster (2013) "Protection in Natural Disasters," available online at: http://www.globalprotectioncluster.org/en/areas-of-responsibility/protection-in-natural-disasters.html.

Government of Colombia (2012) *Official Government Floods Report, No. 21*. Bogota: OCHA, available online at: http://reliefweb.int/report/colombia/21%C2%B0-reporte-oficial-delluvias.

IASC (Inter-Agency Standing Committee) Pakistan Floods Shelter Cluster (2012) "Draft Lessons Learned from Early Recovery Shelter 2010 Response," IASC.

IDMC (Internal Displacement Monitoring Centre) and NRC (Norwegian Refugee Council) (2010) "Colombia Government Response Improves But Still Fails to Meet Needs of Growing IDP Population," IDMC and NRC, available online at: http://www.internal-displacement.

org/8025708F004BE3B1/(httpInfoFiles)/4BCA7DF31521CC16C12577F5002CED98/$file/Colombia_Overview_Dec2010.pdf.

IDMC and NRC (2011) "Research Findings and Recommendations: Flood-displaced Women in Sindh Province," IDMC and NRC, available online at: http://floods2010.pakresponse.info/LinkClick.aspx?fileticket=6PQiEWUqE6A%3D&tabid=208&mid=1776.

IDMC and NRC (2013) "Global Estimates 2012: People Displaced by Disasters," available online at: http://www.internal-displacement.org/publications/2013/global-estimates-2012-people-displaced-by-disasters.

IDMC and OCHA (United Nations Office for the Coordination of Humanitarian Affairs) (2009) "Monitoring Disaster Displacement in the Context of Climate Change," available online at: http://www.internal-displacement.org/publications/2009/monitoring-disaster-displacement-in-the-context-of-climate-change.

IFRC (International Federation of Red Cross and Red Crescent Societies) (2012a) "A Study on Legal Preparedness for International Disaster Assistance in Colombia: Towards the Application of the IDRL Guidelines—Summary Version," available online at: http://www.ifrc.org/FedNet/Resources%20and%20Services/IDRL/IDRL%20reports/IDRL%20in%20Colombia%20-%20Summary%20version.pdf.

IFRC (2012b) "Two-year Consolidated Report: Pakistan: Monsoon Flash Floods," December 20, available online at: http://www.ifrc.org/docs/Appeals/10/MDRPK006%202YR.pdf.

IOM (International Organization for Migration) (n.d.) "Enhancing Development in Colombia through Temporary and Circular Labour Migration to Spain," available online at: http://www.iom.int/cms/en/sites/iom/home/what-we-do/labour-migration/enhancing-development-in-colombi.html.

IOM (2010) "Pakistan Flood Response 2010, Shelter Cluster Evaluation," internal report.

IPCC (Intergovernmental Panel on Climate Change) (2012) *Special Report on Managing the Risks of Extreme Events and Disasters to Advance Climate Change Adaptation*, New York: Cambridge University Press.

Islamic Relief (2011) "Flooded and Forgotten: The Ongoing Crisis Threatening Lives and Livelihoods in Pakistan," available online at: http://www.curtisresearch.org/FINAL.001_Flooded_and_Forgotten.pdf.

Kellett, J. and Sparks, D. (2012) "Disaster Risk Reduction: Spending Where it Should Count," Global Humanitarian Assistance, available online at: http://www.globalhumanitarianassistance.org/wp-content/uploads/2012/03/GHA-Disaster-Risk-Report.pdf.

Koskinen, P. (2006) "Human Rights-based Approach to Humanitarian Assistance: A Tool to Empower Internally Displaced Women?" International Conference on Refugees and International Law: The Challenge of Protection, December 15–16, Oxford, Refugee Studies Centre, available online at: http://www.cihc.org/members/resource_library_pdfs/2_Law_and_Protection/2_6_Rights_based_Assistance/HR_approach_to_Humanitarian_Assistance.pdf.

Mannon, T. (2011) "Colombia Not Prepared for New Rainy Season," Colombia Reports, available online at: http://colombiareports.com/colombia-news/news/18464-colombia-not-prepared-for-new-rainy-season.html.

McLoughlin, S. and Münz, R. (2011) "Temporary and Circular Migration: Opportunities and Challenges: Working Paper No. 35," European Policy Centre, available online at: http://www.epc.eu/documents/uploads/pub_1237_temporary_and_circular_migration_wp35.pdf.

Moloney, A. (2012) *Colombia Flood Victims at Risk As Rainy Season Looms*, Bogota: Alertnet, available online at: http://www.trust.org/alertnet/news/colombia-flood-victims-at-risk-as-rainy-season-looms.

NDMA (National Disaster Management Authority), Government of Pakistan (2011) *Pakistan 2010 Flood Relief: Learning from Experience: Observations and Opportunities*, Islamabad: NDMA, available online at: http://reliefweb.int/sites/reliefweb.int/files/resources/F_R_11.pdf.

OCHA (United Nations Office for the Coordination of Humanitarian Affairs) (2010) "Emergency Response Fund: Colombia," available online at: https://docs.unocha.org/sites/dms/Documents/Colombia%20ERF%202010.pdf.

OCHA (2011a) "Colombia Floods 2010–11, Situation Report No. 40," available online at: http://www.colombiassh.org/site/IMG/pdf/Inundaciones_Colombia_Sit_Rep_40.pdf.

OCHA (2011b) "Colombia Floods 2010–11, Situation Report No. 19," available online at: http://www.colombiassh.org/site/IMG/pdf/OCHA_Colombia_SitRep_19_-_Floods_final_v2.pdf.

OCHA (2011c) "Colombia Floods 2010–11, Situation Report No. 5," available online at: http://reliefweb.int/sites/reliefweb.int/files/resources/Sit_Rep_05_29.12.2011.pdf.

OCHA (2011d) "Pakistan Floods Relief and Early Recovery Plan," available online at: https://docs.unocha.org/sites/dms/CAP/Revision_2010_Pakistan_FRERRP_SCREEN.pdf.

OCHA (2011e) "Colombia: Humanitarian Trends January–June 2011," available online at: http://reliefweb.int/report/colombia/humanitarian-trends-january-%E2%80%93-june-2011.

OCHA (2011f) "Pakistan Floods: Timeline of Events (as of 21 of July 2011)," available online at: http://reliefweb.int/sites/reliefweb.int/files/resources/map_564.pdf.

Oxfam (2011) "Ready or Not: Pakistan's Resilience to Disasters One Year On from the Floods," available online at: http://reliefweb.int/sites/reliefweb.int/files/resources/pakistan-ready-or-not.pdf.

Polastro, R., Nagrah, A., Steen, N. and Zafar, F. (2011) "Inter-agency Real Time Evaluation of the Humanitarian Response to Pakistan's 2010 Floods Crisis," DARA, available online at: http://daraint.org/wp-content/uploads/2011/03/Final-Report-RTE-Pakistan-2011.pdf.

RI (Refugees International) (2011a) *Pakistan: Flood Survivors Still Struggling to Recover*, Washington, DC: Refugees International, available online at: http://www.refugeesinternational.org/policy/field-report/pakistan-flood-survivors-still-struggling-recover.

RI (2011b) *Surviving Alone: Improving Assistance to Colombia's Flood Victims*, Washington, DC: Refugees International, available online at: http://www.refugeesinternational.org/policy/in-depth-report/surviving-alone-improving-assistance-colombias-flood-victims.

RI (2012) *Colombia: Flood Response Improves, But Challenges Remain*, Washington, DC: Refugees International, available online at: http://refugeesinternational.org/sites/default/files/032712_Colombia_Response%20letterhead.pdf.

Thomas, A. (2012) *Poisoned Climate: Still Submerged in Colombia*, Washington, DC: Refugees International, available online at: http://www.refugeesinternational.org/blog/poisoned-climate-still-submerged-colombia.

Thomas, A. and Rendon, R. (2010) *Confronting Climate Displacement: Learning from Pakistan's Floods*, Washington, DC: Refugee International, available online at: http://refugeesinternational.org/sites/default/files/ConfrontingClimateDisplacement.pdf.

UNHCR (UN High Commissioner for Refugees) (2011a) Protection Thematic Working Group, "RapidProtectionAssessmentReportPakistanFloods2011," available online at: http://pakresponse.info/LinkClick.aspx?fileticket=S2FrJAwh35U%3D&tabid=112&mid=780.

UNHCR (2011b) "UNHCR Steps Up Assistance to Pakistan Flood Victims," available online at: http://www.unhcr.org/4e788b759.html.

United Nations News Centre (2010) "Pakistan: UN Issues Largest-ever Disaster Appeal at Over $2 Billion for Flood Victims," available online at: http://www.un.org/apps/news/story.asp?NewsID=35983&Cr=pakistan&Cr1,last%20accessed%20September%2021,%2 02012#.USva6utetM4.

Warraich, H., Zaidi, A.K.M. and Patel, K. (2010) *Bulletin of the WHO: Floods in Pakistan: A Public Health Crisis*, World Health Organization, available online at: http://www.who. int/bulletin/volumes/89/3/10-083386/en/index.html.

WFP (World Food Program) (2011) "Market Monitoring Bulletin, January–April," available online at: http://documents.wfp.org/stellent/groups/public/documents/ena/ wfp246214.pdf.

White, S. (2011) *The 2010 Flooding Disaster in Pakistan: An Opportunity for Governance Reform or another Layer of Dysfunction?* Washington, DC: Center for Strategic and International Studies, available online at: http://csis.org/publication/2010-flooding-disaster-pakistan.

World Bank and GFDRR (Global Facility for Disaster Reduction and Recovery) (2013a) "Colombia Dashboard: Overview," available online at: http://sdwebx.worldbank.org/ climateportalb/home.cfm?page=country_profile&CCode=COL.

World Bank and GFDRR (2013b) "Pakistan Dashboard: Natural Hazards," available online at: http://sdwebx.worldbank.org/climateportalb/home.cfm?page=country_ profile&CCode=PAK&ThisTab=NaturalHazards.

Yonetani, M. (2011) "Displacement Due to Natural Hazard-induced Disasters: Global Estimates for 2009 and 2010," IDMC and NRC, available online at: http://reliefweb.int/ sites/reliefweb.int/files/resources/Full_Report_1079.pdf.

Zafar, N. (2011) "Women and Children Brave Death and Disease: Pakistan Floods 2011," Oxfam, available online at: http://blogs.oxfam.org/en/blog/11-09-23-pakistan-women-children-brave-death-disease.

4

RECURRENT ACUTE DISASTERS, CRISIS MIGRATION

Haiti has had it all

Elizabeth Ferris

Introduction and overview

There is probably no country in the world where migration has been so directly linked to recurrent crises as Haiti. Political instability, poverty, environmental degradation, disasters and the intersections between them have led Haitians to leave their homes in search of protection and survival almost since the country's independence in 1804. Most of these movements have been internal, but large numbers of Haitians, over a long period of time, have sought safety and survival in nearby countries, particularly the Dominican Republic and the United States. In fact, efforts to limit potential Haitian migration flows have been a key determinant in shaping international policies toward Haiti.

Haitian migration, particularly during the past four decades, illustrates the complexity and interconnections between different drivers of migration. When forced—and able—to leave their country either because of a disaster or a political crisis, Haitians have tended to follow established labor migration routes. This seems to be a global pattern in which refugee flows tend to follow existing economic migration patterns. The ability to make it to a given destination country has depended on the policies of the would-be host country. In this respect, Haitians—whether fleeing the effects of a natural disaster or a political conflict—have found limited choices. Both the United States and the Dominican Republic have often closed their borders in response to the perceived threat of large numbers of Haitians arriving on their territories. With Haiti, there are different patterns of movement resulting from acute disasters and from conflicts. International migration has increased most dramatically in response to political events. The major disasters of the last decade—the storms of 2008 and the earthquake of 2010—did not lead to an increase in the number of Haitians taking to their boats for the United States. In these contexts, most of those affected were displaced elsewhere within Haiti.

The case of Haitian migration in response to recurrent acute disasters thus may offer lessons for other situations in the future. This chapter begins, as all accounts of Haiti must, with an acknowledgment of the country's volatile political history and corresponding economic poverty and social inequality. These are the factors that have made Haitian history a series of recurrent crises. The chapter then discusses the political crises of the 1990s, the hurricanes of the 2000s, and the earthquake and cholera epidemic of 2010, with a particular emphasis on the way in which Haitians have used migration as a response to crises. The chapter concludes by drawing out the lessons of the Haitian migration experience for understanding crisis migration more generally.

The sad historical context

Most observers of Haiti begin with a lament for the country's tortured political history, its economic poverty and a centuries-long record of foreign intervention in its affairs. While at the time of its independence, Haiti was the wealthiest French colony in the Americas, within a hundred years it had become the poorest country in the hemisphere.

Political stability proved elusive in Haiti's first tumultuous decades of existence, with repeated interventions by foreign interests. Setting a pattern that would endure for three centuries, Haitians migrated in response to the political turmoil. For example, following independence, many white Haitians and their slaves went to Cuba. In 1809 some 10,000 Haitians went to New Orleans, doubling that city's population when they were expelled from Cuba by the Spanish authorities (In Motion 2005). Political intervention was tied to foreign economic interests. As the price for recognizing its independence, Haiti was forced to pay France for its loss of slave profits—debts that were not paid off until 1947 (*Timelines of History* n.d.). Foreign interests financed Haitian political factions, which competed with each other for power. In 1915, the US Marines occupied the country to protect US economic interests—an occupation that was to last until 1934. During that time, the Marines ran the government, developed infrastructure throughout the country, built an army and formalized the boundary between the Dominican Republic and Haiti. Also during the occupation, Haiti borrowed extensively from the National City Bank of New York (in order to pay its indemnity to France) and vast sums of money were transferred from Haiti to New York to repay these obligations. Meanwhile Haitians migrated to other Caribbean countries, particularly to the Dominican Republic, in search of work—and survival. Haitian elites sought protection by temporarily moving out of the country when their political factions were out of power (primarily to the United States, France and Canada) while Haitians without economic means migrated to the Dominican Republic and Cuba to work in the cane fields. For both the elites and poor agricultural laborers, migration was often temporary and circular, but many Haitians ended up putting down roots in their host countries, creating large Haitian diaspora communities.

Haitian agricultural migrants in the region often worked in appalling conditions and were victims of repressive actions by the governments that hosted them. For example, 25,000 Haitians were rounded up and deported from Cuba in the space of a few months in 1937 (Ferguson 2003, 6). Around the same time, Rafael Trujillo, dictator in neighboring Dominican Republic, ordered a campaign against Haitian migrant workers in that country, culminating in the massacre of some 15–20,000 Haitians in 1937. While migration represented a survival strategy for tens—perhaps hundreds—of thousands of Haitians over the years, migrants were often exploited and vulnerable.

François Duvalier (Papa Doc) was elected president of Haiti in 1957, ruling until 1971 when his son François-Claude (Baby Doc) Duvalier came to power and stayed until 1986. The Duvalier years were marked by the use of a brutal para-military force, the *tontons macoutes*, who terrorized the population into submission (an estimated 30,000 Haitians were killed for political reasons during Papa Doc's tenure) as well as economic mismanagement and corruption on all levels, from the Duvaliers themselves down to rural section chiefs.

While it is impossible to sum up 150 years of Haiti's history in a few paragraphs, certain common themes stand out. Migration—crisis migration—has been part of Haiti's experience since its independence. Political instability, repression and poverty have led Haitians to see migration as a way of escaping adverse conditions. Tens of thousands of Haitians sought work opportunities in neighboring Dominican Republic and nearby Caribbean countries. Those with economic means left for the United States (and later for Canada). Political repression and foreign intervention went hand in hand with economic exploitation as a small elite controlled the country's economy.

In addition to the political turmoil, increasing inequality and economic poverty, Haiti has been vulnerable to natural hazards since the colonial period, particularly hurricanes and earthquakes. In 1751, the city of Port-au-Prince was destroyed by an earthquake (US Geological Service 2010). In 1780, a hurricane killed more than 22,000 people in the Caribbean (*Timelines of History* n.d.). In 1842, Cap-Haïtien and other Haitian cities were destroyed by earthquakes (*Science Daily* n.d.). In 1946 an 8.1 magnitude earthquake struck the island of Hispaniola, producing a tsunami and killing 1,790 people (*Telegraph* 2010). In 1954, Hurricane Hazel destroyed 40 percent of the country's coffee crop and killed 100 people. In 1963, Hurricane Flora killed between 5,000 and 8,000 people and again wiped out the coffee crop (*This Day in History* 1996–2013). Hurricane Gordon killed more than 1,000 in 1994 (*Telegraph* 2010).

As many have pointed out—and as becomes even clearer in looking at Haiti's more recent experience—the disastrous effects of natural hazards in Haiti are largely the result of poverty and poor governance. Most obviously, charcoal is the main source of energy for Haiti's eight million people, which has led to massive deforestation. Deforestation increases the risk of flooding and landslides when storms occur. In 1980, Haiti reportedly still had 25 percent of its forests, allowing the country to get through Hurricane Emily in 1993 without loss of life. But, by

2004, less than 1.4 percent of the forest remained, which meant that tropical storms Jeanne and Gordon (which were not even hurricanes) killed thousands. Moreover, Haiti's extreme vulnerability to flooding, as a result of deforestation and the lack of mitigating measures, meant that it did not even take a tropical storm for deaths to occur. In May 2004, three days of heavy rains from a tropical depression resulted in massive flooding and landslides that killed 2,600 people (Masters n.d.). In comparison, neighboring Dominican Republic, which still has about 25 percent of its tree cover as well as higher economic development and lower population density than Haiti, has much lower disaster casualty rates even though geographically it is equally vulnerable to natural hazards (Carroll 2008).

The combination of poverty and poor governance in Haiti not only meant that more people died from disasters, but that recovery has been slow. As will be discussed below, recovery from the 2008 hurricanes was made more difficult by price hikes in food and fuel, which had taken place in the months before the hurricanes hit. People were already struggling to get by; the effects of the hurricanes were to make it even more difficult for people to eke out an existence (Environment News Service 2008).

The political crises of the 1990s

Migration

It has been almost thirty years since the Duvalier dictatorship came to an end; these three decades have been turbulent ones. A deadly combination of political instability, repression and chaos coupled with economic poverty and recurring natural hazards created a situation where the majority of Haiti's eight to ten million people lived in precarious and vulnerable conditions. In this context, displacement and migration continued to be survival strategies.

While the number of Haitians migrating (both short and long term) to the Dominican Republic was far higher than those seeking to enter the US, US fears of large-scale Haitian migration began to surface in the 1980s. Although Haitians probably emigrated in large numbers to the US before that date, the first reports of Haitians arriving by boat and without documentation stem from 1972. The US Immigration and Naturalization Service (INS) reported that more than 55,000 Haitian "boat people" arrived in Florida between 1972 and 1981, although also cautioning that this figure probably represented only about half of those who entered and that the actual number was likely to have exceeded 100,000 (Haggerty 1989).

In 1981, US President Reagan responded to the flow of Haitian migrants by creating an interdiction program in which the Coast Guard was ordered to intercept vessels carrying undocumented aliens and return them to their point of origin. "The Haitian government at that time agreed to the interdiction program and promised not to prosecute returnees" (Zimmerman 1993, 385). This policy was justified by the Reagan administration in security terms, emphasizing that illegal immigration had become a "serious national problem detrimental to the United States" (Zimmerman 1993, 392). While the interdiction program was based on the assumption that the Haitians leaving their country were doing so for economic reasons, there was a provision that those deemed to qualify for refugee status would not be sent back (Zimmerman 1993, 393f.). From 1981 through 1990, 22,940 Haitians were interdicted at sea. Of this number, INS considered 11 Haitians qualified to apply for asylum in the United States (Wasem 2005, 2f.).

Following the coup of 1991, Haitians were displaced both internally and internationally. Human Rights Watch reports that some 300,000 people were internally displaced, as part of a deliberate strategy by the Haitian military to neutralize opposition to its regime. By keeping people on the move, their capacity to organize and to challenge the regime would be limited (Human Rights Watch 1994).

Haitians fled the country in large numbers, triggering various policy shifts on the part of the United States. It was more difficult for the US—which after all had condemned the coup, as had the Organization of American States—to make the case that Haitians were simply economic migrants and could be safely returned to Haiti. This led to some of the most bizarre experiments in US immigration history: over the course of a few years, US policy included interdictions and forcible returns, the search for other countries to be temporary safe havens, screening on Coast Guard ships and later processing at the US Naval Base at Guantanamo, in-country processing of Haitians fearing persecution, and special parole provisions. Approximately 10,490 Haitians were paroled into the United States in 1991/92 after a prescreening interview at Guantanamo determined that they had a credible fear of persecution if returned to Haiti. But the numbers continued to increase and on May 24, 1992, citing the surge in the number of Haitian arrivals, then-president Bush ordered the Coast Guard to intercept all Haitians in boats and immediately return them without interviews to determine if they were at risk of persecution (Wasem 2005, 3f.). Those who eventually were allowed to remain in the United States faced discrimination and uncertainty; in fact it was not until 1998 that Congress passed the Haitian Refugee Immigration Fairness Act, which enabled Haitians to become legal permanent residents—like others who were paroled into the US and recognized as refugees.

As demonstrated in Figure 4.1, the spikes in numbers of Haitians interdicted by the US Coast Guard (presumably also an indication of the number of Haitians leaving the country) corresponded to political developments in the country, peaking after the 1991 coup that deposed President Aristide, and again in 1994 when Aristide's return was preceded by a particularly turbulent time. It is interesting to note the somewhat different pattern in Haitian immigrants admitted to the US or granted legal residency. For example, the number arriving by legal channels spiked in 1991—the year before the surge in the number of interdictions, suggesting that those with the means to do so left the country before Aristide's overthrow. Interestingly the number of immigrants also increased in 1988, perhaps in response to the coup, and again at the time of Aristide's election, perhaps indicating concern with the consequences of a government led by populist Aristide.

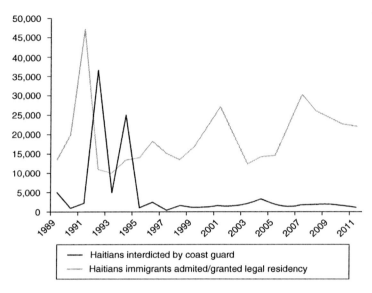

FIGURE 4.1   Haitian migration to the US, 1989–2010

The Haitian diaspora

One of the consequences of Haiti's long-standing history of international migration is the existence of a large and politically active diaspora. About 10 percent of Haiti's population lives outside Haiti (Bhargava, Docquier and Moullan 2011) although some sources estimate the percentage is as high as 25 percent. As is evidenced in Table 4.1, there are large numbers of Haitians living in the United States. Fagen *et al.* report figures of 400,000 in New York, 350,000–400,000 in South Florida and an additional 50,000–100,000 in Boston (Fagen *et al.* 2009, 19, 32, 39). Some reports put the number of Haitians in the US at more than two million. A study by the National Organization for the Advancement of Haitians estimates that of the 2,023,000 Haitians living in the US, 60 percent were US born, 19.4 percent were naturalized citizens, 19.9 percent residents and 1 percent had illegal status (International Crisis Group 2007, 3). In addition, an estimated 102,000 lived in Canada in 2006 (an increase from 82,000 in 2001) (Fagen *et al.* 2009, 32), between 500,000 and 1 million in the Dominican Republic, and 60,000 in France (Ferguson 2003). Most statistics on the Haitian diaspora are rough estimates and many numbers are more than a decade old; Table 4.1 is thus offered as a general indication only and the figures therein should be treated with caution.

Large numbers of Haitians, possibly up to one million, live in the Dominican Republic, many without documentation (Ferguson 2003, 8). The Haitians have faced particular problems with discrimination and statelessness, and most do not have work permits or valid visas. In many cases, the children of Haitian migrants have been unable to obtain residency papers, preventing them from attending school or

TABLE 4.1 Estimates of Haitian diaspora in selected countries

Country	Year of estimate	Estimate[1]
US	2007–10	400,000–2 million[2]
Canada	2006	102,000[3]
France	n/a (before 2003)	80,000[4]
Dominican Republic	1999, 2001	500,000–1,000,000
Bahamas	n/a (before 2003)	80,000
Guyana	n/a (before 2003)	30,000–40,000
Guadeloupe	n/a (before 2003)	15,000
St. Martin	n/a (before 2003)	15,000
Turks and Caicos Islands	n/a (before 2003)	10,000
Martinique	n/a (before 2003)	5,000

Source: author
1 All estimates, except those with separate footnotes, from Ferguson (2003).
2 See Terazzas (2010); Fagen et al. (2009); ICG (2007).
3 Fagen et al. (2009).
4 Haitian diaspora (n.d.).

holding jobs, leaving them essentially stateless (*Bloomberg Businessweek* 2012). The situation of Haitians in the Dominican Republic has received considerable attention from the international human rights community over the years, with UN reports comparing it to modern-day slavery (General Assembly, 2008).

Remittances from the US to Haiti are estimated at between $1.5 and 2 billion annually, almost 20 percent of Haiti's GDP (Ratha 2010), and have become an important survival strategy for Haitians facing poverty, unemployment and disasters. In fact, as Fagen et al. note, "Haiti is estimated to be the world's most remittance-dependent country as measured by remittances' share of household income and of GDP" (Fagen et al. 2009, 17). Moreover, Haitian associations abroad mobilize collective remittances that provide important social services, often substituting for state action (Fagen et al. 2009, 41, 48). Remittances tend to increase following disasters and after Hurricane Jeanne in 2004, affected communities in Gonaïves depended on remittances for survival. But, as Fagen has shown, they yielded only small improvements in the quality of life and were unable to decrease long-term poverty or address the larger problems confronting the country (Fagen 2006).

The natural disasters of the 2000s

The 2008 hurricanes

> The whole country is facing an ecological disaster. We can't keep going on like this. We are going to disappear one day. There will not be 400, 500 or 1000 deaths. There are going to be a million deaths.
>
> (Haitian Prime Minister, Michèle Pierre-Louis,
> after the 2008 hurricanes, in Carroll 2008)

The 2008 hurricane season was a cruel one for Haiti. Within thirty days, four storms—Fay, Gustav, Hanna and Ike—dumped heavy rains on Haitian hillsides stripped bare of virtually all of their forest cover by deforestation. Flood waters surged over large areas of the country. Particularly hard hit was Gonaïves, the country's fourth largest city. The rains from the year's four storms killed 793, left 310 missing, injured 593, displaced 151,072 persons, destroyed 22,702 homes and damaged another 84,625. More than 800,000 people were affected—8 percent of Haiti's total population (USAID 2008). The floods wiped out 70 percent of Haiti's crops, resulting in the deaths of dozens of children due to malnutrition in the months following the storms. Damage and losses were estimated at around $900 million, the economically most costly disaster in the country in more than a century. A joint assessment by the World Bank, the UN, the European Commission and other partners showed that losses were equivalent to about 15 percent of Haiti's GDP (World Bank 2008). In comparison, Cuba, which was hit by three of the four same hurricanes that devastated Haiti and which also sustained major economic damage, reported only seven casualties (*USA Today* 2008).

The devastation from the storms was tremendous but their impact had as much to do with the political and economic situation in the country as with the force of their winds. The World Bank's Country Social Analysis (CSA) for Haiti in 2004 reported that half of all Haitians were living in extreme poverty, 40 percent of the population was illiterate, and 80 percent lacked access to clean water. Moreover, the country had experienced very rapid urbanization, with some 40 percent of Haitians living in urban areas in 2003, compared with 25 percent in 1982. Although the unemployment rate in the metropolitan area already stood at 49 percent, the Bank reported that 100,000 new job-seekers were entering the urban labor market every year, far more than can be absorbed. In terms of security, there were more private security contractors than police in Haiti and the country was recognized as having one of the world's weakest police forces. The CSA concluded that:

> very poor urban neighborhoods are explosive points of conflict in the country's development crisis, combining demographic, socioeconomic, institutional, and political risk factors. Violence and insecurity in the Port-au-Prince slums in particular have undermined the political process, fuelled conflict, and negatively affected development and reconstruction efforts.
>
> *(World Bank 2006, viii)*

In April 2008, rising food and fuel prices had led to economic hardships for Haitians, and to protests in which stores were looted and buildings set afire. UN Security Forces fired into mobs; five Haitians and one UN peacekeeper were killed. Following the violence, Haitian Prime Minister Jacques-Edouard Alexis received a no-confidence vote in parliament and was eventually succeeded by Michèle Pierre-Louis (Perito 2008, 2). Persistent poverty, political instability and widespread deforestation intensified the impact of the four storms.

The international community was slow to respond to the disastrous storms in Haiti (Schneider 2008), with only $45 million of the UN's appeal for $108 million funded by late October 2008. Although members of Congress called on the US government to grant Temporary Protected Status (TPS) to Haitians currently residing in the US, the administration refused to do so, fearing it would create a magnet effect. However, the US government did suspend deportations of Haitians after the storms (Perito 2008, 3).

In May 2009, former US president Bill Clinton was named by the UN Secretary-General to be a special envoy on Haiti (BBC News 2009). He clearly signaled that the international community had a responsibility to address the long-term structural issues confronting Haiti, emphasizing that:

> [l]ast year's natural disasters took a great toll, but Haiti's government and people have the determination and ability to build back better, not just to repair the damage done but to lay the foundations for the long-term sustainable development that has eluded them for so long.
>
> *(BBC News 2009, no page)*

Clinton tried to engage businesses to invest in Haiti by hosting a trade delegation in September 2009 and also appointed the renowned humanitarian Paul Farmer as his deputy (Office of the Special Envoy for Haiti 2009a, 2009b). In 2009 prospects seemed more promising as the economy was growing, and many high-level visitors, from the UN Secretary-General to US Secretary of State Hillary Clinton, visited the country, pledging their support. In April 2009, an international donors' conference, hosted by the Inter-American Development Bank, reaffirmed previous commitments and pledged $324 million in new economic assistance. The UN commissioned a report by economist Paul Collier on how to achieve sustainable development progress in Haiti (Perito 2009). But, long before these plans could be brought to fruition, a much worse disaster occurred in the country.

The earthquake, hurricane, disease and internal displacement

The 7.0 earthquake that struck Haiti on January 12, 2010 left some 223,000 people dead, 300,000 injured and 2 million homeless. Centered in Leogane, near Port-au-Prince, the earthquake killed nearly 30 percent of the country's civil servants, destroyed most of the government's ministries, and knocked out the capital's airport and most of the city's communications infrastructure. By any measure, the Haitian earthquake was a mega-disaster. As most observers note, the effects of the earthquake were more severe because of the country's precarious economic situation, its history of bad governance and its limited governmental capacity. The fact that the earthquake struck so close to a major urban center where most of the population lived in poverty meant—as is always the case—that people who were already poor and marginalized were most affected. Moreover, the earthquake occurred at a time when the political situation was particularly volatile, as 2010 was an election year marked by a crowded political field and widespread protests.

The international community mobilized massively and rapidly to respond to the earthquake. UN and other international agencies deployed staff, thousands of non-governmental organizations (NGOs) rushed to the scene, military personnel were quickly deployed and some $3 billion was pledged in relief and recovery efforts. Indeed almost two-thirds of all international funds mobilized for natural disaster response in 2010 went to Haiti (Ferris and Petz 2011).

Displacement was massive. Haitians responded immediately to the destruction of their homes and their fears of further tremors by setting up spontaneous settlements. The international humanitarian community tried to respond to the needs of the displaced but in some cases, camps became magnets for the urban poor seeking to access services and relief, and it became increasingly difficult to distinguish between who was displaced by the earthquake and who was not. At the peak of the displacement, some 1.5 million individuals lived in more than 1,500 sites around the country—1,200 in Port-au-Prince alone, a city of three million. By mid-2012, close to 500,000 Haitians were still living in tents and under tarpaulins and, by early 2013, the number of displaced was still more than 300,000. Efforts to provide permanent housing to Haiti's displaced were stymied by complex land and property issues, and most international actors focused on providing first temporary and then transitional housing. By 2012, donors had disbursed only 53 percent of the $4.5 billion pledged for 2010–11 (Office of the Special Envoy for Haiti 2012). In 2012 the government shifted from a policy focused on reconstruction and repair of housing to one emphasizing rent subsidies, but the rental housing market was extremely tight, in part because of the substantial international presence. The painfully slow reconstruction process meant that Haitians continued to be displaced—and vulnerable to other disasters.

It has been difficult to get a handle on internal displacement since the earthquake. While the number of Haitians living in camps has decreased, their fate remains unclear. Some who had camped on public lands received rent subsidies and left the settlements, but it is unclear whether they have found more permanent living arrangements or just have been displaced elsewhere. Others were forcibly evicted from private settlements and there have been no reports about where they have gone (Amnesty International 2013; IOM 2013).

During the first ten months after the earthquake, a main fear of international and national responders alike was the upcoming hurricane season, particularly as more than a million people were living under tents. As warnings were issued of the approach of Hurricane Tomás in November 2010, the government urged people to take shelter with family and friends. In some cases, people evacuated although many did not want to leave their temporary homes for fear of losing their possessions (Archibold 2010). Hurricane Tomás bypassed Port-au-Prince, passing instead over the western part of the country, leaving 6,610 people homeless and causing 21 deaths. Almost 50,000 people were evacuated with about half seeking refuge in 75 temporary shelters (OCHA 2010).

But the hurricane caused far fewer casualties than the developing cholera epidemic. By the time Hurricane Tomás hit, 500 people had already died from cholera, and the flooding and displacement of the population raised fears that the outbreak could sweep through communities already made vulnerable by the earthquake and hurricane. Crowded and insanitary living conditions, known risk factors for cholera, were widespread in Haiti although the country had never before experienced an outbreak of cholera. On October 17, 2010, the first hospitalization was reported (McNeill 2012); within weeks, cholera had spread to all ten of Haiti's departments and, by July 2012, more than half a million cholera cases had been reported with more than 7,400 deaths (OCHA 2012; see also WHO 2011). By January 2013, the death toll was 7,900; 6 percent of Haiti's population had had the disease. Although the toll was by far the worst in Haiti, cholera spread to other countries; in 2011 outbreaks were reported in the Dominican Republic, Venezuela and Florida, and in 2012 in Cuba.

Centered in Artibonite, there was initial speculation that the disease had spread rapidly as a result of the influx of IDPs following the earthquake. Within a month of the first outbreak, demonstrations occurred over suspicions that the cholera had been brought to Haiti by Nepali peacekeepers, members of the MINUSTAH (Mission des Nations Unies pour la Stabilisation en Haïti) mission. While the UN denied this, six months later a research report by the Centers for Disease Control (CDCs) confirmed that the cholera outbreak was likely to have had its origins in the Nepali contingent, which had arrived only two weeks before the outbreak and where sewage from the troops' camp emptied into a river (Katz 2013).

The humanitarian community mobilized to prevent the spread of cholera in the IDP camps and by and large was successful, as evidenced by the fact that those living in camps were much less likely to get cholera than those living elsewhere. Haiti was a poor country before the earthquake, with deplorable sanitation systems, but the presence of MINUSTAH troops and the wide-scale displacement were both factors in the introduction of cholera and its rapid spread.

Haitian migration: United States

The earthquake also led to fears of a large-scale migration of Haitians to the United States, but this simply did not happen. On the one hand, the location of the earthquake meant that the roads to the northern coast where boat migration usually begins were impassable. The expenses of the boat trip—estimated by the International Organization for Migration (IOM) to be $600—were prohibitive for Haitians who had lost everything in the earthquake (Murray and Williamson 2011, 8). The US government announced that interdictions of Haitians leaving the island by boat would continue and publicized this fact within Haiti, with messages beamed by the Haitian Ambassador in the United States as well as senior government officials such as Janet Napolitano. Whether the messages got through or Haitians simply lacked the means to take to boats for Florida, US Coast Guard statistics indicate that the number of interdictions of Haitians actually fell in the years following the earthquake: from 1782 in fiscal year (FY) 2009 to 1377 in FY 2010; 1137 in FY 2011; and 977 in FY 2012 (US Coast Guard n.d.).

The US government designated Haiti as eligible for Temporary Protected Status (TPS) on January 15, 2010, making it possible for Haitians living illegally in the US before the earthquake to apply for temporary legal status to live and work in the United States for an eighteen-month period. Since the earthquake, TPS has been extended twice and presently runs until 22 July 2014. Some 200,000 Haitians have benefitted from this provision (Kenny 2012). TPS has been used by the US government in other situations of natural disasters for long periods of time. In the cases of Honduras and Nicaragua, TPS has been repeatedly extended since Hurricane Mitch in 1998 (and for El Salvador since 1999.) TPS provides concrete benefits both to the affected individuals (who do not have to return to their disaster-affected countries) and to the countries themselves. Most immediately, Haitians with TPS status in the US are able to send remittances back to their families—remittances that provide direct support to impacted communities. A World Bank report estimated that remittances to Haiti were likely to increase by 20 percent in 2010, sending an extra $360 million to Haiti; the Bank further projected that an additional $1 billion could be sent by Haitians with TPS over the next three years (World Bank 2010). TPS also reduces pressure on the governments of affected countries, both by providing an inflow of capital through remittances and by relieving the government of the responsibility of having to deal with returnees.

In January 2012, the US government added Haitians to the list of nationals eligible to apply for its H-2 visa programs, which are the country's largest employment-based visa programs. As the Center for Global Development noted, this offers the potential for low-skilled Haitians to work in the United States and thus to ease some of the pressures in their home country by sending remittances back to their communities (Center for Global Development n.d.).

There were other aspects of crisis migration evident in the post-earthquake US response to Haiti. In the immediate aftermath of the earthquake, the US repatriated its nationals and, in the process, a number of Haitians in mixed family situations were admitted. Murray and Williamson report that, of the 27,000 people repatriated to Florida after the earthquake, there were nearly 6,210 Haitians (Murray and Williamson 2011, 9). In addition, some 55,000 Haitians had US immigration applications in process and Haitian rights advocates called for processing of these applications to be expedited. In early 2012, the reports were that 112,000 Haitians had applied for permanent residency on the basis of having a US family member, and again there were calls from the Haitian advocacy community for these applications to be expedited (Kenny 2012).

There is also the messy question of international adoptions in the aftermath of the earthquake. On January 18, less than a week after the earthquake, US Attorney General Janet Napolitano announced that the US would lift visa requirements for those orphans whose adoptions had already been approved by Haitian authorities, and who had been matched with prospective parents in the US, but in practice the standard of proof was reportedly very low (Thompson 2010). In the weeks and months following the earthquake, there was considerable concern that large numbers of Haitian children were being flown out of the country for adoption in the US. This issue was crystallized by the New Life Children's Refuge case, in which a group of ten Baptist missionaries from Idaho were arrested on child trafficking charges when attempting to cross the border into the Dominican Republic with 33 Haitian children. These children had been collected from several orphanages in Haiti, presumably for adoption in the US, but it turned out that most of the children were not orphans. The Americans were arrested and brought to trial inside Haiti. In fact, the numbers of Haitians adopted by US parents declined from 330 in 2009 to 113 in 2010, and to only 33 in 2011 (Bureau of Consular Affairs 2010)—perhaps because of the media coverage around adoptions. The issue of international adoptions is particularly problematic in Haiti, given the endemic poverty that has led to the well-entrenched *restavek* system in which between 250,000 and 500,000 children are placed by their families in positions as domestic servants (Daniel 2012). Haiti also has a history of child trafficking; the government estimated that about 2,000 children a year were being trafficked before the earthquake (Brennan 2012). The number of orphanages in Haiti is large and most have been unregulated. Indeed, of the roughly 30,000 children in Haitian institutions and the hundreds adopted by foreigners each year, the Haitian government estimates that 80 percent have at least one living parent. The decision to turn children over to orphanages is motivated by dire poverty. Since the earthquake, the Haitian government has moved to overhaul its adoption laws for the first time in almost thirty years, in order to bring the country into line with international standards. The Haitian government has also begun inspecting orphanages, but has found it difficult to close those not in compliance because of the lack of alternative institutions for placement of the children (Brennan 2012).

Haitian migration: other countries

Other countries also responded to the Haitian earthquake by easing migration restrictions. In Canada shortly after the earthquake both the federal and the Quebec provincial governments announced that they would speed up processing of family reunification visas, and Quebec announced it would institute a humanitarian sponsorship program to allow petitions for the humanitarian entry of both primary and secondary relatives up to 3,000 persons. In order to qualify for the visa, applicants had to meet criteria for being "seriously and personally affected by the earthquake" (Murray and Williamson 2011, 10) as well as pay applicable fees and, in some cases, demonstrate an ability to support relatives after arrival. During the year after the earthquake, some 3,300 Haitians from all immigration channels were admitted (Murray and Williamson 2011, 11).

The case of Brazil is particularly interesting. Since its deployment in 2004, the military component of the UN peacekeeping mission in Haiti, MINUSTAH, has been led by a Brazilian army contingent, which also suffered serious losses in the 2010 earthquake. Perhaps in recognition of its close relationship with the Haitian people, during the course of 2010, the Brazilian government issued 475 humanitarian visas to Haitians who were already in Brazil or who arrived on its territory without documentation (National Immigration Council 2013). However, the number of undocumented Haitians arriving in its border areas increased dramatically during the course of 2010 and even more so in 2011. Most of the Haitians traveled via Ecuador or Peru, and then through the Amazon to turn up in Brazilian border regions. The Brazilian government reported that many were coming with the assistance of criminal elements. By January 2012, the Brazilian government had issued some 1,600 humanitarian visas but, concerned about border security and about the prospect of more Haitians making the long and dangerous journey, it announced a change in policy. The 2,400 recently arrived Haitians in the country as of January 2012 would receive humanitarian visas, but future visas would be issued only at the Brazilian embassy in Port-au-Prince and would be capped at one hundred per month (Romero and Zarate 2012). The humanitarian visas give the Haitians residence rights and the right to work for five years, after which time they can remain in the country only if they demonstrate that they have a job and a place to live. The Brazilian government reported that 90 percent of the Haitians were employed, but also noted that they had a higher economic status than other Haitians in that they had managed to pay for the journey (National Immigration Council 2012, 13).

While the Brazilian government initially saw the development of the humanitarian visa system as an appropriate response to Haitians suffering from the effects of the earthquake (noting, for example, that the remittances they sent back to Haiti could contribute to the country's recovery), the influx of large numbers of Haitians led the government to limit its visa policy. The Brazilian case also illustrates the importance of economic incentives for disaster-induced migration, as the Haitians settled in areas where jobs were plentiful (Romero and Zarate 2012). More recently, the Brazilian government has taken measures to reduce the number of Haitians being smuggled or trafficked into the country (OJornal 2013).

When the earthquake occurred, the Dominican Republic announced that it would not deport Haitians from the country and initially it opened the border. But as time drew on, there were fears that Haitians would come to the Dominican Republic in large numbers and within several months this generous approach came to an end. One estimate says that the number of undocumented Haitians in the Dominican Republic has risen to one million from approximately 600,000 before the earthquake (BBC News 2011), but there are no official numbers on how many Haitians crossed the border after the disaster. Apart from worries about the economic consequences of illegal laborers at a time of high unemployment in the country, the cholera epidemic added the fear that Haitian migrants could bring cholera to the Dominican Republic, and indeed more than 21,000 cases and 363 deaths were reported there by December 2011 (Kramer 2012). After protests erupted in 2011 the government began deporting Haitians without proper documentation. The situation got so tense that IOM offered Haitians $50 each plus relocation assistance to go home willingly. More than 1,500 Haitians returned

voluntarily through that program. The government also passed a new law denying citizenship to children of illegal immigrants, and in 2012 it passed new and
stricter requirements for work permits for Haitian workers (Archibold 2011; see
also *Bloomberg Businessweek* 2012). Reports in early 2013, however, indicated that
there are new provisions for temporary labor permits, which may be helping some
Haitians regularize their status in the country (Adams 2013).

Conclusions: Haiti and crisis migration

The case of Haiti raises questions about the definition of crisis migration. The 2010
earthquake was clearly a crisis, as were the 2008 storms, and yet the number of Haitians interdicted by the US Coast Guard has remained amazingly stable from 2005
until the present. There is much more variation in the number of Haitians immigrat-
ing to the US during this period, ranging from 14,524 (2005) to 30,405 (2007) per
year. While the image of people clinging to small boats in search of safety seems to
exemplify the term crisis migration (see Kumin, Chapter 15 in this volume), it might
be that more people use traditional immigration channels in response to crises—
particularly given the policies of receiving countries. Moreover, far larger numbers
are displaced internally than those who ever attempt to migrate internationally.

The Haitian case also raises questions about what constitutes a crisis. By most
indicators Haiti has been in crisis almost since its independence in 1804, and migration has been an essential part of Haitians' strategies for coping with their situation.
Poverty and political instability led Haitians to seek survival, and sometimes a better
life, elsewhere. Many moved internally, as part of a massive rural–urban migration,
which accelerated in the past 50 years. Environmental degradation became both a
cause and a consequence of desperate poverty and also increased the risk that natural hazards would become disasters. Different segments of the population migrated
to different countries, with elites tending to go to the United States, Canada and
France, and the poor traveling to nearby countries in search of work. As is the case
in most parts of the world, established migration routes became the means of escape
when political conditions became intolerable.

In looking at both the interdiction and legal immigration statistics over the past
20 years (Figure 4.1), it seems clear that international migration increased most dramatically in response to political events—notably the violence following Aristide's
ouster in 1991, and again in 1994 and in 2004. The major disasters of the last decade—the storms of 2008 and the earthquake of 2010—did not lead to an increase
in the number of Haitians taking to their boats for the United States. Nor is there
evidence that either legal or illegal migration increased following either the 2008
storms or the 2010 earthquake.

Alas, there are no quick fixes to the Haitian situation. For the 1.5 million Haitians displaced internally by the earthquake, there is a lack of monitoring about
what has happened to those who have left the camps. While both the number of
camps and informal settlements, and the number of recorded IDPs have declined,
there is little knowledge about where they have gone and under what conditions

they are living. Better monitoring and evaluation systems would perhaps provide clearer insights into ways of finding solutions for Haitian IDPs.

In terms of international migration responses, in the case of the US, further extensions of TPS would provide concrete assistance to those Haitians currently living in the US and to their families back home through the remittances they send. This is a helpful policy and one that should be extended. Increasing other avenues of legal migration for Haitians, to the United States and elsewhere, is also a helpful measure. As others have noted, increasing migration possibilities does more to help destitute Haitians than sending aid (Kenny 2012).

Yet, given the many entrenched problems in Haiti, it probably is not realistic to think that temporary humanitarian entry programs would do more than help a small percentage of the country's population. Moreover, it is unlikely that temporary entry programs would be temporary, as the incentives to remain in countries offering more opportunities would be strong.

Perhaps a more promising area is to seek ways to strengthen diaspora groups to enable them not only to send money home to family members but to support community-led development efforts. And yet the record of such contributions—important as they are in terms of magnitude—is not all that positive, perhaps because of the reluctance or inability of such groups to build the capacity of local governmental authorities who are, after all, ultimately responsible for the welfare of their citizens. There are some signs that there may be means for doing so, although historically relations between the Haitian diaspora and their counterparts in Haiti have been marked by difficulties and mutual suspicion (Forman, Lang and Chandler 2011).

In the longer term, support for establishing the rule of law and political accountability must be central to international aid programs. Even good development programs cannot make up for incompetent or corrupt leadership and entrenched bureaucracies. Disaster risk reduction measures are urgently needed, but are almost impossible to implement effectively when poverty and inequality leads communities to burn remaining trees for charcoal and to live in areas that are dangerous. Working with local civil society and community-run associations offers some promise—and, indeed, Haiti has a long tradition of such groups—but their possibilities for bringing about needed change are also limited by the political context.

Acknowledgment

With thanks to Daniel Petz for his research assistance.

References

Adams, G. (2013) "Haitian Migrants Get Temporary Work Permits in the Dominican Republic," Report from UN Radio, available online at: http://reliefweb.int/report/dominican-republic/haitian-migrants-get-temporary-work-permits-dominican-republic.

Amnesty International (2013) "Nowhere to Go: Forced Evictions in Haiti's Displacement Camps," January, available online at: http://www.amnesty.org/en/library/asset/AMR36/001/2013/en/894f378f-19ef-4ee0-b8bd-f305fe8225b5/amr360012013en.pdf.

Archibold, R.C. (2010) "Facing New Crisis, Haitians Prove Skeptical," *New York Times*, November 4, available online at: http://www.nytimes.com/2010/11/05/world/americas/05haiti.html?_r=1&ref=world.

Archibold, R.C. (2011) "As Refugees From Haiti Linger, Dominicans' Good Will Fades," *New York Times*, August 30, available online at: http://www.nytimes.com/2011/08/31/world/americas/31haitians.html?pagewanted=all.

BBC News (2009) "Bill Clinton to be UN Haiti Envoy," May 19, available online at: http://news.bbc.co.uk/2/hi/8056762.stm.

BBC News (2011) "Dominican Republic Resumes Deportation of Haitians," January 6, available online at: http://www.bbc.co.uk/news/world-latin-america-12132514.

Bhargava, A., Docquier, F. and Moullan, Y. (2011) "Modeling the Effects of Physician Emigration on Human Development," *Economics and Human Biology*, 9: 172–183.

Bloomberg Businessweek (2012) "Haitian Migrants Must Get Dominican Work Permits," June 11, available online at: http://www.businessweek.com/ap/2012-06/D9VB3S580.htm.

Brennan, E. (2012) "Trying to Close Orphanages Where Many Aren't Orphans at All," *New York Times*, December4, available online at: http://www.nytimes.com/2012/12/05/world/americas/campaign-in-haiti-to-close-orphanages.html?pagewanted=all.

Bureau of Consular Affairs, US Department of State (2010) "Haiti," October, available online at: http://adoption.state.gov/country_information/country_specific_info.php?country-select=haiti.

Carroll, R. (2008) "We are Going to Disappear One Day," *Guardian*, November 7, available online at: http://www.guardian.co.uk/world/2008/nov/08/haiti-hurricanes.

Center for Global Development (n.d.) "Migration as a Tool for Disaster Recovery: H-2A/H-2B Resources," available online at: http://www.cgdev.org/section/initiatives/_active/migration_tool_disaster_recovery/h2ah2bresources.

Daniel, T. (2012) "Haiti Seeks to Fix Broken Adoption System," Associated Press, November 20, available online at: http://bigstory.ap.org/article/haiti-seeks-fix-broken-adoption-system.

Environment News Service (2008) "Deadly Hurricanes Devastate Impoverished Haiti," September 8, available online at: http://www.ens-newswire.com/ens/sep2008/2008-09-08-02.html.

Fagen, P.W. (2006) "Remittances in Crises: A Haiti Case Study," *Humanitarian Policy Group Background Paper*: 1–9.

Fagen, P.W., Dade, C., Maguire, R., Felix, K., Nicolas, D., Dathis, N. and Maher, K. (2009) *Haitian Diaspora Associations and their Investments in Basic Social Services in Haiti*, Georgetown University, prepared for the Inter-American Development Bank, January, 1-97, available online at: http://issuu.com/georgetownsfs/docs/haitian_diaspora_-_fagen/1.

Ferguson, J. (2003) *Migration in the Caribbean: Haiti, the Dominican Republic and Beyond*, London: Minority Rights Group International.

Ferris, E. and Petz, D. (2011) *A Year of Living Dangerously: A Review of Natural Disasters in 2010*, Washington, DC: Brookings-LSE Project on Internal Displacement.

Forman, J.M., Lang, H. and Chandler, A. (2011) "The Role of the Haitian Diaspora in Building Haiti Back Better," Center for Strategic and International Studies, June 14, available online at: http://csis.org/publication/role-haitian-diaspora-building-haiti-back-better.

General Assembly (2008) "Report of the Special Rapporteur on contemporary forms of racism, racial discrimination, xenophobia and related intolerance, Addendum—Mission to the Dominican Republic," Human Rights Council, Geneva, March 18, available online at http://www2.ohchr.org/english/bodies/hrcouncil/docs/7session/A.HRC.7.19.Add. 5.doc.

Haggerty, R.A. (ed.) (1989) "Haiti: The Society and its Environment," *Dominican Republic and Haiti: Country Studies*. Washington: GPO for the Library of Congress: 241–277: 247, available online at: http://countrystudies.us/haiti/.

Haitian Diaspora (n.d.) "About," available online at: http://haitiandiaspora.com/about/.

Human Rights Watch (1994) "Fugitives From Injustice: The Crisis of Internal Displacement in Haiti," available online at: http://www.unhcr.org/refworld/docid/3ae6a7eb4. html.

In Motion (2005) "Haitian Immigration: Eighteenth and Nineteenth Centuries," *In Motion: The African-American Migration Experience*. New York Public Library, available online at: http://www.inmotionaame.org/print.cfm;jsessionid=f8308845313437462218007 migration=5&bhcp=1.

International Crisis Group (2007) "Peacebuilding in Haiti: Including Haitians from abroad," Latin America/Caribbean Report No 24–14, December, available online at: http://www.crisisgroup.org/~/media/Files/latin-america/haiti/24_peacebuilding_in_haiti___including_haitians_from_abroad.ashx.

IOM (2013) "Press Briefing Notes," April 12, available online at: http://www.iom.int/cms/en/sites/iom/home/news-and-views/press-briefing-notes/pbn-2013/pbn-listing/number-of-haitians-living-in-pos.html.

Katz, J.M. (2013) "In the Time of Cholera: How the UN Created an Epidemic—Then Covered it Up," *Foreign Policy*, available online at: http://www.foreignpolicy.com/articles/2013/01/10/in_the_time_of_cholera.

Kenny, C. (2012) "The Haitian Migration," *Foreign Policy*, January 9, available online at: http://www.foreignpolicy.com/articles/2012/01/09/the_haitian_migration.

Kramer, A. (2012) "Cholera in the Dominican Republic: The Outbreak and Response," Center for Strategic and International Studies, March, available online at: http://www.smartglobalhealth.org/blog/entry/cholera-in-the-dominican-republic-the-outbreak-and-response/.

Masters, J. (n.d.) "Hurricanes and Haiti: A Tragic History," *Weather Underground*, available online at: http://www.wunderground.com/resources/education/haiti.asp.

McNeill, D.G. (2012) "Haiti: Cholera Epidemic's First Victim Identified as River Bather Who Forsook Clean Water," *New York Times*, January 9, available online at: http://www.nytimes.com/2012/01/10/health/haitian-cholera-epidemic-traced-to-first-known-victim.html?_r=1.

Murray, R.B. and Williamson, S.P. (2011) "Migration as a Tool for Disaster Recovery: A Case Study on US Policy Options for Post-Earthquake Haiti," Center for Global Development Working Paper 255, June, available online at: http://www.cgdev.org/publication/migration-tool-disaster-recovery-case-study-us-policy-options-post-earthquake-haiti.

National Immigration Council (2012) "Haitian Immigration to Brazil," 1–17, available online at: http://www.iom.int/jahia/webdav/shared/shared/mainsite/microsites/IDM/workshops/moving-to-safety-complex-crises-2012/speeches-presentations/Session-3-Paulo-De-Almeida-Haitian-Migration-to-Brazil.pdf.

National Immigration Council (2013) "Haitian Refugees in Brazil," *Knowledge Migration*, January 28, available online at: http://refugeeresearch.net/ms/km/2013/01/28/haitian-refugees-in-manaus-brazil/.

OCHA (2010) "Haiti—Cholera—Hurricane Tomas, Situation Report # 16," November 8, http://reliefweb.int/report/haiti/haiti-cholera-hurricane-tomas-situation-report-16-8-november-2010.

OCHA (2012) *Haiti Humanitarian Bulletin* Issue 19, 2 July [online]. Available at: http://reliefweb.int/report/haiti/haiti-humanitarian-bulletin-issue-19-2-july-2012

Office of the Special Envoy for Haiti (2009a) "President Clinton Leads Trade Delegation to Haiti," October 1, available online at: http://www.lessonsfromhaiti.org/press-and-media/press-releases/president-clinton-leads-trade-de/.

Office of the Special Envoy for Haiti (2009b) "Paul Farmer Visits Haiti as UN Deputy Special Envoy," September 8, available online at: http://www.haitispecialenvoy.org/press-and-media/press-releases/paul-farmer-visit/.

Office of the Special Envoy for Haiti (2012) "Assistance Tracker: Summary," December, available online at: http://www.lessonsfromhaiti.org/assistance-tracker/.

OJornal (2013) "Brazil Seeks Help to Curb Immigration of Haitians," May 16, available online at: http://ojornal.com/portuguese-brazilian-news/2013/05/brazil-seeks-help-to-curb-immigration-of-haitians/#axzz2UIooopaL.

Perito, R.M. (2008) "Haiti After the Storms: Weather and Conflict," *USIPeace Briefing*, Washington: USIP, November 1–6, available online at: http://www.usip.org/files/resources/USIP_1108.PDF.

Perito, R.M. (2009) "Haiti: Is Economic Security Possible if Diplomats and Donors Do Their Part?" United States Institute of Peace, May, available online at: http://www.usip.org/publications/haiti-economic-security-possible-if-diplomats-and-donors-do-their-part.

Ratha, D. (2010) "Helping Haiti through Migration and Remittances," *People Move: A Blog About Migration, Remittances, and Development*, World Bank, January 19, available online at: http://blogs.worldbank.org/peoplemove/helping-haiti-through-migration-and-remittances.

Romero, S. and Zarate, A. (2012) "Influx of Haitians into the Amazon Prompts Immigration Debate in Brazil," *New York Times*, February 7, available online at: http://www.nytimes.com/2012/02/08/world/americas/brazil-limits-haitian-immigration.html.

Schneider, M.L. (2008) "In the Aftermath of Hurricanes, Haiti Situation is Critical," *World Politics Review*, October 20, available online at: http://www.worldpoliticsreview.com/articles/2792/in-the-aftermath-of-hurricanes-haiti-situation-is-critical.

Science Daily (n.d.) "Great Hurricane of 1780," *Science Daily*, available online at: http://www.sciencedaily.com/articles/g/great_hurricane_of_1780.htm.

Telegraph (2010) "Haiti Earthquake: History of Natural Disasters to Hit the Country," January 13, available online at: http://www.telegraph.co.uk/news/worldnews/centrallamericaandthecaribbean/haiti/6978919/Haiti-earthquake-history-of-natural-disasters-to-hit-the-country.html.

This Day in History (1996–2013) "Oct. 2, 1963: Hurricane Devastates Haiti," History.com, available online at: http://www.history.com/this-day-in-history/hurricane-devastates-haiti.

Thompson, G. (2010) "After Haiti Quake, the Chaos of US Adoptions," *New York Times*, August 3, available online at: http://www.nytimes.com/2010/08/04/world/americas/04adoption.html?pagewanted=all&_r=0.

Timelines of History (n.d.) "Timeline: Haiti," available online at: http://timelines.ws/countries/HAITI.HTML.

USAID (2008) "Haiti—Storms Fact Sheet # 5 (FY) 2009," November 14, available online at: http://reliefweb.int/report/haiti/haiti-storms-fact-sheet-5-fy-2009.

USA Today (2008) "Hurricane Ike Kills 7 in Cuba," September 12, available online at: http://www.usatoday.com/weather/storms/hurricanes/2008-09-12-Ike-Cuba_N.htm.

US Coast Guard (n.d.) "Alien Migrant Interdiction: US Coast Guard Maritime Migrant Interdictions," available online at: http://www.uscg.mil/hq/cg5/cg531/amio/FlowStats/currentstats.asp.

US Geological Service (2010) "Poster of the Haiti Earthquake of 12 January 2010—Magnitude 7.0," November 23, available online at: http://earthquake.usgs.gov/earthquakes/eqarchives/poster/2010/20100112.php.

Wasem, R. (2005) "Social Resilience and State Fragility in Haiti, A Country Social Analysis," US Immigration Policy on Haitian Migrants, CRS Report for Congress, Order Code RS21349, January 21: 1–6.

WHO (2011) "Cholera and Post-Earthquake Response in Haiti," *Health Cluster Bulletin* 29, November 7, available online at: http://www.who.int/hac/crises/hti/sitreps/haiti_health_cluster_bulletin_7november2011.pdf.

World Bank (2006) "Social Resilience and State Fragility in Haiti, A Country Social Analysis," Caribbean Country Management Unit, ESSD Sector Management Unit, Report No. 36069–HT, April 27.

World Bank (2008) "World Bank Supports Haiti's Rebuilding Efforts after Recent Hurricanes," December 12, available online at: http://www.worldbank.org/en/news/2008/12/12/world-bank-supports-haitis-rebuilding-efforts-after-recent-hurricanes.

World Bank (2010) "Haiti Remittances Key to Earthquake Recovery," May 17, available online at: http://www.worldbank.org/en/news/feature/2010/05/17/haiti-remittances-key-to-earthquake-recovery.

Zimmerman, J. (1993) "United States Department of State v. Ray: The Distorted Application of the Freedom of Information Act's Privacy Exemption to Repatriated Haitian Migrants," *American University International Law Review*, 9(1): 385–416.

5

HEALTH CRISES AND MIGRATION

*Michael Edelstein, Khalid Koser and
David L. Heymann*

Introduction

In contrast to almost every other crisis and case considered in this volume, this chapter demonstrates that on the whole health crises do not result in mass migration. Focusing on infectious diseases, the chapter shows that when people move as a result of such health crises, they tend to move over short distances and for relatively short periods of time, and often because of misunderstandings and panic. Where cross-border movements have taken place, it is often difficult to discern health from other factors such as poverty and state collapse as an explanation. International Health Regulations (IHR) contribute instead to an orderly and collective public health response that generally precludes the need for large-scale cross-border movements. Although restricting population movement is a largely ineffective way of containing disease, migration policies worldwide tend still to be predicated on the risk of international migration where health crises emerge, for example focusing on isolation and non-admission for nationals of affected countries.

The emergence of an international regime for disease control

The emergence and spread of disease has been a concern since the early days of organized society. The Bible, in Leviticus 13–14, describes the isolation and decontamination rituals of infected individuals. This concept of isolation shifted from individuals to populations in response to the threat of large-scale mortality, during the plague pandemic in fourteenth-century Europe. One of the earliest recorded government health policies was to isolate communities affected by disease and restrict population movement in response to the threat of a health crisis, for example through the quarantine law put in place in several Mediterranean port cities in the fourteenth and fifteenth centuries. By the late eighteenth century these principles became firmly embedded in the doctrine of isolation and restriction, and had become the norm at international borders, sometimes with a highly disruptive outcome.

The beginning of international governance for infectious diseases was marked by the 1851 international sanitary conference held in Paris, focusing primarily on governance around the importation and exportation of cholera, plague, smallpox and yellow fever. By the mid-twentieth century, a further sixteen International Sanitary Conventions focusing on the same diseases had been held (Aginam 2002) and produced a variety of treaties on topics such as pilgrimage to Mecca, notification procedures for infectious diseases and the inspection of ships. These treaties were not always enforced (Aginam 2002).

The beginning of the twentieth century also saw the emergence of international health organizations, culminating with the creation of the World Health Organization (WHO) in 1948, with a mandate to facilitate international cooperation on matters related to the spread of infectious disease, as well as responsibility for international disease surveillance. In 1951, WHO adopted the International Sanitary Regulations (ISR), which superseded the treaties adopted by the successive sanitary conventions, but continued to focus on four diseases: cholera, yellow fever, plague and smallpox.

The ISR became the IHR in 1969, and had as a goal: maximum prevention of the spread of infectious diseases with minimal disruption of travel and trade. Initially the IHR remained focused on four specific diseases (cholera, plague, yellow fever and smallpox). The IHR were largely based on the assumption that there was a narrow spectrum of diseases that caused a threat to international travel and trade, that migration was unidirectional, and that these diseases could be stopped at international borders (Gushulak and MacPherson 2010).

In 1995 WHO acknowledged that countries did not often report these four diseases because of the risk of decreased travel and trade, and that it did not have a mandate to enforce the reporting requirement. The IHR disease coverage was furthermore too limited: diseases causing high mortality or spreading rapidly, such as Ebola and Marburg hemorrhagic fevers, severe acute respiratory syndrome (SARS) or pandemic influenza, did not require notification. WHO also pointed out that these diseases could not effectively be stopped at international borders because travelers could cross borders while in the asymptomatic incubation period for the disease they carried, and thus appear healthy.

Two further concerns clearly illustrated the need for a paradigm shift in the IHR and global infectious disease control governance. The first was the speed of international travel. The second was the lack of a functional network allowing rapid communication between member states: political boundaries and border posts had gradually become a less important component in the global control of infectious diseases (Davies 2010).

The IHR were eventually revised by the World Health Assembly in 2005 (WHO 2005) and came into operation in 2007. The revised IHR have moved away from specific diseases, and focus on "public health events of international concern" (PHEICs), with the same aim of maximizing control of disease spread while minimizing travel and trade restrictions. PHEICs are not limited to infectious diseases and include contaminated food, chemical contamination of products or the environment, release of radio nuclear material, or other toxic release. The IHR are thus flexible, and adaptable to future, unknown threats (Edelstein 2012).

The revised IHR moved toward a preventive approach to international spread of disease that emphasizes the importance of detection and containment at source, and have a requirement that all countries develop core public health capacity to detect, report and respond to PHEICs where and when they occur (Wilson, von Tigerstrom and McDougall 2008).

The revised IHR take a stepwise approach to managing PHEICs, from monitoring events at the national level to global response. The revised IHR contain no formal enforcement mechanism or penalty for failing to comply with recommendations and there are no sanctions against states for non-compliance with binding resolutions (Fischer, Kornblet and Katz 2011). Despite their adherence to the IHR, countries remain sovereign and sometimes revert to the doctrine of isolation and restriction, threatening or deciding to close borders or impose travel restrictions in an attempt to prevent infections from entering their territory. During the H1N1 pandemic, for example, several countries, such as Slovakia (Gurniak 2009) and China (Huang 2010), imposed travel restrictions, in spite of repeated WHO statements that such restrictions were not recommended.

Case studies

Migration and cholera: Zimbabwe, 2008–2009

Zimbabwe has been considered by some writers a failed state, which saw a gradual collapse of the public health system in the ten years preceding one of the largest outbreaks of cholera ever recorded (WHO 2009a). The years leading to the outbreak in Zimbabwe had seen key health personnel leaving the country and, at the time of the outbreak, the main hospitals in the country had closed, as well as the Medical University of Zimbabwe and many local hospitals and clinics, resulting in ordinary Zimbabweans being unable to access healthcare (Amnesty International 2009). In addition to a health system collapse, a breakdown in the distribution of clean water contributed heavily to the emergence of cholera in Harare.

Although the initial outbreak in August 2008 was rapidly controlled, more infections were reported in the following months. By December, there had been more than 16,000 cases, with 15 percent of cases dying of the illness—a very high proportion. The infection was spreading rapidly in a population already affected by hunger and a high prevalence of HIV/AIDS (MSF 2009). By December 2008, all ten provinces had reported cases, with further spread in South Africa, Mozambique, Botswana and Zambia. Although the Ministry of Health declared an emergency on December 3, there was no formal notification via the IHR framework.

The health crisis contributed to a large-scale population movement out of Zimbabwe into South Africa (MSF 2009). By January 2009, before the outbreak had reached its peak, an estimated 38,000 Zimbabweans had fled into South Africa as a result of the outbreak, with some migration to Botswana as well.

From July 2008 the South African Department of Home Affairs had established an office to process Zimbabwean asylum claims at the "showground" in the border town of Musina, a refugee camp in a large open space where thousands of Zimbabweans were living in precarious conditions (MSF 2009). The South African government, however, considered most Zimbabwean immigrants to be economic migrants not eligible for refugee status, and was therefore deporting large numbers of Zimbabweans before, during and after the cholera outbreak (UNHCR 2009).

The South African government responded to the acute health emergency by providing clean water and medical facilities at the border. An outbreak control team was dispatched locally, as well as additional medical personnel (Hogan 2008). Water samples from the Limpopo River were taken regularly and an emergency preparedness plan was put in place (Hogan 2008). Non-governmental organizations (NGOs) such as Médecins Sans Frontières (MSF) also organized mobile clinics at the border area (Hogan 2008). In early March 2009, however, the South African government decided to close its Musina reception office and ordered Zimbabweans to clear the area, taking down and burning all temporary shelters, leading many recent immigrants to flee into hiding for fear of deportation or arrest (MSF 2009).

Other bordering countries also reacted to the cholera outbreak: Mozambique sent outbreak control teams to border areas, Zambia screened individuals entering from Zimbabwe at its border posts and Botswana dispatched outbreak management teams to the border town of Matsiloje (Berger 2008). No border countries closed their borders. A WHO statement from December 2008 clarified that "WHO does not recommend any special restrictions to travel or trade to or from affected areas" (WHO 2008, no page). Neighboring countries were however encouraged to "strengthen their active surveillance and preparedness systems" (WHO 2008, no page).

The United States government issued a travel warning for Zimbabwe in December 2008, citing the cholera outbreak among other security reasons (US Department of State 2008). The warning was lifted in April 2009. By June 2009, the outbreak was coming to an end, the number of cases reported having dropped from 8,000 a week in February 2009 to one-hundred a week in May of the same year (WHO 2009a). By the end of the outbreak, in June 2009, there had been 98,424 suspected cases in Zimbabwe, including 4,276 deaths (WHO 2009a), with an additional 12,000 cases and fifty-nine deaths in South Africa (AFP 2009).

The precise impact of the outbreak on migration from Zimbabwe into South Africa and Botswana is hard to estimate due to a high level of background migration, with thousands of Zimbabweans crossing every day. Attribution of mass migration to this medical emergency alone is therefore not possible. Similarly, distinguishing deportation as a regular element of managing migration from safe-guarding against return to situations of health crises can be hard to implement in practice.

The SARS outbreak, 2003

SARS was a new viral infection that caused respiratory symptoms and was associ-ated with a very high mortality. Available evidence suggests that SARS emerged in Guangdong Province, southern China in November 2002. In February 2003, a physician incubating SARS traveled from Guangdong Province to Hong Kong, and stayed at a hotel where he infected several other guests, who became ill and transmitted the disease to others when they returned to Vietnam, Singapore, Canada and Taiwan (Tsang *et al.* 2003), starting a worldwide outbreak of more than 8,000 cases and 800 deaths in thirty-two countries. Although local public health officials started reporting a new illness to superiors in Guangdong as early as December 2002, Guangdong health officials did not make a public announce-ment about the disease until February 2003. A coordinated and effective cam-paign to combat SARS in China began in mid-April (Brahmbhatt and Dutta 2008).

The first recorded case in Beijing occurred on March 5, 2003. By the end of April, 1,000 cases had been reported in the city (Brahmbhatt and Dutta 2008), leading to mass attempts to flee the city. Up to one million people had left by April 26 (Pomfret 2003). Earlier on, on March 15, WHO issued a rare emergency travel advisory that urged people not to travel if they developed symptoms. The advisory also included guidance to airlines and airline crew (WHO 2003). On March 27, WHO issued more stringent advice to international travelers and airlines, includ-ing recommendations on screening travelers at certain airports. Some airlines in affected countries began screening departing international travelers (WHO 2003). The global progression of the SARS epidemic, particularly in South East Asia and in Canada, led national governments to put exceptional containment measures in place. In Taiwan, from March 18, anyone who came into contact with a SARS patient was quarantined for ten to fourteen days, either at home or in a healthcare facility, depending on the degree of exposure (CDC 2003). Quarantined individu-als were not allowed to leave the quarantine sites unless authorized by the local authority. On April 28, the quarantine was extended to anyone arriving by air from a WHO-designated SARS affected area, who had to be isolated for ten days. By the end of the epidemic, approximately 130,000 persons had been placed in quarantine in Taiwan. The Taiwanese government also screened all persons enter-ing public buildings and restaurants for fever and required masks for all persons working in restaurants, entering hospitals, and using public transportation systems (CDC 2003).

The city of Toronto, Canada, experienced the largest outbreak of SARS outside Asia, with 225 cases (Institute of Medicine Forum on Microbial Threats 2004). The city adopted a voluntary ten-day home quarantine strategy for individuals with close contact with a case. In total 23,103 individuals were quarantined, of whom twenty-seven were issued a legally enforceable quarantine order (Svoboda *et al.* 2004). Toronto also closed hospitals and required healthcare workers to wear masks to limit the spread of the disease, whilst Hong Kong and Singapore also implemented quarantine measures (Institute of Medicine Forum on Microbial Threats 2004). While some countries decided to give a legally binding quarantine order to non-compliant individuals, others decided on stricter enforcement measures such as isolation in a guarded room, the use of security ankle bracelets, video monitoring, fines and jail sentences. These were the exception rather than the rule, and voluntary quarantine was effective in the majority of cases (Institute of Medicine Forum on Microbial Threats 2004).

Additionally to containment strategies, governments also implemented measures at international borders, such as pre-departure temperature screening, post arrival disembarkation screening, maintaining "stop lists" of people with suspected SARS to prevent such individuals from traveling, and isolation of ill travelers with suspected or probable SARS (Institute of Medicine Forum on Microbial Threats 2004).

A few countries also decided to restrict population movement in order to prevent SARS from entering their territory. On May 8, Kazakhstan closed its 1,700 km border with China to all air, rail and road traffic, as well as repatriating Kazakh nationals from China (Yermukanov 2003). Russia also closed the majority of its border crossings with China and Mongolia in May 2003, as well as suspending flights from China, Hong Kong and Taiwan (Vassileva 2003). Several other countries decided to suspend flights from and toward SARS affected areas.

The SARS outbreak also elicited a global response from WHO which in a rare move, issued a travel advisory on March 15 and April 2 recommending that persons traveling to Hong Kong and Guangdong Province consider postponing all but essential travel until further notice (WHO 2003). This was the most stringent travel advisory issued by WHO in its fifty-five-year history, which was eventually extended to Beijing and Shanxi Province, China, and Toronto, Canada, on April 23 (WHO 2003).

By May 23, 2003, the total number of cases reached 8,000, but the epidemic started to show signs of peaking and travel advisories were gradually removed (WHO 2003). In addition to bringing unprecedented attention on emerging diseases, SARS had a deep impact on global travel, with volume of travel to and from China down 45 percent in June 2003 compared to June 2002, and down 69 percent between Hong Kong and the US (BBC News 2003), and a cost of close to US$40 billion to the global economy (Institute of Medicine Forum on Microbial Threats 2004).

During the SARS outbreak, though there was initially internal migration in some countries, notably China, quarantine, restriction of travel, health communication and other containment measures implemented by national governments contained the epidemic and may have provided the assurance necessary to prevent real or attempted mass internal migration.

The H1N1 pandemic, 2009

In March 2009, human cases of infection with a novel strain of influenza A virus (H1N1) emerged in Mexico, the United States, and Canada. Throughout March and April, Mexico experienced outbreaks of a respiratory illness of unknown origin and reports of patients with influenza like illness came from throughout the country. By mid-April, several cases of severe respiratory illness in Mexico were confirmed as infection with what was first referred to as swine-origin influenza A H1N1 virus. From March 1 to April 30, 1,918 suspected cases and eighty-four deaths were reported in Mexico. By April 28, seven countries on four continents had reported confirmed cases. By June 3, the first case of H1N1 influenza was reported in Africa, the last continent that had remained unaffected by the virus (Sekkides 2010).

It was clear from the early days of the pandemic that the rapid spread of the disease and its high transmission rate could not justify a containment strategy (Huang 2010). On June 11, WHO declared the start of a worldwide pandemic by raising the pandemic alert level to six—its highest level. By then, more than 30,000 cases in seventy-four countries had been reported (Chan 2009). At the global level, H1N1 was the first, and to date the only, event to be declared a Public Health Event of International Concern (PHEIC) under IHR, on April 25, 2009. Such a declaration requires the WHO Director-General to issue temporary recommendations on how countries should respond to the PHEIC. The Director General, on advice of the IHR Emergency Committee, proposed that nations increase their active surveillance for unusual outbreaks of influenza-like illness (Katz 2009). Throughout the pandemic WHO regularly communicated with all member states through the National IHR Focal Points and the WHO public website to inform them of recommendations for actions to mitigate the consequences of the epidemic (Katz 2009), such as vaccine development and distribution, use of antiviral medications, social distancing via school closures, work pattern adjustment, self isolation of symptomatic individuals and advice to their caregivers, cancellation of mass gathering events, and screening at international transit points in some circumstances. WHO explicitly stated that it did not recommend travel restrictions related to the virus but did recommend that persons who were ill delay international travel (WHO 2009b).

National responses to the H1N1 pandemic showed a range of actions, from compliance to WHO's advice to measures taken against WHO recommendations. The United States broadly followed WHO guidelines and focused its response on vaccine and antiviral medication distribution, ensuring sufficient capacity for required medical care and non-medical interventions to mitigate the impact of the disease (PCAST 2009). About twelve million courses of antivirals and vaccine were allocated from the federal reserve to the most affected states; hospitals were allowed to plan for additional sites for treatment and triage of a potential surge of patients; the Centers for Disease Control (CDCs) recommended that people with influenza-like illness remain at home; and the government issued guidelines for schools with one or more cases of H1N1 to close for fourteen days (Huang 2010). Despite pressure from the Congress, the border with Mexico remained open; the US State Department however issued a travel warning to Mexico, leading to major airlines curtailing flights into Mexico (Huang 2010).

By contrast, China's containment policy focused on attempting to prevent the importation of H1N1 and to prevent it from spreading internally (Huang 2010). It screened inbound passengers from countries that had reported H1N1 cases and on May 1, 2009, suspended direct flights from Mexico. By the end of May, China was screening every inbound international flight and quarantined the whole flight if any passenger was found to have a temperature above 37.5 degrees Celsius. Tens of thousands of people were being held in government quarantine facilities by the end of July 2009. In light of mounting evidence of the mild nature of H1N1 influenza and the failure of containment, China formally abandoned this strategy in September 2009 and focused on mitigation by controlling outbreaks and reducing severe cases and fatalities (Huang 2010).

Other countries also put measures in place that contravened WHO recommendations as well as the spirit of the IHR. In Hong Kong, 300 guests and employees of a hotel where an infected man stayed after he arrived in Hong Kong from Mexico were confined for a week under police guard; in Singapore, anyone who had recently visited Mexico was placed in home quarantine (Gostin 2009). In Europe, the European Centre for Disease Prevention and Control (ECDC) released guidelines in line with WHO recommendations (ECDC 2009). Although most European countries adopted a mitigation approach similar to the American strategy, there were some instances of national decisions not in line with WHO and ECDC guidance, such as Slovakia closing its border with Ukraine in November 2009 (Gurniak 2009). By the end of the pandemic in August 2010, 214 countries had been affected and over 18,000 people had died, although this is likely to be an underestimate: in the United States alone, the 2009 H1N1 virus caused an estimated fifty-nine million illness episodes, 265,000 hospitalizations, and 12,000 deaths (Writing Committee of the WHO Consultation on Clinical Aspects of Pandemic 2010). Global travel to and from Mexico was reduced by 40 percent during the pandemic (Bajardi *et al.* 2011).

As in the SARS case, H1N1 is thought to have caused some internal migration, particularly in Mexico soon after its identification. Internal efforts to control the pandemic under the IHR, including open and transparent communication, however, provided an orderly global framework for pandemic response to which most countries adhered, and mass migration did not occur.

HIV-related travel restrictions

Placing entry, stay, or residence restrictions on non-national people living with HIV (PLHIV) has been an early and persistent response to the HIV/AIDS epidemic from governments (Amon and Todrys 2008). The restrictions include restriction on long-term residence, compulsory HIV status disclosure, or an absolute ban on entry (Rushton 2012). These restrictions are generally justified on two arguments:

public health security, arguing that allowing PLHIV to enter the country exposes the domestic population to a public health risk, and the economic argument that allowing PLHIV to enter on a long-term or permanent basis, imposes significant economic costs on the domestic health system (Rushton 2012).

The fear of immigration to seek more advanced healthcare became even more prominent from the mid-1990s when antiretroviral therapy (ART) started to become available in the developed world but remained inaccessible in developing countries (Rushton 2012). In 2010 however, President Barack Obama, announcing the end of the HIV-related travel restrictions in the United States, conceded that "the US (HIV) policy was based on fear, not science" (Franke-Ruta 2009, no page). Indeed from the very early stages of the HIV/AIDS epidemic, evidence showed that such travel restrictions were not justified, mainly owing to the fact that HIV is not transmissible through casual contact (Amon and Todrys 2008).

As early as 1985, several member states sought WHO's advice on the possibility of issuing travel restrictions for HIV infected individuals. WHO advised that testing and certification of international travelers were not warranted, on the basis that it was not justified from a public health point of view and not required under the IHR (WHO 1986). WHO reiterated in 1988 that screening international travelers was not an effective strategy to prevent the spread of HIV (WHO 1988). In 2006, the Office of the United Nations High Commissioner for Human Rights (OHCHR) and the Joint United Nations Programme on HIV/AIDS (UNAIDS) stated that "any restrictions on the rights to liberty of movement and choice of residence based on suspected or real HIV status alone cannot be justified by public health concerns" (UNAIDS 2004, 4).

The economic argument has not been substantiated by evidence either, as countries without HIV travel bans did not find an increase in HIV-positive immigrants (Nieburg *et al.* 2007). Skepticism about the economic argument grew further with the plummeting cost of ART (Rushton 2012). While there is a cost to treating HIV-positive migrants, it has been argued that HIV treatment is not different than other chronic illnesses and as such HIV-specific legislation is not justified.

HIV travel restrictions may not only be ineffective but could even be harmful to public health, by creating a false sense of security in countries where HIV travel restrictions are in place and by discouraging migrants from undergoing testing or seeking treatment (Ganczak *et al.* 2007). A 2006 study showed that a majority of HIV-positive travelers to the US did not comply with the legally mandated disclosure of their HIV status at the time and that a significant minority, fearing deportation if ART medication was found in their luggage, would stop taking medication for the duration of their stay, increasing the risk of developing drug resistance (Mahto *et al.* 2006). Another consequence of the US travel ban has been the refusal to hold the International AIDS Society conference there (or in any other country with HIV-related travel restrictions) due to HIV positive delegates not being able to attend (IAS 2009). Beyond public health implications, HIV-related travel restrictions reinforce stigma and discrimination against PLHIV and strengthen the idea that immigrants are a danger to the national population.

Beyond the theoretical shortcomings of HIV-related travel restrictions, these migration policies have affected individual lives in a very concrete way, such as in the case of a Ukrainian national who immigrated to the Russian Federation in order to be reunited with his partner, and who had to travel back to the Ukraine and re-enter Russia every three months to avoid mandatory HIV testing, as a positive test would lead to his permanent deportation (UNAIDS 2009). Documented individual stories are numerous, and the individual consequences of the HIV-related travel restrictions include irregular immigration, loss of income or livelihood, family breakups, loss of dignity and in some anecdotal reports death in confinement (UNAIDS 2009).

Furthermore, immigrants who are found to be HIV positive in countries with travel restrictions in place often face suboptimal care in government facilities while waiting for deportation (HRW 2007). The United Nations High Commissioner for Refugees (UNHCR) has clearly stated that refugees and asylum-seekers should not be targeted for special measures regarding HIV infection and that there is no justification for screening being used to exclude HIV-positive individuals (Rushton 2012). Nevertheless the impact of HIV travel restrictions has been strongly felt by asylum-seekers, discouraging HIV-positive asylum-seekers from using legal immigration channels (Amon and Todrys 2008) and leading to inappropriate treatment of asylum-seekers. In 1991, the United States denied entry to 115 HIV-positive Haitian political refugees and their families who otherwise would have been eligible for refugee status. UNAIDS has also noted the dilemma of some families with an HIV-positive member having to decide whether to forgo seeking asylum or to leave a family member behind (UNAIDS 2009).

Although the number of countries imposing HIV-related travel restrictions has remained more or less stable between 1989 and 2008 (Rushton 2012), an International Task Team on HIV-related Travel Restrictions was set up by UNAIDS with support from the Global Fund and WHO in 2008, in order to spearhead a major global effort for the elimination of such restrictions (UNAIDS 2009). Since then, some high profile countries such as China, the US and South Korea have started removing travel bans, potentially signaling the beginning of a broader trend (Rushton 2012). Nevertheless, as of July 2012, 45 countries are still maintaining total or partial travel and immigration bans on HIV-positive individuals.

Conclusions

While the health consequences of migration are well documented, it is difficult to attribute collective migration directly to health crises, especially migration across international borders. In cases where population migration occurs, it is generally within a wider humanitarian crisis, either man-made (such as conflict or nuclear disasters) or natural (such as earthquakes or floods). These situations are often an immediate threat to life and are more likely to trigger population movement. Even when the underlying event is not as sudden or catastrophic, such as the gradual collapse of the state in Zimbabwe, migration due to health crises occurs against a background of continuous emigration to bordering countries, with populations displaced by the health crisis using the same mode of movement as those migrating for other purposes, making it difficult to attribute migration directly to health or quantify the health-related population movements. Migration occurring in the context of a health crisis is therefore best described as a mixed pattern of migration.

When there is population movement as a result of a health crisis, migration tends to be internal, to regions directly outside the immediate crisis zone, and early on in the health crisis when information is often scarce, contradictory or erroneous. This movement is usually not sustained. Recent examples in India and China have shown populations leaving large urban centers to go back to their family villages (Pomfret 2003).

At the individual level, migration in search of better healthcare does occur. This can lead to a perceived threat of infection and of economic burden for countries where treatment is available, although evidence on both is weak. This in turn can lead to travel restrictions, deportations, and violation of human rights (Amon and Todrys 2008; Rushton 2012).

One specificity of health crises is, in many instances, the ability of individuals or communities to cope with, or to mitigate the effect of the crisis. The gradual improvement of the understanding of infectious diseases, their causative agents, modes of transmission and evidence-based ways to control their spread have empowered individuals, populations and governments to adopt preventive behavior, pre-empting in many cases voluntary or forced migration (Svoboda *et al.* 2004). Such preventive behavior empowers individuals to take an active stance against the disease—for example, by practicing good personal hygiene or drinking from a safe source such as bottled water, reducing the risk. These possibilities offer an alternative to fleeing, and may explain in part why people often choose not to leave an area where a health crisis is occurring.

While such responses may not be available in resource and infrastructure-poor countries where the majority of health crises occur, they are often provided by international partners and thus contribute to the prevention of mass emigration. In the H1N1 pandemic, social distancing, voluntary isolation, quarantine or mass vaccination were offered to the population of most countries as a pragmatic and evidence-based approach to deal with the health crisis (ECDC 2009; PCAST 2009). It is not, however, possible to predict whether the absence of these measures would have led to larger population movements. Additionally, health crises may lead to individuals or groups being too sick or frail to migrate and being trapped in crisis zones. In July 2012, an Ebola outbreak in Western Uganda led to patients fleeing a hospital where some of the infected patients had died (Chonghaile 2012). In such a context, sick or elderly patients may not be physically able to leave, increasing their chances of contracting the virus, which can kill up to 90 percent of those in contact with it (Chonghaile 2012).

In addition, current understanding of transmission dynamics has made outdated the idea that diseases can be stopped at borders. Modern outbreaks such as SARS or H1N1 have shown that diseases can be disseminated worldwide in a matter of days, the volume and speed of global travel making it impossible to stop infections at borders. Mathematical models provide little evidence that travel restrictions reduce the spread of disease (Bajardi *et al.* 2011), the exception being perhaps when trying to contain localized outbreaks. This evidence is reflected in the IHR, which focus less on control measures at borders and more on detection and response at source, with public health surveillance and response capacity building. As a principle, the IHR attempt to keep restrictions on population movement to a minimum.

The IHR have been amended over the years to enable the international community to respond to cross-border health crises in a rapid and efficient way by enabling global communication channels and encouraging local public health capacity building, both in the detection and management of health crises. The regulations allow for a tailored response to be advocated as and when crises arise, focusing on limiting the spread of diseases while keeping travel and trade restrictions to a minimum. While the IHR encompass travel-related public health measures to limit the spread of disease, such as control measures at points of entry by air, sea or ground, they are not designed to make recommendations on migration-related issues relating to health crises, such as the status of individuals or populations leaving a health crisis area. Individuals leaving purely to escape a health crisis are unlikely to be recognized as refugees pursuant to the 1951 Convention relating to the Status of Refugees. They are more likely to be considered migrants rather than refugees, as for Zimbabweans entering South Africa during the cholera outbreak.

Outside of health crises, infectious diseases can impact travel and migration at the individual level, as seen in the context of the worldwide HIV epidemic. While there are legal precedents for successful health-related asylum claims, particularly for HIV-positive individuals, asylum was granted on the basis of the fear of persecution associated with HIV status or sexual orientation rather than health status. The reverse, individuals qualifying as refugees who are denied asylum and deported because of their HIV status, has been more commonly seen. UNAIDS have stated that HIV-related migration restrictions have regularly violated the human rights principle of *non-refoulement* of refugees (UNAIDS 2004). These cases fall outside of the remit of the IHR.

Nevertheless, the flexibility extended in much national legislation to people who may not satisfy the legal criteria for refugee status, but may be in danger if they return to their country of origin, could be extended to people from countries undergoing health crises. Similar provisions already exist for example for people whose countries have been affected by natural disasters (such as US policy toward Montserrat and Haiti). As this chapter has shown, there is often an interaction between natural disasters and health consequences, and so such a policy understanding should be relatively easy to achieve. The policy challenge would be to know when deportation bans on the basis of health crises may be lifted, and it would seem sensible that these would be aligned with WHO declarations.

In a world of rapid travel, trade and climate change, where the frequency of emerging infectious diseases and other health problems is on the rise, the potential for increased health-related migration makes it a necessity to better define its status. Greater efforts should be made to encourage governments, and organizations that work with migration and migrating populations, to understand and abide by the IHR as a means of strengthening the potential to prevent migration related to health crises while ensuring the best possible protection against disease.

Recommendations

- More research is required on the impact of health crises on migration. While there is limited evidence both historically and more recently, this evidence tends to be anecdotal and hard to verify. Empirical challenges involve identifying and accessing affected populations; conceptual challenges include attribution and distinguishing health from other motivations to migrate.
- Greater coherence is required between the IHR and migration policies and practices at the national and international levels in order to inform government responses during health crises that help populations to avoid migration, and potentially pre-empt unwarranted decisions to close borders or restrict entry, as have been witnessed in the case of HIV.
- Greater efforts are required to encourage states to abide by the IHR, including the need for maintaining strong core capacity in public health, and for organizations that work on migration and/or with migrating populations to fully understand the IHR framework and its potential to prevent migration related to health crises.
- At the national level, greater coordination is required between government agencies separately tasked with migration and health mandates.
- National migration policies should accommodate the assistance and protection of migrants arriving from, or faced with the prospect of returning to, areas affected by health crises, including by suspending deportation orders until the health crisis has subsided.
- Special effort should be made at the national and global levels to ensure that populations are empowered to protect themselves from diseases that have the potential to spread internationally.
- Efforts should be made to ensure that the mass media have the knowledge and understanding to contribute to health protection and understanding of risks and their management.
- Greater efforts are required to continue to promote and increase access to healthcare by strengthening developing country capacity to deliver health services and procure medicines and vaccines in order to address needs that, if unmet, could lead to individual migration for healthcare.

References

Aginam, O. (2002) "International Law and Communicable Diseases," *Bulletin of the World Health Organization*, 80(12): 946–951.

Amnesty International (2009) *Zimbabwe's Health System in Chaos*, available online at: http://www.amnesty.org/en/news-and-updates/news/zimbabwes-health-system-chaos-20081121.

Amon, J. and Todrys, K. (2008) "Fear of Foreigners: HIV-related Restrictions on Entry, Stay, and Residence," *Journal of the International Aids Society*: 1–6.

Associated Foreign Press (AFP) (2009) "US Ends Zimbabwe Travel Warning, Sanctions Stay," available online at: http://www.reuters.com/article/2009/04/17/us-zimbabwe-usa-idUSTRE53G5LP20090417.

Bajardi, P., Poletto, C., Ramasco, J.J., Tizzoni, M. and Colizza, V. (2011) "Human Mobility Networks, Travel Restrictions, and the Global Spread of 2009 H1N1 Pandemic," *PLoS One*, 6(1): e16591.

Berger, S. (2008) "Zimbabwe's Neighbours Fight Cholera Outbreak," *Telegraph*, available online at: http://www.telegraph.co.uk/news/worldnews/africaandindianocean/zimbabwe/3660742/Zimbabwes-neighbours-fight-cholera-outbreak.html.

Brahmbhatt, M. and Dutta, A. (2008) *On SARS Type Economic Effects during Infectious Disease Outbreaks*, Washington, DC: World Bank.

BBC News (2003) "SARS Hit Airlines 'More than War'," available online at: http://news.bbc.co.uk/1/hi/business/2986612.stm.

CDC (2003) "Use of Quarantine to Prevent Transmission of Severe Acute Respiratory Syndrome–Taiwan 2003," *Morbidity and Mortality Weekly Report*, 52(29): 680–683.

Chan, M. (2009) "World Now at the Start of 2009 Influenza Pandemic," available online at: http://www.who.int/mediacentre/news/statements/2009/h1n1_pandemic_phase6_20090611/en/index.html.

Chonghaile, C. (2012) "Uganda Ebola Outbreak: Patients Flee Hospital amid Contagion Fears," *Guardian*, available online at: http://www.theguardian.com/world/2012/jul/29/ebola-uganda-outbreak-patients-flee.

Davies, S.E. (2010) *Global Politics of Health*, Cambridge: Polity.

Edelstein, M. (2012) "Validity of International Health Regulations in Reporting Emerging Infectious Diseases," *Emerging Infectious Diseases*, 18(7): 1115–1120.

European Centre for Diseases Prevention and Control (ECDC) (2009) *Guide to Public Health Measures to Reduce the Impact of Influenza Pandemics in Europe*, Stockholm: European Centre for Diseases Prevention and Control.

Fischer, J., Kornblet, S. and Katz, R. (2011) *The International Health Regulations (2005): Surveillance and Response in an Era of Globalization*, Washington, DC: Stimson Center.

Franke-Ruta, G. (2009) "White House Announces End to HIV Travel Ban," *Washington Post*, available online at: http://voices.washingtonpost.com/44/2009/10/30/obama_to_announce_end_to_hiv_t.html?wprss=44.

Ganczak, M., Barss, P., Alfaresi, F., Almazrouei, S., Muraddad, A. and Al-Maskari, F. (2007) "Break the Silence: HIV/AIDS Knowledge, Attitudes, and Educational Needs Among Arab University Students in United Arab Emirates," *Journal of Adolescent Health*, 40(6): 572.

Gostin, L.O. (2009) "Influenza A(H1N1) and Pandemic Preparedness under the Rule of International Law," *Journal of the American Medical Association*, 301(22): 2376–2378.

Gurniak, V. (2009) "Slovakia Closes Border Crossings with Ukraine amid A/H1N1 Fears," available online at: http://en.rian.ru/world/20091107/156749426.html.

Gushulak, B. and MacPherson, D. (2010) "Migration, Mobility and Health," in Institute of Medicine Forum on Microbial Threats, *Infectious Disease Movement in a Borderless World: Workshop Summary*, Washington, DC: National Academies Press.

Hogan, B. (2008) "Statement by Minister of Health Barbara Hogan on the Outbreak of Cholera in Zimbabwe and South Africa," available online at: http://www.info.gov.za/speeches/2008/08112711451003.htm.

Huang, Y. (2010) "Comparing the H1N1 Crises and Responses in the US and China," NTS Working Paper Series No. 1, Singapore: RSIS Centre for Non-Traditional Security Studies.

Human Rights Watch (HRW) (2007) *Chronic Indifference: HIV/AIDS Services for Immigrants Detained by the United States*, Washington, DC: Human Rights Watch.

Institute of Medicine Forum on Microbial Threats (2004) "Learning from SARS: Preparing for the Next Disease Outbreak, Workshop Summary," available online at: http://www.ncbi.nlm.nih.gov/pubmed/22553895.

International Aids Society (IAS) (2009) *HIV-specific Travel and Residence Restrictions*, Washington, DC: International AIDS Society.

Katz, R. (2009) "Use of Revised International Health Regulations During Influenza A (H1N1) Epidemic," *Emerging Infectious Diseases*, 15(8): 1165–1170.

Mahto, M., Ponnusamy, K., Schuhwerk, M., Richens, J., Lambert, N. and Wilkins, E. (2006) "Knowledge, Attitudes and Health Outcomes in HIV-infected Travellers to the USA," *HIV Med*, 7(4): 201–204.

MSF (2009) "Forced Closure of Refugee Reception Office Further Endangers Zimbabweans," available online at: http://www.msf.org.uk/article/forced-closure-%E2%80%9C refugee-reception-office%E2%80%9D-further-endangers-zimbabweans.

Nieburg, P., Morrison, J.S., Hofler, K. and Gayle, H. (2007) *Moving Beyond the US Government Policy of Inadmissibility of HIV-infected Noncitizens*, Washington, DC: Centre for Strategic and International Studies.

Pomfret, J. (2003) "SARS: The China Syndrome," *Washington Post*, May 17: 56.

President's Council of Advisors on Science and Technology (PCAST) (2009) *Report to the President on US Preparations for 2009 H1N1 Influenza*, Washington, DC: The White House.

Rushton, S. (2012) "The Global Debate over HIV-related Travel Restrictions: Framing and Policy Change," *Global Public Health*, published online October 31, DOI:10.1080/17441692.2012.735249.

Sekkides, O. (2010) "Pandemic Influenza: A Timeline," *Lancet Infectious Diseases*, 10(10): 663.

Svoboda, T., Henry, B., Shulman, L., Kennedy, E., Rea, E., Ng, W., Wallington, T., Yaffe, B., Gournis, E., Vicencio, E., Basrur, S. and Glazier, R.H. (2004) "Public Health Measures to Control the Spread of the Severe Acute Respiratory Syndrome During the Outbreak in Toronto," *New England Journal of Medicine*, 350(23): 2352–2361.

Tsang, K.W., Ho, P.L., Ooi, G.C., Yee, W.K., Wang, T., Chan-Yeung, M., Lam, W.K., Seto, W.H., Yam, L.Y., Cheung, T.M., Wong, P.C., Lam, B., Ip, M.S., Chan, J., Yuen, K.Y. and Lai, K.N. (2003) "A Cluster of Cases of Severe Acute Respiratory Syndrome in Hong Kong," *New England Journal of Medicine*, 348(20): 1977–1985.

UNAIDS (2004) "Statement to the 60th Session of the United Nations Commission on Human Rights, Agenda Item 10: Economic, Social and Cultural Rights."

UNAIDS (2009) *The Impact of HIV-related Restrictions on Entry, Stay and Residence: Personal Narratives*, Geneva: UNAIDS.

UNHCR (2009) "South Africa: End Strain on Asylum System and Protect Zimbabweans," available online at: http://www.unhcr.org/refworld/docid/49670ba4c.html.

US Department of State (2008) "Travel Warning: Zimbabwe," available online at: http://djibouti.usembassy.gov/em_dec12_08.html.

Vassileva, R. (2003) "Russia Tries to Stop SARS Express," available online at: http://articles.cnn.com/2003-05-20/world/sars.russia_1_sars-russian-academy-ill-passenger?_s=PM:asiapcf.

Wilson, K., von Tigerstrom, B. and McDougall, C. (2008) "Protecting Global Health Security through the International Health Regulations: Requirements and Challenges," *Canadian Medical Association Journal*, 179(1): 44–47.

WHO (1986) "AIDS: International Travel," *Weekly Epidemiological Report*, 61(4): 27.

WHO (1988) "Statement on Screening of International Travellers for Infection with Human Immunodeficiency Virus," WHO/GPA/INF/88.3.

WHO (2003) "World Health Organization Issues Emergency Travel Advisory," available online at: http://www.who.int/csr/sars/archive/2003_03_15/en/.

WHO (2005) "International Health Regulations," available online at: http://www.who.int/ihr/en/.

WHO (2008) "Cholera in Zimbabwe," available online at: http://www.who.int/csr/don/2008_12_02/en/index.html.

WHO (2009a) "Global, National Efforts Must be Urgently Intensified to Control Zimbabwe Cholera Outbreak," available online at: http://www.who.int/mediacentre/news/releases/2009/cholera_zim_20090130/en/index.html.

WHO (2009b) "No Rationale for Travel Restrictions," available online at: http://www.who.int/csr/disease/swineflu/guidance/public_health/travel_advice/en/index.html.

Writing Committee of the WHO Consultation on Clinical Aspects of Pandemic Influenza (2010) "Clinical Aspects of Pandemic 2009 Influenza A (H1N1) Virus Infection," *New England Journal of Medicine*, 362(18): 1708–1719.

Yermukanov, M (2003) "Kazakhstan Closes the Door to China as SARS Panic Spreads," available online at: http://www.cacianalyst.org/?q=node/1168.

6

CRIMINAL VIOLENCE, DISPLACEMENT AND MIGRATION IN MEXICO AND CENTRAL AMERICA

Sebastián Albuja

Introduction

People moving as a result of or affected by criminal violence in Mexico, Central America and elsewhere often face limited possibilities for protection. First, because criminal violence and its impacts are rarely interpreted as creating humanitarian crises, countries facing these situations are often reluctant to implement a protection and humanitarian assistance approach to assist victims. Responses are often dominated by the prevalent retributive approach to crime, which focuses on punishing criminals, reducing crime, and re-establishing public order, rather than assisting victims.

Second, because criminal violence is difficult to situate within the normative categories of international humanitarian law (IHL), it is not readily clear if IHL protections apply (and how to promote and enforce them). Furthermore, when people fleeing criminal violence cross borders seeking safety, it is unclear if countries of refuge should grant asylum on the basis of refugee law. Additionally, it is often unclear if protection-mandated agencies, notably the United Nations High Commissioner for Refugees (UNHCR) and the International Committee of the Red Cross (ICRC), as well as protection-orientated International Non-governmental Organizations (INGOs), should intervene, as such situations often fall outside the mandates and mission statements of these agencies, even if their intensity reaches or exceeds that of traditional internal armed conflicts where humanitarian agencies normally operate.

Third, not everyone's linkage to place and everyone's mobility are affected equally by criminal violence. Some people see their livelihoods affected by intense criminal violence, and move either anticipating the impact of this violence on their source of income, or because their means of subsistence has already been affected by it. Others flee when friends or family members have been attacked or killed, or after receiving direct threats. Some people have a mixed motivation to flee, kindled both by fear of violence and by the desire for greater economic opportunity. In sum, the diverse ways in which criminal violence relates to human mobility create challenges for protecting those rendered vulnerable by the phenomenon.

In addition to *causing* or influencing migration, intense criminal violence also *affects* people in transit. The vulnerability of migrants affected by criminal violence while transiting through a third country is particularly acute, as they lack information and knowledge about the environment, location, and intensity of threats, as well as support networks.

This chapter examines these issues in three sections. The first section describes the evidence of violence that causes migration, as in the case of Mexicans fleeing drug-cartel violence, and of violence that affects migration, as in the case of Central American migrants crossing through Mexico. The second section discusses the extent to which these situations of violence constitute a humanitarian crisis. The third argues that responses should go beyond the traditional approach to criminal violence, which centers on punishing and/or neutralizing offenders, and should focus on the victims of criminal violence as people in need of protection. To this end, it examines the applicability and potential for protection offered by existing international protection instruments, and the responses of the Mexican and US governments to people migrating as a result of criminal violence and people affected by it while in transit. On the basis of that analysis, it shows that new and broad interpretations of existing legal frameworks may offer protection to people on the move in contexts of criminal violence. However, even if broad legal interpretations of existing frameworks may provide protection, practical implementation is limited and remains the biggest challenge.

The impact of drug-cartel violence on migration in Mexico

Following the launch of a new security strategy by the Calderón administration in 2007, which deployed the military to fight drug cartels in key locations, a wave of intense criminal violence broke out in Mexico. The military intervention drove cartels to fight not only the military and federal police, but also, and more significantly, one another, for the control of drug smuggling routes (Guerrero-Gutiérrez 2011).

According to the latest official information, 47,000 people were killed between 2006 and September 2011 (Government of Mexico 2012). Because the government has not disclosed official figures since then, the exact number to date is unknown, but some estimates put it at 70,000 by the end of Calderón's six-year term (Movimiento por la Paz con Justicia y Dignidad 2012).

Criminal violence that causes migration: forced displacement linked to drug-cartel violence in Mexico

Beyond the most visible impacts of drug cartel violence—its gruesome killings and attacks—one of its outcomes that has remained hidden and under-investigated has to do with migration. Civil society organizations, academic institutions and the media have begun to document cases and patterns of forced displacement caused by drug-cartel violence and how it relates to migration. However, because most of the displacement has taken place individually, except for a few cases of mass displacement in the states of Tamaulipas (2010) and Michoacán (2011), it has been difficult to document at best or neglected at worst.

Efforts have been made to disentangle or distinguish migration that is forced from migration that is not, in the context of intense criminal violence. Using data from the latest national census, research by the Internal Displacement Monitoring Centre (IDMC) examined the impact that criminal violence has had on migration in the States most affected by it: Baja California, Chihuahua, Coahuila, Durango, Guerrero, Michoacán, Nuevo León, San Luis Potosí, Sinaloa, Sonora, Tamaulipas and Veracruz. Together these states account for only 38 percent of Mexico's population, but for 68 percent of homicides.

Within those states, the study measured if there is a difference in outbound migration flows between localities with insecurity at the national average level, but which have not been hit by intense violence since 2007—i.e. localities with normal crime conditions—and localities with similar average security conditions, which have suffered intense violence since 2007. Overall, in the 104 municipalities with the highest levels of violence included in the study's analysis, the rate of displacement was fifteen times higher than in municipalities without high levels of violence. When the effect of other drivers of migration, including economic and demographic conditions and urbanization, was controlled for using statistical tools, the number of people leaving violent municipalities was four to five times higher than those leaving non-violent municipalities (IDMC forthcoming). As it used national census data, this study did not examine cross-border displacement.

Violence has also been associated with loss of livelihoods, at least in certain locations, where the general climate of insecurity and specific threats by the armed groups have been associated with economic decline. A common practice by the drug cartels is to threaten and demand payments from small business owners, including small corner stores or street businesses. For example, the Chamber of Commerce in Ciudad Juárez has reported that roughly 11,000 businesses have closed in the last three years because of the violence there (*El Universal* 2010a). This decline in livelihoods and subsistence associated with violence has in turn also forced people to move.

Ciudad Juárez: along the border, an epicenter of violence

Ciudad Juárez, located in the State of Chihuahua on the Mexico–US border, has experienced high levels of violence for more than two decades, but there was an unprecedented rise after 2007, following the military intervention, when the Cartel de Sinaloa began to challenge the dominance of the Cartel de Juárez and its control of trafficking routes. Gun battles became common on the city's streets even during the day, and residents said they were afraid go out at night. The homicide rate rose from an average of 234 per year between 2000 and 2006 to 316 in 2007, before leaping to 1,600 in 2008 and 2,600 in 2009 (Universidad Autónoma de Ciudad Juárez 2009, 247). In 2009, the rate was equivalent to 47 per 100,000 people according to official sources, and up to 191 per 100,000 according to unofficial sources, making Juárez the most violent city in the world. Some 28 percent of Mexico's homicides took place there. Since that point, violence in Juárez has decreased and in 2012 even unofficial sources counted only 56 homicides per 100,000 inhabitants, a 70 percent reduction in violence (Consejo Ciudadano para la Seguridad Pública y Justicia Penal AC 2013). Even as violence decreases, displacement persists.

Displacement from Ciudad Juárez and its surrounding areas has gone largely unnoticed, in part because it has taken place alongside economic migration within Mexico and across the border. However, in 2009 and 2010, the Universidad Autónoma de Ciudad Juárez, a large research university, carried out a comprehensive survey in Ciudad Juárez and its surrounding areas. The survey asked a representative sample of families across the city if family members had left in the last three years, and if so how many had left and why. The results showed that as many as 230,000 people had fled because of the violence in the city since 2007. Respondents said that roughly half crossed the border into the United States, while the other half took refuge in other Mexican states, becoming internally displaced persons (IDPs). Survey results showed that those who have fled within the country have gone predominantly to the States of Durango, Coahuila and Veracruz, or to other locations within Chihuahua.

While this research did not inquire into people's choice of destination, anecdotal information shows that people fled within the country because of strong links to their state of origin—many people living and working in Ciudad Juárez had come from southern states to work there—and strong family networks elsewhere. On the other hand, people moved across the border because they had the financial capital and support networks to help them move across.

The high number of abandoned homes complements evidence gathered in this survey. The Municipal Planning Institute of Ciudad Juárez has reported that there are up to 116,000 empty homes in Juárez (CNN 2011b). Of those homes, 5,000 were built by the Mexican National Institute for Workers' Housing (INFON-AVIT) and sold to workers with low-interest loans. It is hardly imaginable that workers would abandon their homes and lose all payments made toward ownership, if staying was an option.

Criminal violence that affects migration: Central American migrants' vulnerability as they transit through Mexico

The safety of Central and South American migrants making their way to the United States through Mexico has become greatly threatened as a result of increased insecurity and drug-cartel violence. It has been estimated that 70,000 Central and South American migrants have disappeared while crossing through Mexico in the last six years (Sherman 2012).

Most Central and South American migrants travel through Mexico using three main routes and unsafe means of transportation (Chávez Galindo and Landa Guevara 2011), including double-bottomed trucks and cargo trains. The so-called Beast, or Train of Death, which goes from southern Mexico to Mexico City, was used by an estimated 20,000 migrants in 2010 (Chavez 2011). Cargo trains do not stop, and migrants have to jump on while the train is moving, which leads to very high rates of injury from falls or getting run over by the train, which causes migrants to lose limbs. While on the train, migrants face dehydration, sexual abuse and theft by gangs running the trains (Chavez 2011). Additionally, the police and/or private security firms frequently carry out operations in which they use violence to get migrants off the trains, including throwing them off.

Many of the abuses against migrants are systematic attempts by criminal groups to generate profit from migrant flows and control their paths. Cartels have increasingly set up a structure of extortion for those who travel on the trains. Criminal operations typically extort $100 for each migrant to travel from the southern border city of Palenque to Coatzacolcos, which is 310 km, and $400 to travel to Tierra Blanca (556 km away) (*La Jornada* 2013). Those who do not pay the charge are thrown off the train or kidnapped.

Kidnapping of migrants by cartels is "widespread and systematic" (Comisión Nacional de los Derechos Humanos 2011). Far from being opportunistic, kidnappings often take place en masse and are perpetrated by the drug cartels. An estimated 20,000 migrants are kidnapped each year (Brodzinsky 2012); the Ombudsman says at least 11,333 migrants were kidnapped in the six months between April and September, 2010 (AlertNet 2011). In the final months of 2011 and the beginning of 2012, there was a wave of kidnapping. A number of high-profile cases, such as the sixty-one migrants rescued in Coahuila in October of 2011 and the seventy-three rescued in Tamaulipas in February of 2012, raise questions about whether the overall numbers have actually increased since the 2011 National Commission for Human Rights (CNDH) study (Organización Internacional para las Migraciones 2012). Kidnapping events tend to be concentrated along the train routes in the south, though some also occur at the border. The CNDH estimated that 67.4 percent of kidnappings occurred in the southeast, whereas 29.2 percent occurred in the north and just 2.2 percent in the center of Mexico (Comisión Nacional de los Derechos Humanos 2011).

State officials, bus drivers for private companies, and state authorities have been accused of corruption and are allegedly involved in organized crime. Members of each group have been known to sell migrants' travel information to criminal groups, facilitating kidnappings (Cote-Muñoz 2011). Further, the CNDH collected evidence of instances when Federal Police and Instituto Nacional de Migración (INM) officials noticed kidnapping events but did not intervene to support victims or detain perpetrators (Comisión Interamericana de Derechos Humanos 2010). Of the 238 victims and witnesses that CNDH officials interviewed in 2009, ninety-one said that their kidnapping was the direct responsibility of state officials and ninety-nine said that state officials were complicit in their kidnapping (Amnistía Internacional 2010).

Kidnapping on this scale requires an intricate network of safe houses, weapons, communications equipment, and vehicles. The Zeta cartel is responsible for a large portion of kidnapping operations since it operates in most of the areas traversed by the train where migrants travel (Comisión Interamericana de Derechos Humanos 2010). Criminal organizations benefit in different ways from kidnapping migrants. Kidnapped migrants are forced to reveal telephone numbers of family members who are asked for ransom (Talsma 2012). An estimated nine out of ten migrants are threatened with death, and typically shown firearms as part of the threat, if they do not reveal their family's phone numbers or if their families do not pay ransom (Comisión Nacional de Derechos Humanos 2009). The CNDH estimated that the 9,758 kidnappings carried out between September of 2008 and February of 2009 generated 25 million US dollars in revenue for the cartels (CNDH 2011). Migrants whose families do not pay are also at times forced to join the criminal groups or to smuggle illegal weapons or drugs (Jimenez 2012; see also CIDEHUM 2012).

Although only 14 percent of migrants travel by train (Chavez 2011), they constitute the most vulnerable group because they tend to be the poorest (Chávez Galindo and Landa Guevara 2011). Those on the train are disproportionately Honduran: 54 percent of Honduran migrants travel by train, compared to 5 percent of Salvadorans and 7 percent of Guatemalans (El Colegio de la Frontera Norte 2013). Youth, women and children who travel by train are especially vulnerable, particularly as sexual abuse of migrants is endemic: it has been estimated that six in every ten female migrants are subjected to some form of sexual abuse during their journey through Mexico (Chávez Galindo and Landa Guevara 2011).

Kidnapped migrants have repeatedly been mass murdered in the state of Tamaulipas, along the northern border. A total of 177 bodies of migrants have been found in at least a dozen mass graves in the locality of San Fernando (CNN 2011a). In 2010, mass graves of Central and South American migrants were found in San Fernando, Tamaulipas, when an Ecuadorian immigrant managed to escape from the Zetas Cartel (El Universal 2010b). In 2011, another grave with 32 bodies identified as those of migrants was found in the same locality. Investigators believe these migrants had been kidnapped while traveling by bus toward the Mexico–US border. In one of the latest massacres, 49 decapitated bodies with their feet and hands cut off were found far from the US border in the state of Tamaulipas. It is believed that these migrants were kidnapped in small groups (Rodriguez 2012).

It is difficult to establish if violence against Central American migrants has reduced the number of migrants crossing through Mexico. From 2005 to 2010, there was a downward trend in the number of Central Americans traveling through Mexico: from a peak of 433,000 in 2005 to just 140,000 in 2010. However, that trend seems to be reversing. Both Mexican and United States officials reported more arrests of Central Americans in 2012 than 2011. (Mexico apprehended 85,100 Central Americans in 2012 compared to 63,072 in 2011 (Centro de Estudios Migratorios 2012), while US officials apprehended 99,013 non-Mexicans on the Mexican border in 2012 compared to just 54,098 in 2011 (Customs and Border Protection 2013).)

Some of this increase in migration could be related to increasing violence in Central America. San Pedro Sula, Honduras, replaced Ciudad Juárez as the most violent city in the world in 2012, with 169 homicides for every 100,000 inhabitants (Consejo Ciudadano para la Seguridad Pública y Justicia Penal AC 2013).

When does criminal violence become a humanitarian crisis?

Humanitarian crises exist when there is massive suffering and loss of life. This book's conceptual framework defines humanitarian crises as "any situation in which there is a widespread threat to life, physical safety, health or subsistence that is beyond the coping capacity of individuals and the communities in which they reside" (Martin, Weerasinghe and Taylor, Chapter 1 in this volume). The indicators and data described above suggest that the constitutive elements of this definition exist in the current situation confronted by Mexico.

First, the intensity and pervasiveness of the violence poses a widespread threat to life: anywhere between 50,000 and 70,000 people killed in a six-year period is by any measure an enormous loss of life, and a threat to people's physical safety and health. The systematic and large-scale kidnapping of migrants—20,000 kidnappings per year—as well as the mass murders of migrants—multiple mass graves of migrants uncovered in recent years—add to the violence's toll and make its threat to life and physical security widespread and constitutive of a humanitarian situation.

Second, as mentioned above, the violence has also been associated with loss of livelihoods and subsistence, which pushes people beyond their coping capacities, forcing them to leave.

Is Mexico a country at war?

While violence and insecurity need not happen in the context of an internal armed conflict to constitute a humanitarian crisis, the existence of a conflict would reinforce the view that Mexico's situation of violence in fact amounts to a humanitarian crisis.

The intensity of the violence has prompted many to assert that the country confronts an internal armed conflict. Some have referred to Mexico's wave of violence as an insurgency, including former US Secretary of State Hillary Clinton, whose remarks in 2010 prompted an angry reaction from the Mexican government. In 2009, the Pentagon released a study arguing that Mexico was in danger of becoming a failed state because of drug cartel violence. Some commentators have talked about the growing "Colombianization" of Mexico (*El País* 2010), and others have drawn comparisons between present-day violence in Mexico and Colombia's violence two decades ago (Beittel 2011). Furthermore, the media and the public in general refer to this violence variously as the "drug war," drug violence and cartel violence, among others.

In this context of politically charged, vague, and imprecise assessments, there is a need for objective standards with which to evaluate and assess Mexico's violence. While establishing the nature of violence in Mexico is irrelevant to protecting people—the nature of the warring parties and the fairness or legitimacy of the violence are all irrelevant when it comes to protection—establishing the status of this violence is relevant in terms of the legal grounds for an intervention by the ICRC and in terms of the obligations of the groups implicated in the violence.

Ultimately, the answer to this question comes from examining Mexico's violence under IHL. Common Article 3 of the Geneva Conventions does not establish a definition of a non-international armed conflict. Article I of Additional Protocol II provides elements to determine when the lower threshold for an internal armed conflict has been exceeded, namely the intensity of the violence and the organization of the groups. Case law, particularly jurisprudence by the International Criminal Tribunal for the Former Yugoslavia (ICTY) has provided further elements to determine when these standards are met (see further Vité 2009).

Criteria established by the ICTY to assess the intensity element are duration of violence, type of weapons used, the use of military rather than police forces to combat the armed group, and existence of displacement of civilians. First, violence in México is hardly sporadic—it has been going on for four years and it includes frequent acts of violence. Second, the weapons used are advanced, sophisticated, and high caliber—financed with the dividends of the illegal drug trade, and brought illegally from the United States.

As regards the criterion of the means used by the government to combat the groups, the Mexican military was sent in early 2007 to combat the cartels in various states. In fact, violence between the cartels and between the cartels and the state exploded after President Calderón, who came to power in December 2006, sent thousands of troops to combat the cartels, which up until then had been operating under tacit agreements to respect one another's turf.

The military intervention has failed by all standards. The Calderón administration regularly boasted of its "victories" against the cartels: the killing or capture of some twenty kingpins, including the leader of the Cartel del Golfo, Tony Tormenta; the capture of the leader of the Cartel de Sinaloa, Edgar Valdez; and the seizing of hundreds of tons of cocaine. However, the impact of the intervention on security as a whole vastly outweighs the gains made. Violence increased dramatically after the military crackdown, and human rights violations by the military have become widespread, including acts of torture, forced disappearance, and sexual harassment (Meyer 2010; Human Rights Watch 2011).

At the beginning of 2010, under mounting pressure from the population in Ciudad Juárez, most of the military forces were withdrawn from the city and were replaced by federal police. Nevertheless, soldiers still remain in various hotspots, including Ciudad Juárez, and military presence has also recently been registered in Ciudad de México, prompting criticism and fear amongst the population in the country's capital (*La Nación* 2011).

Of the criteria established by the ICTY, perhaps the hardest to describe and measure is displacement of civilians because this is an outcome of the violence that is invisible and has been scarcely documented. This consequence of the violence constitutes the bulk of the next section, but for now suffice it to say that displacement of civilians has been a significant if unseen effect of the drug war in Mexico.

The second element established by Additional Protocol I and further specified by case law is the organization of the groups taking part in the hostilities. This is the criterion that is hardly met in the Mexican case. On the one hand, Mexican drug cartels have historically operated as organizations with well-defined leadership and hierarchy—often based on family links guaranteeing extreme loyalty—and have had functional and well-defined structures, including firepower, transportation, money laundering, and intelligence cells (Alvarez 2006). Further, under criteria used by reputable sources such as the Uppsala University UCDP database, these are "formally organized groups" (UCDP n.d.) However, since 2007 when the violence broke out, cartels have increasingly become devoid of leadership and led by thugs with no allegiances (Pacheco 2009). Furthermore, it has been argued that the recent government offensive that has killed or imprisoned the cartel leaders has resulted in atomized and horizontal structures, fleeting alliances and internal conflict (Loyola 2009, 36–38).

Beyond the two core elements of intensity of violence and organization of the groups, the relevance of the armed groups' intentions or motives has remained less clear. The ICTY determined in the *Limaj* case that the motives of armed groups participating in non-international struggles are irrelevant for a determination of the application of IHL (ICTY Prosecutor v. Limaj 2005). Under this unequivocal precedent, the non-political nature of violence in Mexico would be irrelevant for a determination of whether it constitutes an internal armed conflict.

In conclusion, a prima facie analysis of Mexico's violence under criteria established by IHL shows that the situation meets all the criteria set out in Additional Protocol II and its jurisprudential interpretation, except the one regarding organization of the groups. Nevertheless, Common Article 3 of the Geneva Conventions, which does not provide a definition or a specific lower threshold for the existence of a non-international armed conflict, may still apply.

In terms of a potential ICRC intervention, this means that it could exercise the right of humanitarian initiative recognized by the international community and contained in Common Article 3. As regards the obligations for the warring parties that arise under IHL, given the nature of the drug cartels and their use of violence, significant efforts would be required to get the groups to respect IHL distinctions of combatants and civilians, particularly since drug cartels do not have concerns about legitimacy or acceptance by the people, as insurgent groups do.

Other analytical tools

In addition to an IHL legal analysis, other tools are also useful for breaking down Mexican drug-cartel violence analytically. Using an established set of criteria, the

Heidelberg Conflict Barometer categorized the Mexican situation as a war for two consecutive years (2010 and 2011), upgrading it from 2009 when it was labeled a "severe crisis," and putting it in the company of the five other wars fought in the world in 2010 (Afghanistan, Iraq, Pakistan, Somalia and Sudan) and the twenty wars fought in 2011 (University of Heidelberg 2010, 2011). Under the Heidelberg Barometer's definitions,

> a war is a violent conflict in which violent force is used with a certain continuity in an organized and systematic way. The conflict parties exercise extensive measures, depending on the situation. The extent of destruction is massive and of long duration.
>
> *(University of Heidelberg 2008, 1)*

A host of other typologies of armed violence also provide useful tools to analyze the violence currently taking place in Mexico. One such particularly useful typology suggests considering the following elements of armed violence: whether the armed groups have political intentions, whether these groups seek change of the status quo, whether they use violence for psychological or physical ends, and whether they seek territorial control (Vinci 2006; Williams 2008; Schneckener 2009).

Mexican drug cartels clearly do not have a political agenda or an ideology; they are and historically have been entirely profit-driven organizations (Alvarez 2006). Likewise, they do not pursue other political goals such as change in the status quo; again, profit is their overarching motive.

The second two elements in the typology are harder to establish in the case of the Mexican drug cartels. While these groups' motivation for, and means of, using violence, are primarily psychological—gruesome acts of violence intended to scare off rival groups and intimidate the population have been the norm—each group also seeks to strengthen their position and domination over other groups and over the population in general, forcibly recruiting people to work and fight for them.

Likewise, even though profit seems to be their overarching goal, there is evidence that the groups have intentions to control territory beyond what is necessary to pursue their economic goals. Their dominance in some parts of Tamaulipas where there is a near-absolute lack of state presence has become so pervasive that the cartels effectively control territories as the sole authority. The lack of state presence and control in these areas means that the government has limited possibility to provide protection.

A study prepared for a Committee in the Mexican Senate showed that up to 195 municipalities in Mexico are under control of the drug groups, and another 1,536 are infiltrated by organized crime, out of roughly 2,500 municipalities in the country (Beittel 2011). Journalistic sources have also exposed the control of drug groups over entire populations. Further, the infamous Wikileaks cables revealed that the Mexican government fears that it might have lost territorial control in some of Mexico's northern States. A US Embassy cable quotes a high-ranking Mexican official saying that the Government was losing control of some regions to the drug gangs (BBC News 2010).

Moreover, their acts of violence have increasingly targeted the state. The cartels have increasingly attacked public officials, judges and investigators—they have assassinated as many as eleven mayors of small towns, and have also targeted and killed journalists (*Christian Science Monitor* 2010)—in attacks that were not essential to carry out their illegal business, but aimed to make statements about their presence and dominance. In this sense, these groups' mode of operating differs from other organized criminal networks that operate silently and covertly and wish to remain that way. The Mexican cartels seem to have no qualms about making their presence and domination visible.

But the drug cartels do not exercise their power only through violence. Corruption and co-optation of authorities is equally if not more pervasive. The "political–criminal nexus," or collaboration and accommodation between criminal and political actors at all level of government (Godson 2003), is a hallmark feature of the way in which the cartels operate in Mexico.

A case that epitomizes this took place in a prison in Gómez Palacio, Durango, where prison administrators let inmates out during the night, and provided them with guns and vehicles so that they could kill and then return (*Guardian* 2010). There is no question that the drug trade and its associated violence have prospered because they have long been allowed or tolerated by political authorities (Universidad Autónoma de Ciudad Juárez 2009, 231). Finally, regarding the spatial dimension of the violence, it takes place in both rural and urban spaces, but not all cities are affected equally. Important cities that have been affected are, most notoriously, Ciudad Juárez, but the increase of violence in prosperous and heretofore safe towns, notably Monterrey, has captured increasing attention.

In Mexico, violence is certainly not *contained* within well-defined urban spaces such as the walled *favelas* in Río de Janeiro, Brazil. In Tamaulipas, the state that follows Chihuahua in terms of number of killings, violence perpetrated by these groups has taken place in rural areas with very little state presence. Thus, what distinguishes this violence is not necessarily its geographic location—although this certainly matters at the time of planning responses—but the nature of the groups involved, their reasons for using violence, their territorial control intentions, their position in relation to government institutions, and the resources they have at hand.

Legal frameworks and practical responses

Deviant acts that violate criminal law (including robbery, assault, battery, rape and murder) are a common occurrence in every society—or, as Durkheim put it, crime is normal, inevitable, and serves a social function—and are predominantly dealt with through retributive justice focusing on punishing offenders in proportion to the gravity of their violations.

The prevalent approach to crime, focusing on punishment of offenders, largely neglects crime's impact on victims, and has been widely criticized on these grounds.

Restorative justice has long proposed shifting the focus of the response to crime to give greater importance to its impact on victims: "Traditional criminal justice seeks answers to three questions: What laws have been broken? Who did it? and What do the offender(s) deserve? Restorative justice instead asks: Who has been harmed? What are their needs? Whose obligations are these?" (Zehr 2002, 20).

These arguments take on a whole new dimension when crime reaches levels of pervasiveness and violence that go beyond normal conditions of criminality, as described above. In such contexts when the state response focuses exclusively on punishing offenders and combating criminal behavior, people moving as a result of criminal violence or affected by it while on the move—and who therefore have protection needs—are left with no recourse.

The existing international protection framework—the various universal, regional, binding and non-binding instruments of refugee law, IHL and human rights law—provides the desired emphasis on the rights, needs and vulnerabilities of victims, including those who move as a result of criminal violence or are affected by it. But assessing the potential means for protecting people migrating as a result of criminal violence must take into account the diverse circumstances of mobility and vulnerability in the context of intense criminal violence.

The protection structure's focus on forced or coerced movement as its trigger does not adequately respond to the complex mobility circumstances of people in contexts of criminal violence. It provides coverage for some people moving in the context of criminal violence, but is insufficient in others.[1]

People displaced internally seeking safety, for instance after receiving direct threats, or following violence with which they cannot cope, are protected under the Guiding Principles on Internal Displacement, whether their displacement happens as a consequence of violence or in anticipation of its effects. At the other end of the divide, people moving solely to *improve* their economic situation are not IDPs.

The Guiding Principles' descriptive identification of people who may be IDPs includes persons who flee "situations of generalized violence." The situation in certain localities in Mexico may be understood to be one of generalized violence. However, in addition, under the Guiding Principles, *coercion must exist* for people to be IDPs.

But in some situations, people move after their source of income has declined or become less sustainable as a result of or in connection with the pervasive climate of violence and insecurity. For example, people in Ciudad Juárez have moved because their small business became less profitable or threatened to fold when due to violence and insecurity, people stopped shopping or eating in the neighborhood where it is located, or because the market, shop, or corner store where they worked closed. Other people have moved anticipating a looming economic decline linked to violence, or in other words, predicting that their business or source of income will also collapse as a result. In these cases, while people have not been directly coerced to move, their choice to move is not entirely free either—it is linked to the fact that the sustainability of their livelihood or their quality of life in their current place of residence has declined because of the effects of violence and insecurity.

A broad interpretation of the Guiding Principles may provide relief to people in this situation. People who move seeking a source of income, but who would have not chosen to move were it not for the negative impact of insecurity and violence on their livelihoods—in other words, they do not move solely to *improve* their current economic circumstances as a free choice—would require protection as IDPs because they were forced to move because of the climate of insecurity.

For their part, people crossing borders seeking safety and security as a result of criminal violence, whether as a direct consequence of it or anticipating threats, are specifically covered by the expansive definition of a refugee in the 1984 Cartagena Declaration, which includes people that flee the threat posed by generalized violence or circumstances that seriously disturb public order. Under the 1951 Refugee Convention protection is available on a case-by-case basis to those who can show a well-founded fear of persecution based on one of the five grounds enumerated in the Convention. Both instruments apply whether people move anticipating harm or as a consequence of it. People who move in search of better economic opportunities are not protected by the Cartagena Declaration or the Refugee Convention.

However, as with people moving internally, people also move across borders as a result of criminal violence, being neither directly persecuted for a specific reason, nor moving solely for economic reasons, but rather for a mix of both.

Complementary protection may also offer relief to people who have fled criminal violence, but the threshold above which it is applicable makes it an even narrower avenue for protection. The Convention Against Torture (CAT) prohibits states from removing an individual to a place "where there are substantial grounds for believing that he would be in danger of being subjected to torture" (Article 3(1)). Compared to the requirements under the 1951 Refugee Convention, the threshold is higher for Article 3 in CAT as there must be a "foreseeable, real and personal risk of torture," which goes "beyond mere theory or suspicion" or "mere possibility of torture" (Goodwin-Gill and McAdam 2007).

In addition, the term torture, as defined in the CAT, contains a public requirement, which means that an act of torture has to be carried out by a public official or with their "consent" or "acquiescence." In a similar way, Article 7 of the International Covenant on Civil and Political Rights has been interpreted on certain occasions to contain a principle of non- *refoulement* for cases of risk of torture or cruel, inhuman or degrading treatment or punishment (Mandal 2005). In the cases examined above for the US context, all petitioners' requests for non-removal under CAT were denied.

Finally, asylum-seekers could also benefit from complementary protection under the Convention on Enforced Disappearances. Article 16 prohibits the *refoulement* of individuals by a State Party to a state where there is a risk of them being subjected to enforced disappearances.[2] As with the principle of *non-refoulement* under the CAT, this is tied to a public requirement, as an enforced disappearance is defined as "arrest, detention, abduction or any other form of deprivation of liberty" by state agents or people "acting with the authorization, support or acquiescence of the State," which

is followed by a "refusal to acknowledge the deprivation of liberty or by conceal-
ment of the fate or whereabouts of the disappeared person" (International Convention for the Protection of All Persons from Enforced Disappearance, Art. 2). Again,
in the case of enforced disappearances caused by drug cartels, the applicant has the
burden of proof to demonstrate some kind of public involvement.

While reports of cases of torture inflicted by Mexican military or police officers
have been made public, most such acts are perpetrated by drug cartels, i.e. non-
state actors (Buchanan 2010, 57). In these cases, the applicant would have to prove
that these acts are undertaken with the consent or acquiescence of public officials.
According to some authors, this could be established in the Mexican case even
where there is no direct involvement of government officials because of the large
extent of corruption among police officers (Buchanan 2010, 58).

In addition to these universal instruments, regional human rights instruments
also provide a potential avenue for protection. The European Qualification Directive—the first supra-national instrument that seeks to harmonize complementary
protection (McAdam 2005)—provides for a similar prohibition from return in
Article 2 (e) on subsidiary protection. This article applies to third country nationals
and stateless persons who do not qualify as refugees but who are in need of international protection and who, if returned to their country of origin or residence, are at
a risk of suffering "serious harm." As defined in Article 15 of the Directive, "serious
harm" is (a) death penalty or execution, (b) torture or degrading treatment or punishment, or (c) "serious and individual threat to a civilian's life or person by reason
of indiscriminate violence in situations of international or internal armed conflict."
The extent to which the situation in Mexico can be regarded as an internal armed
conflict, discussed above, thus impacts the applicability of this form of relief.

Table 6.1 summarizes these different situations of mobility in Mexico and the
legal protection offered by existing frameworks.

The Mexican government's limited protection to people fleeing within Mexico

Response to internal displacement in Mexico has been limited predominantly
because of a lack of will to acknowledge the issue and address it systematically. The
census data described in the first section shows that locations with very high levels
of violence have lost population. Of the people that have moved from violent locations, some have been coerced to move, but some have also likely moved because
of livelihood decline linked to the violence. Data also show that people who have
left violent locations seeking safety struggle to resume their lives and have more
limited access to housing, education and employment than people who have not
been displaced (IDMC forthcoming).

The Mexican government has yet to fully acknowledge that cartel violence
is causing people to move (coerced or not) and to set up a response. References
to displacement caused by drug cartel violence have been made by government
officials, but usually to discredit or minimize the existence of the phenomenon

TABLE 6.1 Different situations of mobility in Mexico, and legal protection offered by existing frameworks

Location and temporality \ Motivation	*a. Safety and security*	"Between a and b" "Mix of a and b" "Neither a nor b"	*b. Economic and livelihoods*
1. *Internal movement, anticipation*	Protected under guiding principles and national law	Protected as IDPs under a broad interpretation of guiding principles	Human rights law and national law
2. *Internal movement, consequence*	Protected under guiding principles and national law	Protected as IDPs under a broad interpretation of guiding principles	Human rights law and domestic guarantees
3. *Cross-border movement, anticipation*	– Protection under refugee law on case-by-case basis (persecution + statutory ground) – Cartagena declaration – Prohibition of return under complementary protection regimes on case-by-case basis	Cartagena Declaration	Not protected (except for minimal human rights protection)
4. *Cross-border movement, consequence*	– Protection under refugee law on case-by-case basis (persecution + statutory ground) – Cartagena declaration – Prohibition of return under complementary protection regimes on case-by-case basis	Cartagena Declaration	Not protected (except for minimal human rights protection)

(*Reforma* 2011). The Federal Government has not put in place any mechanisms to respond to displacement since the violence broke out.

An exception is the Office for the Victims of Crime (Províctima), which was created by Presidential Decree in September 2011, with a mandate to assist people affected by kidnapping, forced disappearance, homicide, extortion and human trafficking. The agency has also assisted people displaced by one of the five crimes comprised in its mandate, for example by helping people obtain identity documents and preventing forced evictions and forced sales of property. Additionally, Províctima is mandated to create a specific registry for displacement cases, but to date this has not yet happened.

But there is a degree of uncertainty over the agency's future, particularly now that the Government has not signed the historic "General Law for the Protection and Redress of Victims of Human Rights Violations Caused by Violence," which was approved by congress in May 2011. The text of the Law confirms Províctima's mandate, which would allow it to become a key player in protecting and assisting IDPs.

Separately, the National Human Rights Commission has, since 2011, taken complaints of people displaced by violence, and is in the process of drafting a protocol to guide its attention to IDPs (*El Universal* 2012).

The coexistence of sharply contrasting strong and weak governance structures in Mexico negatively impacts the potential for protection of people displaced by violence, especially with regard to an international intervention reaching people in need. Local governments in whose jurisdictions the violence takes place are resource poor, plagued with corruption and co-opted by the very illegal groups they are supposed to fight. They are thus utterly unable to provide protection to the populations affected by the violence. At the same time, though, the Federal Government is powerful and professionalized, and may have little inclination to request support from foreign and multilateral humanitarian agencies, whose intervention is much needed at the local level. Mexico is a country with highly nationalistic values and a well-developed sense of sovereignty, and it is not insignificant that it is the world's twelfth economic power (in terms of GDP), after Spain, Australia and Canada.

The Mexican government's response to people affected by violence while transiting through Mexico

In 2011, in response to pressure from humanitarian groups and international organizations, Mexico passed a new migration law to increase protection for migrants transiting through the country. The law guarantees access to healthcare, education and legal protection regardless of documentation status. It also increases accountability for Mexican migration officials and makes them the sole enforcement authorities related to migration status, with the hopes of reducing the potential for abuse by other branches of law enforcement. The statute provided for a procedure to denounce abuses and crimes and required migration officials to provide information on rights to all migrants in detention (Sin Fronteras 2012).

Part of the effort to provide protection to migrants involves international and multilateral collaboration. Central American governments have taken steps to protect their migrant citizens, including giving input during the drafting process of the 2011 migration law. El Salvador, Guatemala, and Honduras have also pooled some of their consular resources to create joint offices in Mexico to serve the needs of Central Americans there, such as providing information on migrant rights (Papademetriou, Meissner and Sohnen 2013). In 2011, the Salvadoran government inaugurated a call center where Salvadoran migrants in transit can call for free for consular assistance, to denounce an abuse, and to talk to family in El Salvador. The Salvadoran government disseminates information about this service along with information on the migrants' rights (Zuñíga 2011). Mexico is also starting to collaborate with international organizations. Recently, the Mexican government, on the one hand, and UN agencies and IOM, on the other, signed a cooperation agreement to strengthen the protection of vulnerable migrants crossing through Mexico. The agreement creates a program with a USD3.1 million budget over two years (IOM 2012).

But despite these important improvements, the government's response to Central and South American migrants affected by violence while crossing through Mexico is still insufficient. It is too early to say whether the 2011 law will be effectively implemented and whether international collaboration will be productive. Observers doubt that the new migration law will live up to its promise and are particularly skeptical of the extent to which INM officials will operate effectively within the structure and regulations established by the law (Alba and Castillo 2012). For example, since 2007 Mexico has had legal provisions to guarantee a humanitarian visa to migrants who suffer a crime during their journey north. However, in spite of many crimes committed against transit migrants, only ten such visas were issued in 2007, fourteen in 2008, and eight in the first six months of 2009—all of which were given only in response to extreme pressure from advocacy organizations (Amnistía Internacional 2010).

The United States government's limited response to people crossing the border

The response by US authorities to asylum claims linked to drug-cartel violence in Mexico serves as an example to examine the potential for protection that the refugee regime offers to people fleeing criminal violence across borders.[3] The statistics of successful refugee claims by Mexicans seeking asylum as a result of drug-cartel violence, on the one hand, and the legal reasoning supporting court decisions, on the other, suggests that this avenue is limited as a form of protection for people crossing borders as a result of intense criminality.[4]

Asylum claims from Mexicans in the US have gone up in recent years. In 2011, 4,042 Mexicans sought asylum affirmatively in the US, more than triple the number of applications five years earlier. In the same year, 6,133 Mexicans sought asylum defensively (i.e. to avoid being removed from the US), up from 4,510 the year before, according to US Justice Department figures. At the same time, the rate at which Mexicans are granted asylum is low when compared to other nationalities: from 2007 to 2011, US immigration courts received 21,104 defensive asylum claims from Mexicans. Only 2 percent of those applications were granted.

Because immigration records are not public, the analysis that follows examines 203 appeals cases (which are public), from Mexican citizens who applied for asylum in the United States as a result of criminal violence. Of those cases, forty-five claimed persecution on the grounds of political opinion and fifty under membership of a particular social group. Among the latter, petitioners attempted unsuccessfully to argue that they were part of a cognizable social group persecuted by organized crime (the claims pertained predominantly to cases of defensive asylum claims pending removal from the US).

Petitioners asserted, for example, that as "Mexicans returning home from the United States who are targeted as victims of violent crime as a result,"[5] they were part of a cognizable social group or that being "Mexicans returning from the United States"[6] made them part of a social group targeted because it is perceived as wealthy.

Where petitioners failed to show that their fear was on account of membership in a cognizable social group, courts ruled that fear of "general country conditions" or "indiscriminate violence" was not ground for asylum, unless victims are singled out on account of a protected ground. Likewise, courts held that exposure to criminal violence for financial reasons is not a basis for asylum.[7]

The majority of these cases were rejected for failure to show a well-founded fear of persecution under the social group statutory ground. The cases that were successful had specific evidence (names of cartel or police members, hospital or police reports, and witness testimony). They also could demonstrate and articulate why and how they feared persecution (i.e. who would harm them).

Twenty-two cases that argued fear of generalized violence or unstable country conditions as the reason for fleeing and as grounds for asylum were rejected. This jurisprudence shows that, under the refugee regime (and its interpretation by US courts thus far), people fleeing because they are affected or threatened by drug cartel violence, but not explicitly persecuted, have not gained asylum in the US.

Finally, the United States put in place in 1990 a temporary immigration status (known as temporary protected status, TPS) under which nationals of countries experiencing armed conflict, disasters, or other extraordinary conditions are allowed to remain in the United States. The Department of Homeland Security must designate countries whose nationals are eligible to benefit from TPS. Despite the high levels of violence observed in Mexico, the country has not been designated as a TPS country, and it is unlikely that it will be. Currently, only three countries are TPS countries for reasons related to conflict and violence: Somalia, Sudan and South Sudan. The rest are countries that experienced large-scale natural disasters—El Salvador, Honduras, Nicaragua and Haiti (US Citizenship and Immigration Services n.d.).

Conclusions

Evidence shows that criminal violence causes and affects human mobility in Mexico. This chapter has argued that in the case of Mexico, the intensity and pervasiveness of criminal violence creates a humanitarian crisis. Therefore, there needs to be a fundamental shift in responses by concerned states (and also the international community), from punishing or defeating offenders, to giving full weight to the needs of victims, including migrants, just as we focus on the needs of victims and people displaced by armed conflicts proper.

The existing international protection framework provides such a focus on the needs and vulnerabilities of people moving as a result of criminal violence or affected by it while in transit. However, its existing norms and praxis do not fully respond to the complexity of human mobility in situations of intense criminal violence. As described in the chapter, a hallmark feature of environments where criminal violence is rampant is that it pushes people to move in a variety of ways, from direct coercion and physical threats to the erosion of the general environment and quality of life, to shrinking livelihood opportunities.

The analysis above showed that, through new *interpretations* of existing legal norms, people who do not clearly fall within the existing legal categories could potentially find protection. A broad interpretation of the Guiding Principles could include as IDPs people who flee without direct coercion but who do not move out of free choice either. Likewise, innovative interpretations of the grounds for asylum in the Refugee Convention could provide relief to people in these situations, as well as the application of complementary protection and subsidiary protection mechanisms in certain countries.

However, even if legal interpretation of existing frameworks may in principle offer protection, practical implementation remains the biggest challenge. The chapter showed that, in the case of both internal and cross-border migrants, such interpretations are still limited in practice by relevant governments and/or adjudicating courts (US courts, where the analysis focused). The potential of complementary protection is narrower because of its higher threshold of application and its limitation to the prohibition of return. Regarding migrants who are affected by violence while crossing Mexico, they are in principle protected by universal human rights, but respect and promotion of their human rights in third countries is minimal. Mexico's 2011 migration law was a step in the right direction, but kidnapping and extortion of migrants actually rose in the last months of 2012 and the first months of 2013, which shows that the law's impact is still limited.

Furthermore, decisions by relevant authorities are not free of political calculation. Immigration authorities in the US have stated that, with more than six million Mexicans living in the United States illegally, they fear that granting asylum to people fleeing drug-cartel violence in Mexico would "open the floodgates" for an unprecedented number of applications.[8] In Mexico, despite evidence described above about significant displacement, the government has yet to recognize displacement and set up a comprehensive response.

In the absence of a state response for people displaced by violence in Mexico, humanitarian agencies should engage to protect people affected and displaced by violence. But situations of insecurity caused by criminal violence often fall outside the mandates and mission statements of humanitarian agencies. For instance, among the international agencies currently in Mexico, no agency has thus far set up programs to respond to the impacts of criminal violence on local communities. Most of their focus is on supporting and protecting irregular migrants, but focus also needs to shift to the large number of Mexicans impacted and displaced by violence.

The limited involvement of international agencies is partly due to the fact that up to now, the Mexican Government has not sought cooperation from international agencies in relation to drug-cartel violence, as doing so would acknowledging that the country faces a humanitarian crisis or an armed conflict. The character of the Mexican Government and the situation of the country in the international arena are in themselves limitations on a humanitarian response.

Acknowledgments

The author would like to acknowledge all those who have contributed to this paper in various ways: Susan Martin, Sanjula Weerasinghe and Abbie Taylor, and Nina Schrepfer for comments to the text; participants at the ISIM Crisis Migration Review Workshop (2012) and participants at the PROCAP Technical Workshop (2013) for their feedback; and Johanna Foote, Melanie Wissing and Samantha Howland for excellent research.

Notes

1   Alexander Betts has developed the concept of survival migration, which describes "people who have left their country of origin because of an existential threat for which they have no domestic remedy." Global Governance 16 (2010), 361–382—"Survival Migration: A New Protection Framework" page 1, available online at: http://siteresources.worldbank.org/INTRANETSOCIALDEVELOPMENT/Resources/244329-1255462486411/Betts_Survival_Migration.pdf. Betts' work highlights the inadequacy of the refugee protection regime to address the protection needs of people that flee new drivers of displacement including environmental disaster, livelihood failure or state fragility.

2   With a particular view to the situation of Mexican asylum-seekers in the US, note that the US is not a State Party to the Convention on Enforced Disappearances.

3   Drug cartels are understood here as higher-level, sophisticated, drug trafficking enterprises (Ribando, Clare 2007, CRS Report for Congress "Gangs in Central America" (available online at: http://www.ansarilawfirm.com/docs/CRS-Report-Gangs-in-Central-America-1-11-2007.pdf) differing from gangs, which are understood as "relatively durable, predominantly street-based groups of young people for whom crime and violence is integral to the group's identity" (UNHCR 2010).

4   In the case of people fleeing gang violence in El Salvador, Honduras and Guatemala, in various cases petitioners have argued that their age and sex make them a particular social group targeted by gangs. In several of these cases, courts have ruled that the following are not a cognizable social group: "Boys 8–15 who oppose gangs on moral grounds are not a protected social group"; "Young females who resist gang recruitment is not a cognizable social group"; "Young Salvadoran men who have already resisted gang recruitment and whose parents are unavailable to protect them is not a statutorily protected social group"; "Young women who resisted recruitment by gangs in El Salvador is not a cognizable social group"; "Salvadoran adolescents who refuse to join the gangs because of their opposition to their criminal activities is too broad of a social group." See, *inter alia*, Orellana-Monson v. Holder, No. 11-60147, 2012; Rivera-Barriento v. Holder No. 10-9527, 2012; Mayorga-Vidal v. Holder, 2012 WL 883193, 2012; Mendez-Barrera v. Holder, 602 F.3d 21, 2010. Cases available online at: http://www.refugees.org/resources/for-lawyers/asylum-research/gang-related-asylum-resources/court-of-appeals.html.

5   Delgado-Ortiz v. Holder, No. 09-72993, 2010. Available online at: http://caselaw.
    findlaw.com/us-9th-circuit/1521430.html.
6   Aguilar-Medina v. Holder, 2010, available online at: http://law.justia.com/cases/
    federal/appellate-courts/ca9/10-70130/10-70130-2011-04-18.html.
7   See, *inter alia*, Lara-Castillo v. Holder, No. 11-60710, 2012, available online at: http://
    law.justia.com/cases/federal/appellate-courts/ca5/11-60710/11-60710-2012-10-
    11.html; Aguilar-Medina v. Holder, No. 10-70130, 2010, available online at: http://law.
    justia.com/cases/federal/appellate-courts/ca9/10-70130/10-70130-2011-04-18.html;
    Soria v. Holder, No. 09-71230, 2010, available online at: http://cdn.ca9.uscourts.gov/
    datastore/memoranda/2010/07/01/09-71230.pdf; Fierro v. Holder, No. 09-72517,
    2010, available online at: http://law.justia.com/cases/federal/appellate-courts/ca9/09-
    72517/09-72517-2011-04-18.html; Yanagui v. Holder, No. 09-73674, 2010 available
    online at: http://cdn.ca9.uscourts.gov/datastore/memoranda/2010/09/21/09-73674.
    pdf; Vazquez v. Gonzales, No. 04-75108, 2007; Garcia-Camacho v. Holder, No.
    10-3537, 2011, available online at: https://www.courtlistener.com/ca2/4ack/garcia-
    camacho-v-holder/; Galvan v. Mukasey, No. 08-71222, 2010, available online at: http://
    caselaw.findlaw.com/us-9th-circuit/1309125.html; Mejia Malagon v. Holder, No. 08-
    70979, 2012, available online at: http://law.justia.com/cases/federal/appellate-courts/
    ca9/08-70979/08-70979-2012-10-22.html; Bernal v. Holder, No. 08-73406, 2010,
    available online at: http://cdn.ca9.uscourts.gov/datastore/memoranda/2010/06/04/08-
    73406.pdf; Arreola v. Holder, No. 09-73940, 2010, available online at: http://law.
    justia.com/cases/federal/appellate-courts/ca9/09-73940/09-73940-2011-04-18.html,
    Cartel-based; Delgado-Ortiz v. Holder, No. 09-72993, 2010, available online at: http://
    caselaw.findlaw.com/us-9th-circuit/1521430.html.
8   Grillo (2011); in a somewhat similar case worth noting, the US Seventh Circuit Court
    of Appeals recognized that young men who leave gangs (such as Central American
    gangs) and refuse to return could constitute a "particular social group" for purposes
    of asylum, while noting that there will be other potential hurdles for these appli-
    cants to ultimately obtain asylum. US Seventh Circuit Court of Appeals, Benitez-
    Ramos Decision, December, 2009, available online at: http://www.uscrirefugees.
    org/2010Website/5_Resources/5_4_For_Lawyers/5_4_1%20Asylum%20Research/
    5_4_1_2_Gang_Related_Asylum_Resources/5_4_1_2_1_Court_of_Appeals/Beni-
    tez_Ramos_v_%20Holder.pdf.

References

Alba, F. and Castillo, M.A. (2012) "New Approaches to Migration Management in Mexico
    and Central America," *Migration Policy Institute*, October, available online at: http://
    www.migrationpolicy.org/pubs/RMSG-MexCentAm-Migration.pdf.
AlertNet (2011) "FEATURE: Mass Kidnappings New Cash Cow for Mexico Drug
    Gangs," April 11, available online at: http://www.trust.org/alertnet/news/feature-mass-
    kidnappings-new-cash-cow-for-mexico-drug-gangs/.
Alvarez, V.M. (2006) "The History, Structure, and Organization of Mexican Drug Cartels,"
    University of Texas El Paso doctoral dissertation, available online at: http://digitalcom-
    mons.utep.edu/dissertations/AAI1441328/.
Amnistía Internacional (2010) "Victimas Invisibles: Migrantes en movimiento en México,"
    available online at: http://www.amnesty.org/es/library/asset/AMR41/014/2010/
    es/1345cec1-2d36-4da6-b9c0-e607e408b203/amr410142010es.pdf.
BBC News (2010) "Wikileaks Cables: US Mexico Drugs War Fears Revealed," December
    3, available online at: http://www.bbc.co.uk/news/world-latin-america-11906758.
Beittel, J.S. (2011) "Mexico's Drug Trafficking Organizations: Source and Scope of the Ris-
    ing Violence," Congressional Research Papers, January.

Betts, A. (2010) "Survival Migration: A New Protection Framework," *Global Governance*, 16: 361–382.

Brodzinsky, S. (2012) "Migrant Kidnappings by Criminal Organizations 'Systematic' in Mexico," *In Sight Crime*, May 11, available online at: http://www.insightcrime.org/news-analysis/migrant-kidnappings-by-criminal-organizations-systematic-in-mexico.

Buchanan, H. (2010) "Fleeing the Drug War Next Door: Drug-related Violence as a Basis for Refugee Protection for Mexican Asylum-seekers," *Merkourios*, 27(72): 28–60.

Centro de Estudios Migratorios (2012) "Síntesis 2012: Estadística Migratoria," available online at: http://www.inm.gob.mx/estadisticas/Sintesis_Grafica/2012/Sintesis2012.pdf.

Centro Internacional para los Derechos Humanos de los Migrantes (CIDEHUM) (2012) "Desplazamiento Forzado y Necesidades de Protección, generados por nuevas formas de Violencia y Criminalidad en Centroamérica," *Diagnóstico*, May, available online at: http://www.acnur.org/t3/fileadmin/Documentos/BDL/2012/8932.pdf?view=1.

Chavez, E.R. (2011) "Apuntes sobre migración: Migración centroamericana de tránsito irregular por México. Estimaciones y características generales," *Instituto Nacional de Migración*, available online at: http://www.inm.gob.mx/static/Centro_de_Estudios/Investigacion/Avances_Investigacion/APUNTES_N1_Jul2011.pdf.

Chávez Galindo, A.M. and Landa Guevara, A. (2011) "Migrantes en su paso por México: nuevas problemáticas, rutas estrategias y redes," *Somede*, available online at: http://www.somede.org/xireunion/ponencias/Migracion%20internacional/147Pon%20Ana%20Ma%20Chavez-Antonio%20Landa.pdf.

Christian Science Monitor (2010) "Why So Many Mayors are now Targets in Mexican Drug War," September 28, available online at: www.csmonitor.com/World/Americas/2010/0928/Why-so-many-mayors-are-now-targets-in-Mexican-drug-war.

CNN (2011a) "Los Cadáveres Ubicados en Fosas Clandestinas de Tamaulipas ya son 177," April 22, available online at: http://mexico.cnn.com/nacional/2011/04/22/los-cadaveres-ubicados-en-fosas-clandestinas-de-tamaulipas-aumentan-a-177.

CNN (2011b) "La poca demanda de hogares redujo un 56% la construcción en Ciudad Juárez," October 14, available online at: http://mexico.cnn.com/nacional/2011/10/14/la-poca-demanda-de-hogares-redujo-un-56-la-construccion-en-ciudad-juarez.

Comisión Interamericana de Derechos Humanos (2010) "Secuestros a Personas Migrantes Centroamericanas en Tránsito por México," available online at: http://proteccionmigrantes.org/?p=145.

Comisión Nacional de Derechos Humanos (2009) "Informe especial sobre los casos de secuestro en contra de migrantes," available online at: http://www.cndh.org.mx/sites/all/fuentes/documentos/informes/especiales/2009_migra.pdf.

Comisión Nacional de los Derechos Humanos (2011) "Informe especial sobre secuestro de migrantes en México," available online at: http://www.cndh.org.mx/sites/all/fuentes/documentos/informes/especiales/2011_secmigrantes_0.pdf.

Consejo Ciudadano para la Seguridad Pública y Justicia Penal AC (2013) "San Pedro Sula otra vez primer lugar mundial; Acapulco, el segundo," available online at: http://www.seguridadjusticiaypaz.org.mx/biblioteca/finish/5-prensa/163-san-pedro-sula-otra-vez-primer-lugar-mundial-acapulco-el-segundo/0.

Cote-Muñoz, N. (2011) "Central American Migrants in Mexico: Lost in a Black Hole of Violence," Council on Hemispheric Affairs, September 7, available online at: http://www.coha.org/central-american-migrants-in-mexico-lost-in-a-black-hole-of-violence-2/.

Customs and Border Protection (2013) "US Border Patrol Nationwide Apprehensions from Other Than Mexico by Sector Fiscal Year 2000–2012," available online at: http://www.hsdl.org/.

El colegio de la frontera Norte (2013) "Encuesta sobre migración en la frontera sur: bases de datos," available online at: http://www.colef.net/emif/bases.php.

El País (2010) "La 'Colombianización' de México," October 5, available online at: www. elpais.com/articulo/opinion/colombianizacion/Mexico/elpepuopi/ 20101005elpepiopi_5/Tes.

El Universal (2010a) "Cierran negocios en Juárez, abren en El Paso," August 29, available online at: http://www.eluniversal.com.mx/notas/704901.html.

El Universal (2010b) "Migrantes, 72 muertos de fosa en Tamaulipas," August 25, available online at: http://www.eluniversal.com.mx/notas/704017.html.

El Universal (2012) "CNDH elabora protocolo de atención a desplazados," July 3, available online at: http://www.eluniversal.com.mx/notas/857167.html.

Godson, R. (ed.) (2003) *Menace to Society: Political Criminal Collaboration Around the World*, New Brunswick: Transaction.

Goodwin-Gill, G. and McAdam, J. (2007) "The Refugee in International Law," *Merkourios*.

Government of Mexico (2012) "Base de Datos de Fallecimientos por Presunta Rivalidad Delincuencial," available online at: http://www.pgr.gob.mx/temas%20relevantes/esta-distica/estadisticas.asp.

Grillo, I. (2011) "Will Mexico's Runaway Sheriff Find Asylum in the US?" *Time*, March 11, available online at: http://www.time.com/time/world/article/0,8599,2058298,00.html.

Guardian (2010) "Mexico Prison 'Lets Inmates Out to Kill'," July 26, available online at: http://www.guardian.co.uk/world/2010/jul/26/mexico-prison-accused-inmates-out.

Guerrero-Gutiérrez, E. (2011) *Security, Drugs and Violence in Mexico: A Survey*, Mexico City: Lantia Consultores.

Human Rights Watch (2011) "Neither Rights nor Security," November, available online at: http://www.hrw.org/reports/2011/11/09/neither-rights-nor-security-0.

ICTY, Prosecutor v. Tadic, Case No. IT-94-1-T, Judgment (Trial Chamber), May 7, 1997; Prosecutor v. Limaj, Case No. IT-03-66-T, Judgment (Trial Chamber), November 30, 2005; Prosecutor v. Boskoski, Case No. IT-04-82, Judgment (Trial Chamber), July 10, 2008.

IDMC (forthcoming) "Forced Displacement Linked to Transnational Organized Crime in Mexico".

IOM (2012) "UN Trust Fund for Human Security Backs Effort to Improve Protection for Migrants in Mexico," November 30, available online at: http://www.iom.int/cms/en/ sites/iom/home/news-and-views/press-briefing-notes/pbn-2012/pbn-listing/un-trust-fund-for-human-security.html.

Jimenez, S. (2012) "The Route of Death for Central and South American Illegal Immigrants Can Come to an End with a Change in the United States' Policy," *Global Business & Development Law Journal*, 25: 447–473.

La Jornada (2013) "La ruta de Palenque a Tenosique, con más secuestros y extorsiones para migrantes," May 27, available online at: http://www.avanzada.com.mx/inicio/index. php?option=com_content&view=article&id=18715%3Ala-ruta-de-palenque-a-teno-sique-con-mas-secuestros-y-extorsiones-para-migrantes-mision&catid=88%3Aminutoc l&Itemid=128).

La Nación (2011) "Ciudad de México, en Alerta por la Narcoguerra," February 19, available online at: http://www.lanacion.com.ar/1351199-ciudad-de-mexico-en-alerta-por-la-narcoguerra.

Loyola, M. (2009) "Mexico's Cartel Wars," *National Review*, 61(11), June 22.

Mandal, R. (2005) "Protection Mechanisms Outside of the 1951 Convention ('Comple-mentary Protection')," *Legal and Protection Research Series, Department of International Protection*, UNHCR.

McAdam, J. (2005) "The European Union Qualification Directive: The Creation of a Subsidiary Protection Regime," *International Journal of Refugee Law*, 17(3): 461–516.

Meyer, M. (2010) "Abused and Afraid in Ciudad Juárez: An Analysis of Human Rights Violations by the Military in Mexico," Washington Office on Latin America, available online at: http://idpc.net/sites/default/files/library/Abused%20and%20afraid%20in%20ciudad%20juarez.pdf.

Movimiento por la Paz con Justicia y Dignidad (2012) "A seis meses del gobernio de Enrique Peña Nieto: La guerra sigue . . . y se agrava," April 20, available online at: http://movimientoporlapaz.mx/.

Organización Internacional para las Migraciones (2012) "Guía de buenas prácticas para la asistencia y protección a personas migrantes víctimas de secuestro en México," available online at: http://www.oim.org.mx/pdf/Guia_de_buenas_practicas.pdf.

Pacheco, F.C. (2009) "How has Narcoterrorism Settled in Mexico?" *Studies in Conflict and Terrorism*, 32(12): 1021–1048.

Papademetriou, D., Meissner, D. and Sohnen, E. (2013) "Thinking Regionally to Compete Globally: Leveraging Migration and Human Capital in the US, Mexico, and Central America," Migration Policy Institute, May, available online at: http://www.migrationpolicy.org/pubs/RMSG-FinalReport.pdf.

Reforma (2011) "Minimiza SEGOB cifra de desplazados por el Narco," March, available online at: http://www.reforma.com/nacional/articulo/601/1200968/default.asp?compartir=7b451c97641ad969c9351cfbf24ee7a2&plazaconsulta=reforma.

Rodriguez, O. (2012) "Central American Migrants Flood North Through Mexico to US," *Huffington Post*, July 13, available online at: http://www.huffingtonpost.com/2012/07/13/central-americans-in-the-united-states_n_1671551.html.

Schneckener, U. (2009) "Spoilers or Governance Actors? Engaging Armed Non-state Groups in Areas of Limited Statehood," German Research Foundation (DFG), *Governance Working Paper Series*, No. 21, available online at: http://www.sfb-governance.de/publikationen/sfbgov_wp/wp21_en/SFB-Governance_Working_Paper_No21.pdf?1277900587.

Sherman, C. (2012) "Central American Mothers Look for Missing Migrants," Associated Press, October 23, available online at: http://bigstory.ap.org/article/central-american-mothers-look-missing-migrants.

Sin Fronteras (2012) "Evolución y retos del marco normativo migratorio en México: Una perspectiva histórica," available online at: http://www.sinfronteras.org.mx/attachments/article/1406/informeMigracion_web.pdf.

Talsma, L. (2012) "Human Trafficking in Mexico and Neighbouring Countries: A Review of Protection Approaches," UNHCR Policy Development and Evaluation Service, June, available online at: http://www.unhcr.org/4f070a83540.html.

UNHCR (2010) *Guidance Note on Refugee Claims Relating to Victims of Organized Gangs*, available online at: http://www.unhcr.org/refworld/docid/4bb21fa02.html (last accessed January 25, 2013).

Universidad Autónoma de Ciudad Juárez (2009) "Diagnóstico sobre la realidad social, económica y cultural de los entornos locales para el diseño de intervenciones en materia de prevención y erradicación de la violencia en la región Norte," available online at: http://www.conavim.gob.mx/work/models/CONAVIM/Resource/pdf/JUAREZ.pdf.

University of Heidelberg (2008) "Heidelberg Conflict Barometer," available online at: http://hiik.de/de/konfliktbarometer/pdf/ConflictBarometer_2008.pdf.

University of Heidelberg (2010) "Heidelberg Conflict Barometer," available online at: http://www.hiik.de/en/konfliktbarometer/pdf/ConflictBarometer_2010.pdf.

University of Heidelberg (2011) "Heidelberg Conflict Barometer," available online at: http://www.hiik.de/en/konfliktbarometer/pdf/ConflictBarometer_2011.pdf.
Uppsala University UCDP Database (n.d.) *UCDP Conflict Encyclopedia*, available online at: http://www.ucdp.uu.se/gpdatabase/search.php.
US Citizenship and Immigration Services (n.d.) "What is TPS?" available online at: http://www.uscis.gov/portal/site/uscis/menuitem.eb1d4c2a3e5b9ac89243c6a7543f6d1a/?vgnextoid=848f7f2ef0745210VgnVCM100000082ca60aRCRD&vgnextchannel=848f7f2ef0745210VgnVCM100000082ca60aRCRD#What is TPS?.
Vinci, A. (2006) "The 'Problems of Mobilization' and the Analysis of Armed Groups," *Parameters*, Spring, 36(1): 49–62.
Vité, S. (2009) "Typology of Armed Conflicts in International Humanitarian Law," *International Review of the Red Cross*, March.
Williams, P. (2008) "Violent Non-state Actors and National and International Security," International Relations and Security Network (ISN), Swiss Federal Institute of Technology, Zurich.
Zehr, H. (2002) *The Little Book of Restorative Justice*, Intercourse, PA: Good Books.
Zuñíga, R. (2011) "Consulado de El Salvador lanza la 'Línea que te protege'," June 20, *Diario del Sur*, available online at: http://www.oem.com.mx/laprensa/notas/n2114768.htm.

7

INTRACTABILITY AND CHANGE IN CRISIS MIGRATION

North Koreans in China and Burmese in Thailand

W. Courtland Robinson

Introduction

In April 2011, US Senate Foreign Relations Committee member, Richard Lugar, introduced a resolution expressing concern that the governments of North Korea (or the Democratic People's Republic of Korea (DPRK)) and Burma (or Myanmar) were "expanding their bilateral military relationship" (S. Res. 139 2011, 2). Specifically, the sponsors of the resolution asserted that North Korea was providing military and technical personnel and advanced weaponry, and that the two governments were collaborating on development of Burma's nuclear program. There were further accusations: Burma continued to coerce children, including ethnic minorities, into combat and civilians into portering; and hundreds of thousands had been forced to flee persecution and conflict in Burma.

The resolution was referred to Committee and, while that document lies dormant, Burma has been undergoing sweeping changes. In March 2011, the State Peace and Development Council (SPDC) handed over power to a new government, headed by President Thein Sein. In January 2012, the new government released 651 prominent political prisoners and signed a ceasefire agreement with ethnic Karen rebels; in response to these gestures, the United States, among other democracies, restored diplomatic relations with Burma. In April 2012, Aung San Suu Kyi's National League for Democracy (NLC) won nearly every seat in a by-election, providing the Nobel Prize-winning political leader a seat in Parliament after fifteen years of house arrest. In May 2012, the United States suspended economic sanctions, though it stressed it would maintain its arms embargo as well as its list of companies, tycoons, and generals alleged to be engaged in human-rights violations and corruption.

As if to bring the changes full circle from the Lugar resolution, President Thein Sein took the occasion of a May 2012 state visit by then-South Korean President Lee Myung-bak to announce that Burma would cease buying weapons from North

Korea, a country where the more things changed the more they stayed the same. In December 2011, after more than seventeen years in power, the "Dear Leader" Kim Jong-il died of a heart attack. He had succeeded his father—the "Great Leader" and founder of the Democratic People's Republic of Korea, Kim Il-sung—and was in turn succeeded by his son, Kim Jong-un, an untested twenty-eight year old with little military or political experience.

Of these two Asian countries, once both pariah nations, one seems now to be in rapid transition toward democracy and fundamental reforms, while the other seems locked in what one observer termed "structural intransigence" (*Korea Times* 2012), unwilling or unable to find a path to rejoin the international community. While Burma and North Korea may be charting widely divergent courses going forward, both have created what one might call a living legacy of refugees, asylum-seekers, and vulnerable migrant populations in neighboring countries. Some have moved to flee natural disaster, some to cope with famine and food insecurity, some to escape civil conflict and persecution, some to find family, and some to work. The response to this protracted migration flow has been shaped by local and regional politics (especially the policies of Thailand and China), by international donor and stakeholder priorities, and by the mandates and capacities of international, regional, governmental, and non-governmental organizations (NGOs).

This chapter will examine the migration of Burmese into Thailand from 1988 to 2012, and of North Koreans into China from 1998 to 2012, focusing on how migration patterns changed in response to conditions in countries of origin and destination, how various stakeholders responded, and what lessons might be gleaned to address challenges and opportunities for effective policies and practices in response to crisis migration from Burma and North Korea and beyond. The lessons from Burma, though the situation is still fluid and evolving, suggest that regional associations like the Association of Southeast Asian Nations (ASEAN) and regional frameworks like the Bali Ministerial Conference on People Smuggling, Trafficking in Persons and Related Transnational Crime (Bali Process), may have helped coax the country out of isolation. The lack of such frameworks in East Asia makes the situation in North Korea even more of a seemingly intractable challenge.

North Koreans in China

Though there are a great many unknowns about migration and displacement within and outside of North Korea in the last two decades, some patterns can be identified and estimates made, focusing first on internal migration within the country, and then on the various patterns, phases, and types of populations migrating to (and often through) China.

Internal migration and displacement

While official data from the DPRK suggest limited movement internationally or internally, the unofficial picture is one of a great deal more mobility, most of it without authorization. Studies the author conducted in China in 1998–99 that included nearly 3,000 North Korean refugees and migrants who were asked about movements into and out of the household of more than one month suggested a net migration rate of 18.7 percent (Robinson 2000). The retrospective study covered a four-year period, 1995–98, which included two years (1996–97) of severe famine, characterized by significant malnutrition, a rise in infectious disease, and a dramatic spike in mortality among all age groups (Katona-Apte and Mokdad 1998; KBSM 1998; Robinson et al. 1999; Schwekendiek 2008). In the 1998 study, more than 30 percent said their main reason for moving out of the household was to "search for food," and more than one-fifth of the out-migrants were children under fifteen. Based on the research among displaced North Koreans in China, the author determined "the bulk of this internal movement [in DPRK] to be 'distress migration' as defined in the famine literature,"[1] and moreover, that "these distress migrants should be considered to be internally displaced persons consistent with the *Guiding Principles on Internal Displacement*" (Robinson 2000, 115).

Migration to China

Crisis migration of Koreans into Northeast China dates back to at least the 1880s, when poor farmers crossed the border to escape economic hardship (Liang and Dohm 2006). Cross-border movements accelerated with the Japanese occupation of Korea in 1910, but declined when the Communists came to power in China in 1949 and Korea became a divided country. The more recent surge in cross-border movement of North Koreans into China likely began in the mid-1990s but probably did not peak until 1998 or 1999, several years after the worst of the famine in North Korea in 1996–97 (migration usually is a trailing indicator in famines). Since that time, North Koreans have been crossing the northern border into China, seeking to escape food shortages, economic hardship, and state repression in their own country. Most of these North Koreans have left their own country and entered neighboring China without travel authorization or documentation. Given their undocumented status and the repressive nature of the DPRK, these North Koreans have been labeled refugees and asylum-seekers by those who seek their protection.[2] Conversely, they are called illegal migrants by both China and North Korea.

From 1999 to 2008, the Center for Refugee and Disaster Response at the Johns Hopkins Bloomberg School of Public Health worked with local and international partners to monitor movements of North Koreans crossing into China. Each month for ten years from September 1998 to September 2008, local community-based monitors at ten sites in Northeast China kept track of arrivals and departures of North Koreans in their area (a defined village or urban neighborhood), including some demographic estimates (proportion female, for example) and arrest/deportation events (see Figures 7.1 and 7.2). Key trends over the years included an obvious seasonal spike in arrivals during the winter months when food and fuel were scarce in North Korea (and security might have been relatively more relaxed on

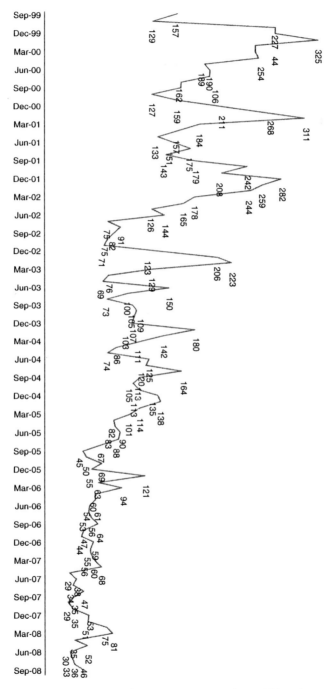

FIGURE 7.1   North Korean arrivals at ten sites in Northeast China, September 1999
– September 2008

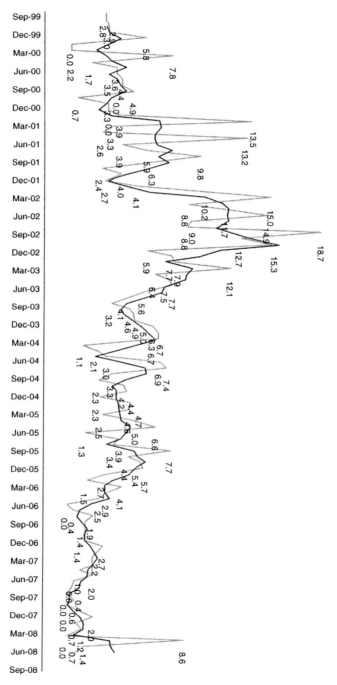

FIGURE 7.2   North Korean monthly arrest rates (with three–month moving average) in ten sites in Northeast China, September 1999 – September 2008

both sides of the border) and an overall (nearly ten-fold) decline in the number of arrivals from 1998 to 2008.

In terms of arrests and deportations, there were also two clear trends. First, arrest and deportation rates tended to be highest in the summer months when more international (particularly South Korean and Korean–American) aid and advocacy groups were more active in China and border security was perhaps correspondingly tightened. Second, the historical high point of arrests occurred in March 2002, when the first group of twenty-five North Korean asylum-seekers pushed their way inside the Spanish Embassy in Beijing and sought international protection. That event precipitated many more such entries into foreign embassies, consulates, even foreign-run schools in Beijing and other large cities in China, and also prompted an immediate crackdown by Chinese authorities on North Korean refugees and asylum-seekers and on foreign groups seeking to help them. Arrivals declined dramatically and—despite continued food insecurity, economic hardship and ongoing human rights abuses in North Korea—never returned to anything close to 1998–99 levels. Rates of arrest and deportation declined somewhat for several years but spiked again in the months just before and after the 2008 Summer Olympics held in Beijing. This, ultimately, also proved to be the end of the ten-year, community-based, sentinel site surveillance project, as authorities cracked down on organizations aiding or documenting protection problems of North Koreans in China.

Estimation of North Korean numbers, population characteristics, and migration trends over time has proved logistically difficult and politically sensitive and the only point on which refugee advocates and critics seem to be able to agree is that North Koreans in China have been difficult to count. As noted by a report from the Congressional Research Service (2007, 4):

> There is little reliable information on the size and composition of the North Korean population located in China. Estimates range from as low as 10,000 (the official Chinese estimate) to 300,000 or more. Press reports commonly cite a figure of 100,000 to 300,000. In 2006, the State Department estimated the numbers to be between 30,000 and 50,000, down from the 75,000 to 125,000 range it projected in 2000. UNHCR also uses the 2006 range (30,000 to 50,000) as a working figure.

In late 2009, Johns Hopkins conducted an estimation of the population of North Koreans in three provinces in Northeast China: Heilongjiang, Jilin (including Yanbian Korean Autonomous Prefecture) and Liaoning (Robinson 2010). The primary objective of this research was to estimate the total population of North Koreans in selected areas in Northeastern China, including the estimated number of children in these provinces who were born to North Korean women in China. Key informant and in-depth interviews with 324 respondents in 108 randomly selected sites in three provinces gave evidence of the following trends.

- **North Korean population decline in China:** Comparing the North Korean population estimates from 1998 to 2009, it is clear there has been a dramatic decline in the North Korean population in Northeast China, from a mid-range estimate of around 75,000 North Koreans in 1998 to around 10,000 as of 2009. When North Koreans first began crossing into China in significant numbers in 1998, it was in the wake of a severe famine and only four years after the death of Kim Il-sung. More than a decade later, hardships continued for the North Korean people in the face of continued food shortages, a moribund economy, periodic natural disasters, and a government unwilling to make the reforms necessary to improve the health, welfare, and human rights of the general population. Factors contributing to the declining population include reduced arrivals (caused by tighter border security and lower expectation of what is available in China) and increased out-migration (including increased migration to South Korea and other countries, and continued deportation).
- **Increase in proportion of North Korean women:** The proportion of the North Korean refugee and migrant population in Northeast China who are female rose from 50 percent in 1998, to 54 percent in 2002, then to more than 77 percent in 2009. This increased proportion of females likely is due to the sustained deportation (or spontaneous return) of North Korean men, and the continued demand, brought on by high male:female sex ratio imbalances in China, for foreign brides. Overall, for roughly four-fifths of North Korean women in Northeast China, marriage to a Chinese man appears to be the primary (though not necessarily the preferred) means of remaining, and staying *relatively* secure, in China. The emphasis here is on relatively, as North Korean women generally still are considered illegal migrants subject to arrest and deportation and, as females and undocumented migrants, carry a double burden of vulnerability to exploitation, particularly to the risk of being trafficked, not so much into commercial sex work or forced labor (though these exist) but into forced marriage.
- **Increase in China-born children:** The Northeast China study also suggested that the number of children born to North Korean women in China rose from around 8,000 in 1998, to 9,000 in 2002, then to more than 10,000 in 2009. The study further estimated that perhaps three-quarters of these children were living without their mothers and often without either parent, creating a variety of stresses for these children who are "left-behind." Children may be left behind due to labor migration dynamics (many absent fathers were working either overseas or in other areas in China) or to the pressures that force North Korean mothers out of these households including arrest/deportation, search for work, family conflict, divorce/separation, and resettlement opportunities in South Korea.

The UN High Commissioner for Refugees (UNHCR) has declared all North Koreans in China to be "persons of concern" (POC) and, as such, encompassed within UNHCR's Framework for Durable Solutions for Refugees and Persons of Concern. Durable solutions, in this sense, primarily refer to voluntary repatriation, local integration, and resettlement, but also development assistance pending a durable solution (UNHCR 2003). Though China is a signatory to the 1951 Convention relating to the Status of Refugees (1951 Refugee Convention), it does not recognize North Korean claims to asylum as valid. Indeed, as recently as March 2012, a Chinese official reiterated "that these North Koreans are not refugees but rather they have entered China illegally for economic reasons . . . As for how to handle this issue, it is entirely within China's sovereignty . . . China is opposed to the attempt to turn the issue into a political and international subject" (Citizens' Alliance for North Korean Human Rights 2012, 111).

In the ongoing leadership transition that has occurred following the death of Kim Jong-il in December 2011, tightened security on both sides of the border contributed to a virtual halt to all movement of North Koreans into China for several months, and reports from knowledgeable sources in Northeast China indicate that clandestine, cross-border arrivals slowed to a trickle in 2012. South Korean government data also indicate that North Korean entries in 2012 were down to around 1,500, about half of the totals in previous years.[3] A recent new study in one Northeastern Chinese province by Johns Hopkins and a partner South Korean research institution (unpublished data) suggests that most of the remaining North Korean refugees and irregular migrants in China are women (numbering possibly around 7,000), while plausible estimates of the number of children born to these North Korean women in China may be 15,000 to 20,000, possibly twice as high as estimates from the 2009 study.

A protection framework for these women certainly includes the 1951 Refugee Convention, particularly to prevent their forced repatriation, which remains an ongoing risk, and to provide an opportunity for international resettlement (though their movement to South Korea results in recognition as nationals of Korea and not as refugees). As mothers of children with a clear and valid claim to Chinese nationality, however, North Korean women in China would be more effectively protected through their inclusion in household registrations and the recognition of their children as Chinese. As possible victims of trafficking, mainly into forced marriage, their best protection may be under national and regional frameworks such as the Bali Process, a voluntary regional forum involving more than forty countries and numerous international agencies. For North Korean men, protection from forced repatriation might only be found through the 1951 Refugee Convention, if they can demonstrate a well-founded fear of persecution on return.

Burmese in Thailand

The waves of conflict and human rights abuse inside Burma over the years—particularly since the student-led uprising in August 1988 and the subsequent crackdown to 2011—have compelled hundreds of thousands to flee their homes and livelihoods and escape to what they hoped would be safer areas inside Burma and

within Thailand. Further internal displacement due to natural disasters (especially
Cyclone Nargis in 2008), sporadic ethnic conflict (including the recent sectarian
violence against the Muslim Rohingya in Rakhine State), and ongoing migration
for labor to Thailand, have added complexity to the protracted displacement and
crisis migration of Burmese.

Internally Displaced Persons (IDPs)

Since gaining independence from Britain in 1948, the state of Burma has been
subject to varied and violent armed conflicts and ethnic insurgencies, particularly
in the eastern areas of the country. A 1962 military coup by General Ne Win
established a military dictatorship that, in various names and forms, governed the
country until political reforms in 2011 led to the establishment of civilian government. In the 1960s, the Burma Army (or *Tatmadaw*) introduced the "four cuts"
policy, aimed at cutting off access to food, money, information and recruits by
the more than thirty ethnic insurgent groups (IDMC 2011). The strategy led to
widespread and ongoing displacement due to forced relocation and destruction
of villages, conflict, use of landmines, and human rights abuses, including forced
labor, confiscation and destruction of food supplies, arbitrary taxation, torture, rape
and extrajudicial execution (BPHWT 2010). From 1996 to 2011, it was estimated
that more than 3,700 villages and hiding sites in eastern Burma were destroyed or
forcibly relocated (TBBC 2011). Estimates of IDP numbers at the turn of the millennium ranged from 600,000 to one million, particularly in Karen and Shan States
of eastern Burma (Caouette and Pack 2002). As of 2012, UNHCR cited a figure
of 339,200 IDPs in Burma, of whom it assisted 239,000 in the southeastern part of
the country with a planned expansion to aid an additional 100,000 more IDPs in
2012, along with 2,000 spontaneous returnees from Thailand (UNHCR 2012a).
UNHCR also noted that it was assisting 808,075 "stateless persons" in 2012, of
whom 800,000 were Muslim Rohingya, residents of northern Rakhine State who
lacked citizenship in Burma.

Refugees in Thailand

Burmese refugees have been crossing the border into Thailand since the 1970s,
with outflows escalating in the wake of the 1988 student revolution and the military regime's subsequent crackdown. A Thailand Border Consortium (TBC) (formerly Thailand Burma Border Consortium (TBBC)) history of Burmese refugees
in Thailand cites 1984 as the year the Burmese launched a dry-season offensive
against Karen National Union (KNU) forces and were able to maintain a front-line
position during the rainy season (TBBC 2012). Around 10,000 Karen refugees
entered Thailand and were unable to return. In 1988, large-scale public demonstrations against the military regime were crushed, sending around 10,000 student
activists across the border into Thailand, where some settled into border camps but
others sought to carry on political activities in Bangkok. By 1995, the number of

Burmese refugees in camps was estimated at around 80,000. Several years of punishing and ultimately successful campaigns by the Burma Army against the armed ethnic insurgencies gave the government forces nearly full control of the border areas by 1997, increasing the refugee numbers in Thailand to 115,000 (IDMC 2011). From 1996 to 2011, the Burma Army launched a massive village relocation program in an effort to consolidate control in the ethnic territories. In the widespread displacement that followed, TBBC estimates that "probably more than 300,000 have fled to Thailand as refugees (the majority being Shan and not recognized [even as "temporarily displaced persons"] by the Thai government)" (TBBC 2012, 172; see also Refugees International 2004).

In 1998, the Thai government for the first time offered UNHCR a more formal role in the border camps (Lang 2001). UNHCR's Global Appeal for 1999 described its role as one in which

> UNHCR field-based protection staff will advise the Government of Thailand in establishing criteria for refugee status determination procedures to ensure that groups of asylum seekers fleeing conflict, or the effects of conflict, will be permitted temporary protection in camps in Thailand. UNHCR will provide assistance as required, to relocate camps at risk of incursion further away from the border, and, in collaboration with the Government, will conduct comprehensive and verifiable registration exercises and monitor the civilian character of the camps.
>
> *(UNHCR 1999, no page)*

In 1998, the Thai government instituted a refugee screening process through the establishment of Provincial Admissions Boards (PAB). UNHCR was given observer status and permitted to submit names of new arrivals to Thai authorities; meanwhile it continued refugee status determinations for Burmese asylumseekers in urban areas, and providing these with POC documents. In a 2005 report, the Jesuit Refugee Service and the International Rescue Committee described the PABs as "never fully functional . . . [with] unclear decision-making guidelines" (JRS and IRC 2005, 3). The system collapsed in 2001, was dormant for several years, and was rejuvenated in 2007, though it has remained inefficient at best and plagued with procedural problems and ethical challenges.

As of 2012, UNHCR cited figures of 92,000 registered refugees from Burma, along with 54,000 unregistered asylum-seekers in nine camps along the border (UNHCR 2012b). UNHCR noted that the international resettlement, begun in 2005, had reduced the camp population by around 70,000 as of 2011 (the majority resettling in the United States) and that an additional 10,000 refugees per year were expected to resettle in 2011 and 2012. "Despite this remarkable burdensharing effort," UNHCR noted, "the camp population is not likely to decrease rapidly," citing the 54,000 unregistered camp residents and "a steady flow of new entrants" (UNHCR 2012b, no page). The needs of the refugees in camps are critical (UNHCR 2012b):

The situation of refugees from Myanmar in camps in Thailand is one of the most protracted in the world. These refugees have been confined to nine closed camps since they began arriving in the 1980s. According to Thai law, those found outside the camps are subject to arrest and deportation. Legally, refugees have no right to employment. The prolonged confinement of Myanmar refugees in camps has created many social, psychological, and protection concerns. The coping mechanisms of refugees have been eroded, and the restrictions imposed on them have increased their dependence on assistance.

Thailand is not party to the 1951 Refugee Convention, nor does it have any domestic laws granting refugee status or protection mechanisms for asylum-seekers. The Immigration Act of 1979, in fact, prohibits entry to those without travel documentation, those without means of support, those without vaccinations, and "those who are dangerous to society or to the peace and security of the Kingdom" (Chapter 2: 7). The Ministry of Interior's 1954 Regulation Concerning Displaced Persons from Neighboring Countries defines a "displaced person" as "he who escapes from dangers due to an uprising, fighting or war, and enters in breach of the Immigration Act" (UNHCR 1980, 167). Thus, even the Burmese in the refugee camps are classified as "temporarily displaced," not refugees. For a time, Thailand accorded differential status and treatment for "students and political dissidents," allowing UNHCR to register them and provide them assistance outside of the border camps. All the remaining Burmese in Thailand are classified simply as "migrants" and, to the extent that they have entered the country without travel documents (and most are undocumented), they are considered illegal migrants subject to arrest and deportation (Caouette and Pack 2002).

Burmese displaced populations and migrants in Thailand

In addition to the 146,000 Burmese living in the border refugee camps, and the small numbers of POC document holders living outside the camps, estimates suggest that as many as 2–2.5 million Burmese migrants are living in Thailand, most of whom have entered the country without documentation. It has been common for many years to refer to the different populations of displaced Burmese as those "in the camps" and those "outside the camps," the implication being that little other than geography distinguished their circumstances (Banki and Lang 2008). Among the latter group who are most clearly "refugee-like" are the estimated 200,000–250,000 Shan (sometimes referred to as Tai Yai) who have fled conflict and forced relocation and are living in northern Thai provinces, where they are classified as "economic migrants" and prohibited access to refugee camps (Refugees International 2004; Banki and Lang 2008; see also Suwanvanichkij 2004). The Shan, both because they are so numerous (some informal estimates run as high as two million) and because they are seen as closer kin, culturally and linguistically, to the Thai people, seem to have been granted the dubious "acceptance" of being recognized as nothing more than ordinary migrants. As of 2011, there were another 10,000 or so displaced Burmese living in various unofficial camps along the border (Refugees

International 2011). Most recently, around 1,700 Muslim Rohingya fleeing religious and ethnic violence in Rakhine State have been held in crowded detention centers in the Thai southern provinces (Human Rights Watch 2013).

Overlapping with the conflict-displaced and more refugee-like populations of Burmese in Thailand are hundreds of thousands of Burmese migrant workers, most of whom arrived in Thailand without travel documentation and face a host of problems that, increasingly, have less to do with the conditions they left behind in Burma and more to do with the conditions they face in Thailand. In the 1990s, as economic growth and demographic stabilization in Thailand created a labor shortage of unskilled workers, larger numbers of workers from neighboring countries began entering Thailand in search of higher wages. In 1992, Thailand informally began to allow migrants workers to be employed in selected jobs in nine sectors. In 1996, migrant worker registration was formally introduced, expanding to 43 provinces and a wider number of sectors and positions. Migrant worker registration policies evolved through a variety of phases or stages to a point where there were (as of 2010) 1.3 million registered migrant workers in Thailand, of whom 1,078,767 were from Burma working in more than 24 types of work, the top three types being agriculture, construction and seafood processing (IOM 2011).

Chantavanich, Vungsiriphisal and Laodamrongchai (2007, 25) noted that migrant registration and permission to work generally "had not changed their illegal status . . . The impact of illegal status includes no protection from arrest, lower wages, poorer working conditions, fewer holidays, exploitation and limited access to social services." Green-Rauenhors, Jacobsen and Pyne (2008, 2) added to this list the "risk of trafficking and exploitation." A 2010 study of Burmese migrant labor in the seafood processing industry in Samut Sakhon Province conducted by Johns Hopkins Bloomberg School of Public Health in conjunction with the Labour Rights Promotion Network (LPN) found that 33 percent of respondents met the international definition of trafficking into forced labor and more than 50 percent reported some form of labor rights violation and/or deception and coercion in recruitment (Robinson, Branchini and Srakaew 2011).

The typology of migrants in Thailand, in addition to the migrant workers and displaced persons (inside and outside of camps), must also include minorities (among the 15 groups of "non-Thai nationals" granted permission to reside in Thailand are included "displaced Burmese nationals, Burmese irregular migrants, [and] displaced Burmese with Thai ancestry") and "Stateless/Nationality-less persons," most of whom are ethnic minorities who were born in Thailand or who have lived in the country for a long time (Archavanitkul and Hall 2011, 65). Research by David Feingold (2008, 5) found that lack of citizenship or lack of legal status was

> the single greatest factor for young hill tribe people in northern Thailand to be trafficked or exploited. It also contributes to increased vulnerability to HIV/AIDS. Without proper identity documents and recognition of permanent residence, ethnic minority people are considered "illegal aliens" in their own country

As the refugee camps for Burmese in Thailand begin to close, and international aid and attention begins to re-focus on the needs inside Burma, it is critical not to forget the needs of the long-term displaced Burmese populations in Thailand, who are living outside of camps and inter-mixed with Burmese migrant workers and Thai communities. Protection frameworks (further outlined below in the discussion on the Thailand Migration Report 2011) would need to include, for the refugees, onward resettlement, safe and voluntary return, or integration in place. For the larger number of long-term displaced and migrant worker populations, the answers lie more in an integrated approach to development of border areas on both sides of the Thailand–Burma boundary line, incorporating enfranchisement of long-standing stateless populations, work permits and migration documents for migrant workers, education for migrant children, and access to health, social services, and basic legal protections for migrant populations in Thailand.

Toward integrated approaches to crisis migration

Caouette and Pack's 2002 report, *Pushing Past the Definitions*, examined "the mixed migration from Burma into Thailand and the ever-blurring nexus between migration and asylum" (2). Their perspective at that time was that

> there is an arbitrary line between the groups that the Thai government categorizes as "temporarily displaced," "students and political dissidents," and "migrants." These faulty distinctions often result in the vast majority of people being denied asylum and protection and the superficial identification of millions as simply economic migrants.
>
> *(Caouette and Pack 2002, 1)*

A decade later, in light of the political and social reforms taking place in Burma, the question needs to be asked: what are the new opportunities, and continuing challenges, to respond to the mixed migration flows between Burma and Thailand? Pushing beyond Burma, what are the opportunities and challenges regarding North Korea and what lessons might the recent breakthroughs in Burma have to offer for changes in and around the DPRK?

The Thailand Migration Report 2011, a collaborative effort by members of the United Nations Thematic Working Group on Migration in Thailand under the leadership of the International Organization for Migration (IOM), notes that "there are more than 3.5 million persons without Thai nationality living in the country, including many long-term residents and children of migrants born in Thailand" (IOM 2011, xii). These included (as of the end of 2009) 2,455,744 migrants from Cambodia, Laos, and Burma; 513,792 (principally ethnic minorities) "residents awaiting nationality," "born in Thailand to non-national parents," and "previously undocumented persons"; and 141,076 displaced persons, refugees and asylum-seekers.[4] Among these 3.5 million, about 11 percent, or roughly 377,000, are children (under 18 years of age), including 113,000 children of registered ethnic minorities, 128,000 children of registered migrant workers, 82,000 children of unregistered workers, and 54,000 children of displaced persons and refugees (IOM 2011, xii). Just under half of these children were born in Thailand, surely an indication that protracted displacement and crisis migration flows have multi-generational effects.

The report notes that, "the degree to which the Government of Thailand envisages that the 3.5 million international migrants in the country will be integrated into the Thai nation varies according to the group" (IOM 2011, xv). A 2005 Cabinet Resolution lays out a strategy to begin to solve the problems of legal status and rights for at least some of the ethnic minorities and highland populations who have been effectively stateless for far too long. On the other hand, temporary migrant workers are expected to return home at the end of their work contracts. As for the Burmese refugees and displaced persons, one Thai scholar notes that, "It is quite clear that the Government of Thailand has no intention to fully integrate displaced persons into Thai society. In fact, the Government announced in April 2011 its intention to close all the shelters along the border" (Chantavanich 2011, 128). Though official dates for camp closings are not available, rumors at the border suggest dates in 2014 or 2015.

The Thailand Migration Report 2011 sponsoring agencies and authors (comprising 18 Thai and international scholars and migration experts) offered a series of recommendations to the Government of Thailand and, indirectly, to other governments, international organizations and NGOs. These include (IOM 2011, xv–xvi):

- Formulate a comprehensive migration policy document in consultation with stakeholders, including migrants' representatives. The policy would state long-term goals of migration policies and link migration with national social and economic development strategies.
- Greater public dialogue on international migration should be promoted. Such a dialogue [should] . . . include the active participation of the mass media, academia, the private sector, and civil society.
- Strengthen strategic planning with the governments of destination countries to develop more efficient and effective migration programs that provide enhanced protection to migrant workers.
- Policymakers should consider a strategy of earned adjustment of immigration status for the integration of some members of groups that have established a long-term presence in Thailand in particular (a) migrant workers who have been registered for several years, (b) displaced persons who have lived in shelters for many years, and (c) ethnic minorities who are long-term residents but remain stateless or without nationality, particularly those born in Thailand.
- A renewed effort should be made to achieve durable solutions for displaced persons from Myanmar. These include (a) safe and voluntary repatriation, (b) partial local integration based on a self-reliance strategy, and (c) continuation of the resettlement program.

The recommendations are ambitious but offer practical and practicable steps for Thailand, and neighboring countries including Burma, to develop a more integrated and cross-cutting approach to responding to labor migration needs and, to a more limited extent, crisis migration issues. On this last point, it should be noted that the Thailand Migration Report 2011 offers solutions to existing and long-standing populations in displaced and/or stateless conditions though it does not explicitly call for solutions to address the issue of *future* migration flows caused by conflict, natural disasters or other crises. This is certainly a gap to be filled and has undoubtedly been the most contentious aspect of migration policy for Thailand and Burma over the past thirty years (Human Rights Watch 2012). Events in late 2012 and early 2013—including threats by Thailand to deport hundreds of thousands of Burmese who lack immigration documents (IRIN 2012) and protests surrounding Thailand's treatment of 1,700 Muslim Rohingya asylum-seekers (Human Rights Watch 2013)—suggest that the issue is likely to remain contentious and challenging.

There is a recent and successful model of regional and bilateral cooperation to draw upon, namely the ASEAN response to Cyclone Nargis. On May 2, 2008, Cyclone Nargis made landfall in the Ayeyarwady Delta region of Burma; in the devastation that followed, an estimated 140,000 people died and 2.4 million people were severely affected (Belanger and Horsey, 2008). The government of Burma initially resisted offers of international assistance, delaying offers of visas to foreign aid workers and media, and obstructing access to affected areas and populations, all in violation of the Guiding Principles on Internal Displacement and the broader Responsibility to Protect (R2P).[5] In the face of government obstruction and inaction, local civil society organizations began to organize private relief efforts, while calling for more help from the regional and international community (Human Rights Watch 2012).

On May 5, ASEAN Secretary-General Surin Pitsuwan called on all member states to offer immediate relief assistance through the framework of the ASEAN Agreement on Disaster Management and Emergency Response (AADMER). Following acceptance of this ASEAN-coordinated aid approach, the ASEAN Secretariat established a two-tiered structure: a diplomatic body, the ASEAN Humanitarian Task Force (AHTF), and a Rangoon-based Tripartite Core Group (TCG), comprising ASEAN, the Burmese government, and the United Nations. Following a visit to cyclone-affected areas, UN Emergency Relief Coordinator John Holmes offered that "Nargis showed us a new model of humanitarian partnership, adding the special position and capabilities of [ASEAN] to those of the United Nations in working effectively with the government." He added that ASEAN leadership was "vital in building trust with the government and saving lives" (Creac'h and Fan 2008, 7; see also ASEAN 2010).

ASEAN has been criticized for its decision in the 1990s to admit Burma and to maintain its policies of "non-interference" and "constructive engagement." It is not too much of a stretch to suggest that this regional engagement strategy, coupled with the apparent success of the pivotal ASEAN role in responding to Cyclone

Nargis, may have helped coax Burma's leaders out of their self-imposed isolation and shift toward civilian rule. It may be hoped that Thailand and its neighbor to the west will be able to build on this evolving trust and cooperation to forge migration policies that will recast a border that has been home to so much conflict and displacement into a bridge for two-way movement of peoples, goods, and ideas. In this bridge-building exercise, it may be that Burmese migrants and displaced persons—though wary of the durability of recent reforms inside their country—can be a force for peace building and development (Brees 2009).

In considering what lessons might be drawn for China and North Korea from a study of Thailand and Burma, it is best to begin by noting all the dissimilarities. First of all, Burma was a military dictatorship while the DPRK was and is a totalitarian state, perhaps unique in modern history in maintaining virtually total control of its citizenry for more than fifty years. There are no opposition parties, let alone a dynamic opposition figure akin to an Aung San Suu Kyi to provide hope for democratic processes within the country and to rally human rights activism worldwide. Second, for all its many gaps and inadequacies in its refugee and asylum policies, Thailand is not China and has provided temporary refuge and international access to refugee and displaced populations. China, meanwhile, has scarcely wavered from its position that North Koreans in China are illegal immigrants subject to arrest, detention and deportation (though it has allowed a small number of North Korean asylum-seekers in Beijing onward passage to third countries). Third, there is no East Asian equivalent to ASEAN to promote regional cooperation on migration, development, or other issues. The key East Asian countries—including North and South Korea, China, Japan, and the Russian Federation—are all participants in the Bali Process as is the IOM and UNHCR. The Bali Process, at least, could bring these parties together to acknowledge that "poverty, economic disparities, labor market opportunities and conflict were major causes contributing to the global increase in people smuggling and trafficking in persons," to commit that States should "provide appropriate protection and assistance to the victims of traffickers of people, particularly women and children" and to affirm that "these problems should be addressed cooperatively and comprehensively" (Bali Process 2002, 2).

A new discussion of North Korean migration must begin by framing an understanding of population mobility within and outside the country as something more than a simple threat to stability. The migration of North Koreans in the last two decades has always encompassed a mix of motives—food, health, shelter, asylum, family formation, family reunification, labor/livelihood and more. The problem is that the discussion of this migration (and the policy/program options that either are or should be available) has been dominated by a dichotomous question: are they or are they not refugees? The two principal countries that say they are not—North Korea and China—control both sides of the border that virtually all of the migrants and asylum-seekers must cross. Those who would argue for refugee status or at least refugee-like protections—the United States and other Western democracies, as well as UNHCR and human rights organizations—have no direct access to these migrant populations at their greatest vulnerability in the border regions.

The South Korean approach, which has provided permanent settlement to around 25,000 North Koreans in the last fifteen years, is to avoid the refugee debate and treat the migration of North Koreans as a fundamentally Korean issue, not an international one. While this ignores some obvious geographic realities—namely that North Koreans generally must enter at least one country of asylum/transit while moving—it has proven reasonably effective in avoiding regional confrontations. The refugee framework, in other words, while it may apply to many North Koreans entering China, has proven impossible to implement due to Chinese intransigence, and is not a particularly good fit for women trafficked into forced marriage, for the effectively stateless children who are a product of those marriages, or for the thousands who cross the border for family reunification or for work (and might wish to return) and might wish not to be politically stigmatized in the process. There are no quick or easy answers to this dilemma but one approach would be to broaden the framework to encourage all the stakeholder countries—and affected local communities—to consider a more complex range of migration categories and, to acknowledge that safer migration options may be in their self-interest.

Though China is signatory to the 1951 Refugee Convention, it has steadfastly refused to acknowledge the claims of North Korean refugees and asylum-seekers or to grant UNHCR access to this population. Though further international pressure on this point is justifiable on the merits of many of the claimants' bona fides, more practical and immediate benefits could be achieved if China could be persuaded to see that its own self-interest, and the interests of local Chinese communities, would be served by offering nationality status to the children born to Chinese fathers and North Korean mothers, and by granting at least temporary protection if not more permanent status to the North Koreans who have been living peacefully for many years in Northeast China. It would also be in China's interests to protect migrant adults and children from the risk of trafficking and to provide access to basic healthcare as a public health benefit, as well as education for school-age children and other core social services.

Similarly, it would be in North Korea's interests to permit households with motives of family reunification, labor and economic betterment, or simply survival to leave without risk of penalty to themselves or their family members left behind. For their part, South Korea should continue to resettle North Korean migrants as nationals of one Korea and broaden migration opportunities for families of mixed-nationality children. Finally, the United States and other resettlement countries should broaden support for North Korean refugees (in terms of humanitarian assistance and resettlement) to include acknowledgment of, and support for, other vulnerable populations, including victims of trafficking, stateless children and other vulnerable populations.

In February 2002, China, the DPRK, and South Korea joined with forty-four other countries (including Thailand and Burma) at a meeting in Indonesia to become members in the Bali Process. In 2011, the member states endorsed a non-binding Regional Cooperation Framework "to support and strengthen practical cooperation among Bali Process Member States regarding refugee protection and international migration, including irregular migration, human trafficking and smuggling" (Bali Process 2011). While the Bali Process is no panacea—among other things, it has been criticized for its Australia-centric agenda, its lack of civil society participation, and a gradual decline of East Asian country participation (Douglas and Schloenhardt 2012)—it provides a framework for regional dialogue and "soft diplomacy" to address the complexity of issues presented by refugee displacement, irregular migration, and human trafficking. A revitalization of commitments within East Asia certainly would be a welcome signal of renewed support for the Bali Process, and offers a venue for member countries to engage in regional dialogue and "soft diplomacy" on, among other things, modes of North Korean migration within the Asian and Southeast Asian region. In the case of the Rohingya, the Bali Process and, more specifically, the ASEAN Ministerial Meetings and discussions with the ten Dialogue Partners (Australia, Canada, China, the EU, India, Japan, New Zealand, Russia, South Korea and the United States) could help provide a regional framework to engage with Burma in seeking a resolution to the sectarian violence and a durable solution to the status of Rohingya within and outside Burma.

Conclusion

In one of those curious coincidences of history, on the same day that North Korea's "Dear Leader" Kim Jong-il died, the Czech dissident and human rights activist, Václav Havel, also passed away. In 2004, Havel had written an appeal for greater attention to the problem of hunger and human rights abuse in North Korea and lack of protection for North Korean refugees in China (*Globe and Mail* 2004). At a January 2012 tribute to Havel's life, Aung San Suu Kyi reflected on a letter she had received from him shortly before his death. He wrote:

> After nearly 50 years of totalitarian rule, the road to free, pluralistic and democratic society will not be an easy one. We also had 50 years of tyranny—first the Germans, and then the Communists, and although 22 years have already elapsed since the Velvet Revolution, we are still not fully there. However, if there is anything we can do to help, for example, and only if you wish, to share some of our transformational experience with you, we shall gladly do it.
>
> *(Suu Kyi 2012, 95)*

Fifty years of dictatorial rule in Burma and more than sixty years of totalitarianism in North Korea have given rise to protracted crisis migration internally and externally, in the form of IDPs, refugees, irregular migration, human trafficking and patterns of mobility that are at once shaped by motives for family reunion and work opportunity but misshapen by state abuse of rights or negligence of core obligations to human security. Forging solutions out of protracted crisis and conflict may no doubt benefit from the transformational experiences and energies of a Václav Havel or an Aung San Suu Kyi. The challenge in responding to ongoing crisis migration in and from North Korea, and in continuing to deal with the consequences of the long-term crisis in Burma, is to build local and regional frameworks that are stronger and more durable than individual leaders, grounded in approaches that protect the right to a durable solution *and* the right to continued mobility as appropriate responses to crisis.

Notes

1   In his analysis of the 1990–91 famine in Sudan, Patel suggested that migration might take place in the early, intermediate or late stages of a famine (caused, in the case of Sudan, by drought). In the early stages, household members might migrate in search of employment. As wages fall in the intermediate stages, employment migration intensifies along with the "migration of adult males to reduce food burden" in the household. In the late stage of a crisis, only after a household has already reduced consumption, sold or traded away its productive or other assets, and sent individual members out in search of work or food, there comes the "distress migration of whole families" (Patel 1994, 323).

2   This chapter uses the terms "refugees," "migrants" and, occasionally, "asylum-seekers" to refer to those North Koreans who have crossed the border into China generally without travel documents and largely without access to proper determination of their status under international law. China is signatory to the 1951 Refugee Convention but has introduced no implementing legislation nor do its policies acknowledge North Koreans generally as entitled to refugee protections under either national or international law. The approach in this chapter follows that of Seymour (2005) to the effect that "Unless otherwise obvious from the content, the use of 'refugee' in this report does not imply any judgment as to whether the people concerned fit any particular definition of 'refugee' under international law. Likewise, the use of a term like 'migrant' does not imply that the person is not entitled to refugee status. The term 'asylum seeker' does not denote any formal application for asylum."

3   According to data made available by South Korea's Ministry of Unification, the numbers of North Koreans resettling in South Korea from 1998–2012 were: (1998) 71; (1999) 148; (2000) 312; (2001) 583; (2002) 1,139; (2003) 1,291; (2004) 1,894; (2005) 1,383; (2006) 2,018; (2007) 2,544; (2008) 2,809; (2009) 2,927; (2010) 2,376; (2011) 2,737; (2012) 1,509.

4   Also included are 106,486 professional, skilled and semi-skilled workers, 121,109 temporary stayers, 92,014 tourist and transit visa extensions, and 19,052 foreign students in higher education.

5   UN Guiding Principles on Internal Displacement ("IDP Principles"), UN Doc. E/ CN.4/1998/53/Add.2 (1998), noted in Comm. Hum. Rts res. 1998/50, principles 3(1) and 25(1). General Assembly Resolution 46/182 concerning the Strengthening of the Coordination of Humanitarian Emergency Assistance of the United Nations of December 19, 1991, emphasizes that "[e]ach State has the responsibility first and foremost to take care of the victims of natural disasters and other emergencies occurring on its territory. Hence, the affected State has the primary role in the initiation, organization, coordination, and implementation of humanitarian assistance within its territory" (cited in HRW 2010, 21).

References

Archavanitkul, K. and Hall, A. (2011) "Migrant Workers and Human Rights in a Thai Context," in Huguet, J.W. and Chamratrithirong, A. (eds.), *Thailand Migration Report 2011*, Bangkok, Thailand: International Organization for Migration.

Association for Southeast Asian Nations (ASEAN) (2010) *A Humanitarian Call: The ASEAN Response to Cyclone Nargis*, Jakarta, Indonesia: ASEAN Secretariat.

Back Pack Health Worker Team (BPHWT) (2010) "Diagnosis: Critical—Health and Human Rights in Eastern Burma," available online at: http://www.backpackteam.org/wp-content/uploads/reports/Diagnosis%20critical%20-%20Eng%20website%20version.pdf.

Bali Process (2002) Bali Ministerial Conference on People Smuggling, Trafficking in Persons and Related Transnational Crime Co-Chairs' Statement, February 26–28, available online at: http://www.baliprocess.net/ministerial-conferences-and-senior-officials-meetings.

Bali Process (2011) "Steering Group Note on the Operationalization of the Regional Cooperation Framework in the Asia Pacific Region (2011)," Fourth Bali Regional Ministerial Conference on People Smuggling, Trafficking in Persons and Related Transnational Crime, Co-Chairs' Statement. Bali, Indonesia, March 29–30, available online at: http://www.baliprocess.net/ministerial-conferences-and-senior-officials-meetings.

Banki, S. and Lang, H. (2008) "Protracted Displacement on the Thai–Burmese Border: The Interrelated Search for Durable Solutions," in Adelman, H. (ed.), *Protracted Displacement in Asia: No Place to Call Home*, Hampshire, UK: Ashgate Publishing Ltd.

Belanger, J. and Horsey, R. (2008) "Negotiating Humanitarian Access to Cyclone-affected Areas of Myanmar," *Humanitarian Exchange*, 41: 2–5.

Brees, I. (2009) "Burmese Refugee Transnationalism: What is the Effect?" *Journal of Current Southeast Asian Affairs*, 2: 23–46.

Caouette, T. and Pack, M. (2002) *Pushing Past the Definitions: Migration from Burma to Thailand*. Washington, DC: Refugees International and Open Society Institute.

Chantavanich, S., Vungsiriphisal, P. and Laodamrongchai, S. (2007) *Thailand Policies toward Migrant Workers from Myanmar*, Bangkok, Thailand: Asian Research Center for Migration, Institute for Asian Studies, Chulalongkorn University.

Chantavanich, S. (2011) "Cross Border Displaced Persons from Myanmar in Thailand," in Huguet, J.W. and Chamratrithirong, A. (eds.), *Thailand Migration Report—2011*, Bangkok, Thailand: International Organization for Migration.

Citizens' Alliance for North Korean Human Rights (2012) "Document: Interactive Dialogue with the Special Rapporteur on the Situation of Human Rights in the Democratic People's Republic of Korea," *Life and Human Rights in North Korea*, 64: 110–120.

Congressional Research Service (CRS) Report for Congress (2007) *North Korean Refugees in China and Human Rights Issues: International Response and US Policy Options*, Washington, DC: Congressional Research Service.

Creac'h, Y.-K. and Fan, L. (2008) ASEAN's Role in the Cyclone Nargis Response: Implications, Lessons and Opportunities, *Humanitarian Exchange*, 41: 5–7.

Douglas, J.H. and Schloenhardt, A. (2012) "Combating Migrant Smuggling with Regional Diplomacy," University of Queensland Research Paper, February, available online at: www.law.uq.edu.au/migrantsmuggling.

Feingold, F. (2008) "UNESCO Promotes Highland Citizenship and Birth Registration to Prevent Human Trafficking," UNESCO Bangkok Newsletter, available online at: http://www.unescobkk.org/fileadmin/user_upload/culture/Trafficking/project/abc/Selected_Articles_and_Publications/Pages_from_DAF_UNESCO_Newsletter_Citizenship_Article.pdf.

Globe and Mail (2004) "Vaclav Havel on Kim Jong-il," June 17.

Green-Rauenhorst, M., Jacobsen, K. and Pyne, S. (2008) *Invisible in Thailand: Documenting the Need for International Protection for Burmese*, Washington, DC: International Rescue Committee.

Human Rights Watch (HRW) (2010) *"I Want to Help My Own People": State Control and Civil Society in Burma after Cyclone Nargis*, New York: Human Rights Watch.

HRW (2012) *Ad Hoc and Inadequate: Thailand's Treatment of Refugees and Asylum Seekers*, New York: Human Rights Watch.

HRW (2013) *Thailand: End Inhumane Detention of Rohingya*, available online at: http://www.hrw.org/news/2013/06/03/thailand-end-inhumane-detention-rohingya.

Internal Displacement Monitoring Centre (IDMC) (2011) *Myanmar: Displacement Continues in the Context of Armed Conflicts*, available online at: http://www.internal-displacement.org/asia-pacific/myanmar/2011/myanmar-displacement-continues-in-context-of-armed-conflicts.

International Organization for Migration (IOM) (2011) *Thailand Migration Report 2011*, Bangkok, Thailand: International Organization for Migration.

Jesuit Refugee Service (JRS) and International Rescue Committee (IRC) (2005) "Nowhere to Turn: A Report on Conditions of Burmese Asylum Seekers in Thailand and the Impacts of Refugee Status Determination Suspension and the Absence of Mechanisms to Screen Asylum Seekers," available online at: http://reliefweb.int/sites/reliefweb.int/files/resources/E80ED524DC5A310A4925703C0006FA14-jrs-tha-11jul.pdf.

IRIN Humanitarian News and Analysis (2012) "Myanmar–Thailand: Burmese Migrant Workers Risk Deportation," available online at: http://www.irinnews.org/printreport.aspx?reportid=96346.

Katona-Apte, J. and Mokdad, A. (1998) "Malnutrition of Children in the Democratic People's Republic of North Korea," *Journal of Nutrition*, 128: 1315–1319.

Korea Times (2012) "Rocket Plan Shows New NK Regime's 'Structural Intransigence': Lee advisor," available online at: http://www.koreatimes.co.kr/www/news/nation/2012/04/113_108312.html.

Korean Buddhist Sharing Movement (KBSM) (1998) "The Food Crisis in North Korea. 6th Phase of Research (30 Sep 1997–15 Sep 1998)," available online at: http://www.goodfriends.or.kr/eng/report/1694e.htm.

Lang, H. (2001) "The Repatriation Predicament of Burmese Refugees in Thailand: A Preliminary Analysis, New Issues in Refugee Research," Working Paper No. 46, UNHCR.

Liang, Z. and Dohm, K. (2006) "Migration and Minority Opportunities in China," abstract prepared for the 2006 annual meeting of the American Sociological Association.

Patel, M. (1994) "An Examination of the 1990–91 Famine in Sudan," *Disasters*, 18(4): 313–331.

Refugees International (2004) *The Shan in Thailand: A Case of Protection and Assistance Failure*, available online at: http://www.unpo.org/article/832.

Refugees International (2011) *Thailand: No Safe Refuge*, Washington, DC: Refugees International.

Robinson, C. (2000) "Famine in Slow Motion: A Case Study of Internal Displacement in the Democratic People's Republic of Korea," *Refugee Survey Quarterly*, 19(2): 113–127.

Robinson, C. (2010) *Population Estimation of North Korean Refugees and Migrants, and Children Born to North Korean Women in Northeast China*, Baltimore, MD: Center for Refugee and Disaster Response, Johns Hopkins Bloomberg School of Public Health.

Robinson, C., Lee, M.K., Hill, K. and Burnham, G. (1999) "Mortality in North Korean Migrant Households: A Retrospective Study," *Lancet*, 354: 291–295.

Robinson, C., Branchini, C. and Srakaew, S. (2011) *Estimating Labor Trafficking: A Case Study of Burmese Migrant Workers in Samut Sakhon Province, Thailand*, Bangkok: United Nations Inter-agency Project on Human Trafficking.

Schwekendiek, D. (2008) "The North Korean Standard of Living During the Famine," *Social Science and Medicine*, 66: 596–608.

Senate Resolution No. 139, 112th Cong., 1st Sess. 2 (2011).

Seymour, J.D. (2005) "China: Background Paper on the Situation of North Koreans in China," a Writenet Report commissioned by UNHCR, available online at: http://www.refworld.org/docid/4231d11d4.html.

Suu Kyi, A.S. (2012) "Vaclav Havel's Last Letter," *Life and Human Rights in North Korea*, 63: 94–96.

Suwanvanichkij, V. (2004) "Displacement and Disease: The Shan Exodus and Infectious Disease Implications for Thailand," *Conflict and Health*, 2(4): 1–5.

TBBC (2012) "Programme Report: 2012, January to June," available online at: http://www.tbbc.org/resources/resources.htm#reports.

Thailand Burma Border Consortium (TBBC) (2011) "Displacement and Poverty in Southeast Burma/Myanmar," available online at: http://www.tbbc.org/idps/idps.htm.

UNHCR (1980) "Roundtable of Asian Experts on Current Problems in the International Protection of Refugees and Displaced Persons: April 1980," Publication Box 4, Section 1.

UNHCR (1999) *Global Appeal 1999—Myanmar/Thailand*. available online at: http://www.unhcr.org/3eaff44138.html.

UNHCR (2003) "Framework for Durable Solutions for Refugees and Persons of Concern," May 1, UNHCR Refworld, available online at: http://www.unhcr.org.

UNHCR (2012a) "Country Operations Profile: Myanmar," available online at: http://www.unhcr.org/pages/49e4877d6.html.

UNHCR (2012b) "Country Operations Profile: Thailand," available online at: http://www.unhcr.org/pages/49e489646.html.

8

ENVIRONMENTAL PROCESSES, POLITICAL CONFLICT AND MIGRATION

A Somali case study

Anna Lindley

Introduction

In 2011, a severe drought combined with intense political violence and general governance failure, causing widespread hardship in south-central Somalia, with famine declared in parts of the territory. The construction of crises—in Somalia as elsewhere—depends critically on the perceptions and pronouncements of dominant actors, and it is a process that is shaped by political agendas often as much as by empirical realities (see Lindley and Hammond forthcoming). However, the situation in Somalia in 2011 certainly fulfilled the criteria set out by the editors of this book in their working definition of a *humanitarian crisis*: there was a widespread threat to life, physical safety, health and subsistence, which exhausted the coping capacity of many individuals and communities. By September 2011 an estimated four million people, around half the Somali population, were thought to be in need of basic humanitarian assistance (Hammond and Vaughan-Lee 2012). It is estimated that some 258,000 people died as a result of the famine, about half of whom were children under five (Checchi and Robinson 2013). This humanitarian crisis generated—and indeed was itself partly constituted by—high levels of forced displacement, with around a quarter of the population living in displacement, within the Somali territories and abroad, in 2011 (UNHCR 2011). This was accompanied by major protection problems, with many people on the move struggling to access humanitarian support and securing very limited protection of their rights in the various Somali areas and foreign states to which they travelled.

Prominent in the political and media hype which ensued were references to so-called "drought displacement," as distinct from movements prompted by conflict and persecution. For example, according to the government of Kenya in July 2011, "The current influx of refugees into Kenya is of Somalis seeking food and not people running away from violence" (Government of Kenya 2011, no page). There are problems with this interpretation, both empirically and in terms of the policy responses it tends to support. As we shall see, political dynamics—in this case, severe structural violence and years of ongoing armed conflict—strongly shaped the experience of drought by different groups in society, and whether they were forced to migrate or not. Famine, and therefore any migration that follows in its wake, is indisputably political.

Bearing this in mind, this chapter seeks to develop a clearer understanding of the environmental dimensions of Somali mobility and displacement, and to draw out the implications for policy responses. Little research has addressed this to date. Investigations of migration have generally focused on the role of political conflict in prompting acute displacement, or on how environmental processes shape more routine nomadic movements (for a rare exception, see Kolmannskog 2010). "Mixed migration" is often referenced in a rather cursory fashion, with little systematic analysis of the complex interplay between political, economic and environmental drivers of migration. Understanding the drivers of crisis-related movement is vital for the development of appropriate policy responses, particularly in a context where distinct drivers may connect with very distinct packages of rights, protection and solutions.

This chapter makes two simple points. First, it argues—against the spirit of the "drought displacement" epithet—that the role of the recent severe drought in producing crisis and migration cannot be understood in isolation from a series of vital historical and concurrent political processes (see Zetter and Morrissey, Chapter 9 in this volume). Second, it argues that the environmental dimensions of recent displacement prompt a series of policy challenges in relation to prevention, response and rights protection, and that these often necessitate working across the policy silos which characterize responses to humanitarian crises and related movement.

The chapter is based on a review of secondary literature on environmental issues (NGO and UN reports as well as academic articles) combined with the author's own primary research on Somali mobilities and policy responses over the last ten years (see also Lindley 2010a, 2010b, 2011, 2013). This is a wide, rich and complex research area, and a detailed study of the contemporary political ecology of mobility in the Somali territories is needed; meanwhile, this chapter brings together some insights that might form a basis for more systematic research.

The first section of the chapter examines the relationship between environmental conditions, rural livelihoods and mobility in the Somali territories, focusing primarily on south-central Somalia, and how political authorities and armed forces have mediated this over the years. The second section presents key aspects of the recent humanitarian crisis in south-central Somalia and related migration issues. The final section of the chapter explores policy responses, highlighting key challenges and areas that might be strengthened.

Environmental conditions, rural livelihoods and mobility

The Somali territories have an arid and semi-arid environment. The main *gu* rains fall during the April–June period, and the minor *deyr* rains fall during the

September–December period; but even during these seasons, rain can be low and sporadic, vary considerably by year, and include long dry periods and intense downpours. The south, particularly the area around and between the Jubba and Shabelle rivers, tends to receive the highest rainfall and the northern territories the least.

Frequent episodes of drought, when rainfall is low for a prolonged period, are a major problem. Droughts are often quite localized, but some affect the whole region. For example, there were severe and widespread droughts in 1974–75, 1991–92, 2000–01 and 2010–11. By definition droughts are slow-onset processes, unfolding over an extended period. Flooding is another hazard, particularly in the southern riverine areas. In the El Niño flooding of 1997–98, some areas received twenty times the typical monthly rainfall, and large shallow lakes appeared between the Jubba and the Kenyan border (Little 2003). Serious river flooding again took place in 2006 and 2010, and there is intermittent flash flooding.

It is broadly accepted that there have been increases in temperatures, which will continue in coming decades, but scientific sources differ somewhat regarding the size and direction of trends in rainfall variability (Lott, Christidis and Stott 2013). There are ongoing studies to determine if these may be reliably linked to a permanent trajectory of human-induced climate change. Nevertheless, the perception in many affected communities is that drought is occurring in intensifying cycles, and that whereas earlier it occurred once every ten years or so, and each was given a specific name, now it seems to have become a "nameless constant" (Kolmannskog 2010, 108).

Prevailing ecological conditions are critical to the rural activities—livestock rearing and crop production—which are a key component of the livelihoods of the majority of Somalis.[1] While this section explores these rural livelihood systems in turn, it also highlights the numerous onward connections to urban areas, as well as the interconnections between pastoralist and crop production. These connections may be visible at the level of the family or at the level of market relationships. Moreover, the livelihood systems described are not static, but crucially mediated by political dynamics in ways that are woven into the account.

Pastoralism is particularly widespread as a livelihood activity, prevalent in the arid northern and central areas, and along the Ethiopian and Kenyan borders. Camels are reared throughout these areas (this is the "prestige" animal, and a major cultural and political symbol: the *mandeeq*, or she-camel, symbolizing Somali autonomy and nationhood). Sheep and goats are also common in various locations, and cattle are reared mainly in the south. More than half the population are either "pure" pastoralists or agropastoralists, deriving food and income from rearing livestock. The marketing of livestock and related products (milk, hides) provides income and employment for many more people (e.g. herders, brokers, health inspectors, truckers) (FSNAU 2011). Livestock and related products typically account for some 80 percent of exports, primarily to markets in the Arab peninsula and neighboring Kenya, Djibouti and Ethiopia (UNOCHA 2006).

Mobility is at the center of this flourishing livelihood system: pastoralism is a nomadic or semi-nomadic activity, involving the seasonal concentration and dispersal of herders and their livestock, according to the availability of forage and water in different places. In times of relative peace and reasonable environmental conditions, most pastoralists have a yearly cycle that entails somewhat similar movements, staying by permanent water sources in the dry season, and grazing where there is temporary water in the wet season (Devereux 2006). The term *deegaan* is used to refer to the collection of landscapes where the sub-clan has the customary right to move around, reside and graze their livestock; this does not traditionally have precise boundaries, may straddle colonially imposed international borders, and an area may host different groups simultaneously or at different times of the year (Little 2003). At the same time, there have long been tendencies toward sedentarization and enclosure of rangeland, which are highlighted below.

There are also many connections between *reer miyi* (people of the bush) and *reer magaal* (people of the town), which are part of a rich array of risk-spreading strategies deployed by pastoralists. Most nomadic pastoralists have some urban-based kin, and some of the family may settle on the edge of town for part of the year, or move to urban areas on a temporary or more permanent basis to work or for schooling (Simons 1995; Luling 2002; see also Fagen, Chapter 16 in this volume). These movements give rise to important flows and exchanges of basic supplies, cash, business transactions, and mutual hospitality and assistance. The international migration of some family members—whether to the Gulf countries during the oil boom or Europe and North America during the civil war—often provides an additional dimension, and an important source of cash remittances (Lindley 2010a).

Drought is a major challenge for pastoralists. When adequate pasture and water are hard to access, the animals' condition deteriorates, and eventually they may die of starvation or disease, depriving pastoralists of their primary means of subsistence and income. Skilled animal husbandry techniques and community support mechanisms prove vital for surviving drought. Another key coping strategy is moving longer distances in search of water and pasture, sometimes even across international borders. This may be seen as displacement in the sense that the pastoralists' usual migratory pattern has been disturbed: it may be a temporary and successful measure, or become a more desperate search (Kolmannskog 2010). Key tools in this process are traditional provisions within customary law, which oblige Somalis to allow access to other groups at times of drought, and the modern mobile phone which helps pastoralists seek information about water availability in other locations. Where water and pasture are found, rapid convergence of herds can lead to environmentally damaging over-grazing and conflict between groups (Kolmannskog 2010). While luckier families are able to call on credit, savings or diaspora remittances to truck water to livestock or livestock to water, others have no such help, or become heavily indebted (Horn Relief 2009; Lindley 2010a). Pastoralists may be forced to migrate to urban areas to find work or access assistance from kinspeople or humanitarian agencies, sometimes on a temporary basis until able to stabilize their rural livelihood, sometimes permanently "dropping out." The poorest pastoralists are unsurprisingly usually the hardest hit by drought as they more often have to pay for water (not having their own reservoirs), their herds are small and livelihood relatively less diversified (Le Sage and Majid 2002).

The vital interconnections between environmental conditions and political context are illustrated by the Somali proverb *nabad iyo caano* (peace and milk), stressing the strong positive association for pastoralists between security and prosperity, with access to pasture and water relying on peaceful cooperation. Conversely, the complementary proverb *col iyo abaar* (conflict and drought) highlights the negative synergies of conflict and drought, which threaten access to pasture and water. The relationship is multi-directional: drought may lead to pressures on resources and spark violent conflict, or extant conflict and insecurity may exacerbate environmental problems and the experience of drought. Pastoral livelihoods were shaped by changing forms of authority throughout the twentieth century. Colonial rule, particularly in the British Somaliland Protectorate (where the administration was keen to secure a meat supply for the Aden garrison across the Red Sea), oversaw a shift from more family-oriented, semi-subsistence pastoralism, to more commercialized production, empowering a range of middlemen and urban actors (Lewis 2003). After independence and unification in 1960, in 1969 General Barre came to power, and nomadic pastoralism came to be viewed as economically backward and politically troublesome, the "scientific socialist" regime officially encouraging sedentarization, large-scale export crop farming and ranching where possible, including through the mass resettlement of northern pastoralists affected by the severe 1974–75 drought (known as *daba-dheer*, or "long-tailed") in agricultural and fishing cooperatives (Bradbury 2008). Land was nationalized and redistributed or sold off in a process which disproportionately favored those with money and regime connections, and privately controlled water points were developed through a corrupt government program and private business initiative (Little 2003). While helping to sustain elite loyalty to the ailing regime, these developments fomented conflict and grievance, and are thought to have contributed in no small way to the civil war and state collapse in 1991.

In 1991, the rebel movements that had emerged during the 1980s ousted Barre from Mogadishu and as the conflict unfolded into clan-aligned factional violence, state institutions collapsed. While the civil war developed differently in the northern territories, in south-central Somalia clashes within, as well as between, armed groups identifying with the Hawiye or Darod clan-families, both of which are associated with predominantly pastoral traditions, devastated the area (Kapteijns 2013). Soon the conflict devolved along multiple, shifting and often more localized axes of violence or cooperation. Many pastoralists were directly affected by political reconfiguration and insecurity, and some were displaced to urban or other rural areas, or to refugee camps, on a temporary or more permanent basis. Many found themselves hosting relatives fleeing the urban violence. As time went on, other challenges emerged or worsened in the absence of state institutions. Clashes over access to rangeland and water continued, made more severe by the ready availability of automatic weapons (Kolmannskog 2010). The further unregulated development of private *berkads* (reservoirs) has reinforced tendencies toward enclosure and sedentarization. In the 1990s and 2000s, Saudi bans on imported Somali livestock due to disease fears were hard to combat in the absence of a public veterinary service (Bradbury 2008). In parts of south Somalia, rural–urban linkages were severed and livelihoods *de*-diversified: the southern city of Kismayo became largely cut off from its economic hinterland by clan-based factional conflict (Little 2003).

However, free from state interference, in many ways many pastoralism thrived, despite the tough ecological conditions and political violence. It has proved a remarkably robust and adaptive livelihood system, albeit with high levels of vulnerability for some actors (Little 2003; Devereux 2006; Bradbury 2008). The movability of the main asset—livestock—is key. Also key is the relative politico-military strength of traditionally pastoralist clans in the civil war context (LeSage and Majid 2002). By contrast, crop production is more closely dependent on local conditions, and has been more vulnerable to environmental hazards and conflict conditions in recent decades (Little 2003). Moreover, it is primarily the livelihood activity of relatively marginalized groups in society: the Somali Bantu farmers and Rahanweyn agropastoralists.

Crop production—mainly sorghum, maize, sesame, cowpeas, vegetables and fruit—is concentrated in the fertile area around and between the Juba and Shabelle rivers in southern Somalia, and in western Somaliland. Irrigated, rain-fed and flood-recession agriculture is practiced, depending on the location and crop. Domestic farming generally accounts for some 50 percent of domestic cereal consumption, and remains an important source of income and labor opportunities (UNOCHA 2006). It is primarily a smallholder activity (Luling 2002; Little 2003).

Although it implies a more sedentary way of life, given the fixed assets involved, crop production is routinely connected to other livelihood systems and to migration in various ways. Agropastoralism is common, with some families split or switching seasonally between moving with livestock and settled crop cultivation (Luling 2002). There are cross-cutting economic interactions: for example, some farmers diversify their income sources by investing in livestock, or allowing herders to graze animals on the stubble of harvested fields (Little 2003; UNOCHA 2006). People may migrate seasonally to supplement food and income gained through crop production with earnings from rural and urban labor opportunities. There has long been some more long-term migration of people seeking informal or formal employment in the major urban centers, and some evidence of remitting. However, historically, international migration was limited, due to the more settled way of life of people involved in farming, and their limited resources (Little 2003; FSNAU 2010; Hammond *et al.* 2011).

While rain-fed agriculture is clearly vulnerable to drought, historically irrigated agriculture had been relatively resistant, until civil war politics undermined access to effective water management systems, with many smallholders struggling to pay for the costs of pumping water or the taxes imposed by armed groups controlling their area who did little to maintain water infrastructure (Luling 2002). The impact of drought on farmers is complex, and depends on the severity and spread of the drought, the market effects, and individual farmers' situations. In the worst cases, drought may lead to total crop failure, depriving farmers of food and income. Too much rain is also a threat for farmers: in 1997 El Niño destroyed the bulk of the Lower Jubba river area agricultural crop, damaged property, and prompted mosquito-spread diseases (Little 2003). At the same time, flooding often damages transport links, making it hard for people to import food or aid, leading to rapid price hikes and compounding food insecurity.

For farming and agropastoral communities, migration is among the key responses to these natural hazards. Recurring drought was associated with displacement throughout the 1990s and 2000s. Flooding displaced some 455,000 people in the southern riverine areas in 2006 and 66,000 in 2010, and there are frequent smaller-scale episodes (FSNAU 2010). Some of this movement is quite localized—for example, flood-affected people often move to higher ground in the same community or local area (FSNAU 2010). Meanwhile, many people are forced to move to urban centers to access humanitarian assistance, or try to find work (LeSage and Majid 2002). Families may split: sending some to town to reduce the mouths to feed at home, to leave some members on the property, or because some are simply too old or disabled to withstand the journey. This may occur on a temporary basis, but sometimes turns into a more permanent situation, depending on the nature of urban opportunities and the extent of damage to rural livelihoods. Some people successfully secure an economic foothold in town, but others end up in dire conditions squatting in and around urban centers. International migration of farmers due to environmental hazards is relatively rare, reflecting the historically limited migration networks of the riverine social groups (Thomas, Chapter 3 in this volume).

The history of farming in the riverine areas of south Somalia, where it is most prevalent, is intertwined with that of the more marginalized social groups. Many of the farming population identify as *Jareer*, now also known as Somali Bantu (associated with historical Bantu expansion, settlement and slavery), a group that has long been ascribed the position of a lower caste by dominant pastoral clans. Meanwhile, the Rahanweyn agropastoralists in this area, who also have more settled communities, have a long history of relative political marginalization. (However, it is important to note that social identifications are flexible, and politically constructed: they were shaped and in some ways hardened during the civil war (see Kapteijns 2013).) The fertile agricultural areas occupied by these groups saw successive "land grabs": first, Italian colonial confiscation for plantation agriculture; second, Barre's land reforms, which gave much nationalized agricultural land over to regime loyalists, poorly performing state enterprises and people with money; and third, land-grabbing during the early years of the civil war by incoming militia from dominant social groups, particularly the Habar Gedir of the Hawiye clan-family (Luling 2002).

When the civil war broke out, largely unarmed and unprotected, people from these marginalized groups were treated with impunity: killing, rape and other abuses were commonplace (Kapteijns 2013). Their livelihoods were also attacked: militia looted their stores and possessions; impeded safe access to farms and disrupted cropping cycles; confiscated farmland, forcing people to live in flood-prone or infertile land, or as laborers on what had been their own land; and deliberately destroyed irrigation infrastructure or assumed control of it in order to extort money from the farming population. In this context, the 1991–92 famine was basically "man-made"; and was most severe in the supposed breadbasket of Somalia (De Waal 1997). The severe political violence disrupted production, trade and aid delivery, and restricted movement, turning drought into a total disaster, killing some 212,000–248,000 people in the main emergency period of 1992–93, and prompting large-scale movement of rural displaced people toward the coastal cities (Hansch *et al.* 1994).

In sum, both routine mobility and displacement in the wake of natural hazards have been recurring features in rural Somali livelihoods, but these dynamics cannot be understood without reference to the political context in which they occur. Before the civil war, state intervention in people's relationship with their natural resource environment was far from benign. Nonetheless, since the collapse of the state in 1991, violent and predatory political actors have exacerbated the impact of environmental hazards on particular groups, and existing governance frameworks often lack the capacity and will to regulate environmentally damaging practices, or respond adequately to changing vulnerabilities in the population.

A clear example is charcoal production, long an important domestic energy source, which more recently became a profitable export, with some 80 percent exported to Gulf countries (Webersik and Crawford 2012). Wood collection and charcoal production has considerably expanded, and is a common survival strategy for the poor, as wood is considered to be a "free" good that anyone can use (Dini 2011; Webersik and Crawford 2012). However, it can cause considerable environmental damage, undermining local livelihoods in other ways, as cutting the slow-growing wood increases soil erosion and reduces vegetative cover (Oduori *et al.* 2011). The real winners are not the producers but local businesspeople and armed actors who—through transport, tax and trade—cream off the profits (Webersik and Crawford 2012). Because of the powerful actors involved in charcoal export, wood collection remains largely unregulated (Dini 2011).

A multi-faceted, multi-layered humanitarian crisis

The humanitarian crisis which reached its apogee in 2011 has often been described as the result of a "perfect storm" of concurrent conflict, drought and poor governance (Hammond and Lindley forthcoming). The northern Somali territories followed a distinct political trajectory, with the declaration of independence in Somaliland and the establishment of a regional administration in Puntland. But, following years of much more localized, lower-intensity conflict, and pockets of peace in south-central Somalia, in 2006, the political conflict entered a new and intense phase (Menkhaus 2007). The Islamic Courts Union (ICU) emerged from local Islamic courts, originally established to combat insecurity in Mogadishu, to become a national political force, and in 2006 brought south-central Somalia under a unified administration for the first time since the collapse of the state, improving security for ordinary people and gaining wide public support. The international response was hostile, in the context of the "global war on terror," and regional fears of Somali irredentism. The ICU was ousted by the internationally sponsored Transitional Federal Government (TFG) (formed in Kenya in 2004) with the help of Ethiopian and African Union forces. The Courts fragmented politically, and the hardline militia group Al Shabaab pitted itself against the government and its allies.

In an earlier study by the author (Lindley 2010b), people displaced from Mogadishu by the conflict in 2007–08 emphasized the dramatic change in the nature of urban violence: the high level of combat-generated insecurity, the disregard for civilian life, the weakening of clan-based protection mechanisms, and the persecution of journalists and civil society activists by both sides in the conflict. The conflict also had economic consequences, disrupting livelihoods, with the wholesale destruction or confiscation/occupation of personal homes and property as well as business premises, equipment and stocks; key infrastructure such as roads or markets damaged or blocked; and mobility heavily circumscribed by violence. Estimates of internally displaced persons (IDPs) escalated from 400,000 in 2006 to one million in 2007 (Lindley 2010b). A relatively wide socio-economic spectrum of people were displaced, with much displacement *out* of war-torn Mogadishu and other urban centers (Lindley 2013). The numbers of Somali refugees in neighboring countries also began to climb substantially year on year.

The Ethiopian forces eventually withdrew, and a moderate Islamist was inaugurated as president. However, far from won over, Al Shabaab gained control of most of south-central Somalia, introducing forced military recruitment, heavy social regulation, severe punishment and bans on many humanitarian agencies on grounds of political bias (HRW 2011). Meanwhile, counter-terror legislation led many aid agencies to draw back from operating in areas controlled by Al Shabaab, now designated a terrorist organization (Hammond and Vaughan-Lee 2012).

It was against this background of conflict that drought emerged in 2010 with the failure of the *deyr* rains in 2010, and compounded by meager *gu* rains in 2011. While the failure of the short rains in late 2010 was related to the weather pattern known as La Niña, the lack of long rains in early 2011 was probably partly related to warming due to greenhouse gas emissions (Lott *et al.* 2013). Unfolding over several months, it was clearly a slow-onset situation: in some areas this dry period came on top of several consecutive seasons of droughts (FSNAU 2011).

The impact on rural livelihoods was severe, and documented in considerable detail by the Food Security and Nutrition Assessment Unit (2011). The harvest was the worst in seventeen years, a quarter of normal production levels, undermining the means of subsistence and income of already poor farming and agropastoralist communities in the areas between and along the Juba and Shabelle rivers. Cereal prices reached record highs, doubling or even tripling in some areas (exacerbated by intermittent trade disruption and drastic decrease in humanitarian food aid). Pastoralists struggled to access water and pasture (particularly those herding cattle, which need more regular access to water than camels), leading to high levels of animal mortality, low prices due to the deteriorated condition of livestock reaching market and over-supply, and localized conflict between pastoralists. The situation was compounded by the restriction of usual risk-spreading and coping strategies used by rural people: due to the widespread nature of the hardship people were dealing with, casual labor opportunities and wages contracted and family and community support mechanisms were eroded. Moreover, the worst-hit Rahanweyn and Bantu groups had limited diasporic networks to tap for assistance (Majid and McDowell 2012).

Thus—by contrast with the post-2007 period, and echoing the famine displacement triggered in 1992—rural people migrated in large numbers toward urban centers, particularly Mogadishu, in the hope of accessing humanitarian assistance. There were reports of pastoralists arriving with their animals as early as January 2011, an early warning of the severity of the impact of the first failed rains (UNOCHA 2011). An estimated 150,000 people arrived in Mogadishu as a result of the famine, mostly thought to be from Rahanweyn agropastoralist and Bantu farming communities (HRW 2013a). The proximity of the capital to famine-affected areas, readier availability of international humanitarian assistance, and fact that Al Shabaab withdrew in August 2011, made it an obvious destination for many affected by the crisis.

Thus a "dual disaster" unfolded, an environmentally induced emergency overlapping with extant political conflict (Hyndman 2010). Available domestic and international governance mechanisms failed to check the situation, despite the numerous alerts issued by the drought Early Warning System, until the UN declared a famine situation in several areas of south Somalia in July 2011.[2] The Somalia situation was thrown into sharp relief by more coordinated policy responses to the regional drought in Kenya and Ethiopia, underlining that natural hazards like drought do not automatically lead to human disasters like famine: questions of governance, accountability and entitlements are key (Sen 1981; De Waal 1997). In Somalia, domestic political actors on all sides failed to address the significant vulnerabilities in the populations under their control, and indeed the way they pursued the conflict often exacerbated the situation for civilians. Al Shabaab restricted population movements, denied the famine, and maintained varying bans and restrictions on the activities of aid agencies. The sheer magnitude of the humanitarian crisis may have eased the intensity of the conflict in the initial aftermath of the famine declaration (FSNAU 2011).

Meanwhile, in the first half of 2011 humanitarian efforts were hampered by the long-standing "normalization" of the humanitarian crisis in Somalia, which made it hard to get donor attention, combined with the acute politicization of aid, which made it hard to reach those in need (Bradbury 1998). The famine declaration had an electrifying "CNN effect" (Lautze *et al.* 2012), triggering an outpouring of donations and relaxation of counter-terror provisions. The drought dimension no doubt helped, as natural disasters are often viewed as blameless, inducing greater generosity than thorny and protracted political conflicts (Hyndman 2010). But aid agencies still struggled to access people outside TFG-controlled areas, achieving access to a variable extent and often as the frontline moved back rather than through negotiation with Al Shabaab. Cash-based interventions increased dramatically, with unconditional grants, vouchers and cash for work, reaching more than 1.7 million people in south-central Somalia, in partnership with Somali money transfer agencies (Ali and Gelsdorf 2012). Meanwhile, Islamic donors, Muslim charities and diaspora associations were increasingly prominent providers of aid, finding it easier to navigate Al Shabaab-held areas (Hammond and Vaughan-Lee 2012).

Thus, from several angles, the famine declaration marked something of a turning point. In October 2011, in response to a range of security and economic interests, Kenya sent troops into southern Somalia, which pushed Al Shabaab northward out of the border area. The 2011/12 *deyr* rains were above average, allowing some livelihood recovery. By the end of 2012, the TFG's successor, the Somali Federal Government (SFG), with the support of various regional and AMISOM forces, controlled all urban centers, with Al Shabaab retreating into rural areas.

This humanitarian crisis was both multi-faceted and multi-layered. The concurrent factors outlined above overlaid historical "stressors": prior political conflict, recurrent drought and chronically poor governance as described in the earlier sections of this chapter, which had significantly eroded the resilience of the worst-affected people. It is perhaps not surprising, then, that the precise beginning and ending of this humanitarian crisis are blurry (Lindley and Hammond forthcoming). The situation pre-2011 was already described as a humanitarian crisis by conventional indicators, and although the famine was declared over in 2012, in early 2013 more than a million people were estimated to still be unable to meet their basic needs without assistance.

As a result of the way this crisis unfolded, a combination of factors was evident in most people's decisions to leave their places of residence. While for some there was a clear primary driver, for many things were more blurry. As one refugee from Mogadishu said, "I cannot say in one story why I wasn't safe, there are too many stories" (Amnesty International 2008, 10). For example, many people might have been able to weather drought using normal coping strategies, without becoming forcibly displaced, were it not for the contracting labor opportunities, restricted mobility and uneven distribution of humanitarian aid, which resulted from the political conflict. Others would have been able to weather conflict better were it not for the drought—for example, the continued extraction of taxes on resources and livestock by Al Shabaab became too much to bear for many households once the drought hit (HRW 2011).

Beyond this simultaneous combination of factors, there is also the culmination of factors over time: the underlying structural factors and the personal histories that shape migration. For example, for some people drought and hunger was the immediate driver of movement, but the groundwork was laid by years of conflict, marginalization and abuse. For some, an upsurge in violence was the straw that broke the camel's back, against a background of long-term strains on rural livelihoods that already predisposed them to migrate. In this context, a useful analytical distinction may be made between structural factors, proximate causes, immediate triggers and intervening factors (Van Hear 1998). We often focus on proximate events and immediate triggers, but less on the structural causes and processes of deprivation, vulnerability and disempowerment that underlie displacement in humanitarian crises (see McAdam, Chapter 2 in this volume).

This nuanced view of causation stands in sharp contrast to the frequent references to "drought displacement" by politicians and the media in the context of the 2011 humanitarian crisis. While host country governments like Kenya have been particularly keen to use such terminology, it has also featured in the announcements of international humanitarian organizations—for example, UN Office for the Coordination of Humanitarian Affairs (OCHA) stated that "Drought, not insecurity, is now the main reason for new displacement in Somalia" (UNOCHA 2011, 1; see also FSNAU 2011). This is also a distinction that exists in the Somali language: people forced to move by drought are traditionally referred to as *daaduun*, whereas the word *qaxooti* refers to people running from war and insecurity, usually translated as "refugee" (Goth 2011). However, the multi-faceted and multi-layered nature of the humanitarian crisis as outlined above suggests that the large displacements in 2011 clearly cannot be viewed as purely "environmentally induced" migration, if such a thing even exists.

It is true that the "drought displacement" terminology sometimes fits with survey evidence—for example, in Food Security and Nutrition Analysis Unit's (FSNAU) (2011) sample of IDPs in south-central Somalia who had been displaced in January–August 2011, 60 percent said that they were displaced by the drought. In a 2012 survey of recent arrivals in Kenya, 43 percent of respondents said that they had come to the camps for drought, livelihood or family reasons, not making reference to additional conflict- or persecution-related reasons for leaving (they were able to specify multiple causes) (RCK 2012). There are multiple methodological challenges, not least as many people escaping Al Shabaab-controlled areas continued to live in fear of speaking out, but more critically, these kinds of surveys tend to capture immediate triggers, and tell us little about the crucial structural context of people's migration. Thus the common label "drought displacement" greatly over-simplifies the nature of Somali mobility in 2011. We need to be wary about the political functions that this serves, of which more below.

The protection of IDPs varies, depending considerably on what social networks, political identity and economic resources they are able to tap into (see Lindley 2013). For those lacking any of these assets, life is extremely difficult and insecure. Both Islamic and customary law contain provisions regarding the treatment of displaced people which are implemented to a variable extent. Many have depended on irregular and frequently diverted distributions of humanitarian assistance. Diaspora fundraising initiatives have reached some IDP camps. Officially, in international law, those displaced by drought and/or conflict qualify as "IDPs" under the non-binding 1998 Guiding Principles on Internal Displacement and the binding 2009 African Union Convention for the Protection and Assistance of Internally Displaced Persons in Africa (Kampala Convention). While all the macro-political authorities have purported protective stances toward displaced people, their actions often fall far short. For example, Puntland has sent back IDPs into war-torn south-central Somalia; Al Shabaab forced back people trying to flee; the TFG and now the SFG's sway is secured in significant part by militia and local powerbrokers, who exploit the presence of displaced people in order to attract and extract humanitarian aid, as has occurred throughout the civil war (HRW 2013a). No formal attempt has been made to differentiate between those displaced internally "by drought" or "by conflict," but by virtue of moving from areas controlled by Al Shabaab (at the time, at least), displaced people have often been treated with suspicion by government and allied forces (HRW 2013a).

Contrary to the received wisdom that most climate-related movements are short-distance and temporary (Tacoli 2009), in this case, because of the difficulty of accessing humanitarian assistance inside the country, and the ongoing insecurity, the 2010–11 drought was associated with high levels of regional movement. In the context of regionalized drought, people migrating from the worst-hit areas inside Somalia found themselves crossing borders into areas that were also under considerable environmental stress. The largest numbers went to Kenya, which has generally admitted people fleeing Somalia as prima facie refugees and settled them in the Dadaab refugee camps in the North Eastern Province. This did not change, even as the official refugee population in Kenya escalated from around 174,000 in 2006 to over half a million in 2011, and security concerns mounted.

However, while African governments have generally admitted people fleeing natural disasters in neighboring countries—indeed preventing such an influx can be impossible—they generally have not considered this to be an obligation under either the 1951 Convention relating to the Status of Refugees or the 1969 Organization of African Unity (OAU) Convention Governing the Specific Aspects of Refugee Problems in Africa, which emphasize persecution and serious public order disturbances respectively (Edwards 2006). This helps contextualize the Kenyan government's pains to distinguish in public statements between long-term refugees displaced by conflict and people more recently displaced by drought—implicitly circumscribing its responsibilities. In this fraught political context, many protection problems have arisen, overlaying earlier ones (see Horst 2006; Lindley 2011). Many people fleeing Somalia in 2011 experienced problems entering Kenya and reaching Dadaab camps safely and registration was suspended for long periods (RCK 2012). While additional camps were eventually opened, conditions in the Dadaab complex deteriorated as the government's Department for Refugee Affairs, United Nations High Commissioner for Refugees (UNHCR) and partner agencies struggled to deal with a rapidly growing population with major needs, alongside mounting security incidents (RCK 2012). While there had been a climate of increasing openness to urban refugees, the atmosphere changed markedly in the context of the surge in displacement. In December 2012 government efforts to enforce the encampment approach, although subsequently successfully challenged in court, effectively unleashed a wave of police and criminal harassment of Somalis in urban areas (HRW 2013b). Meanwhile, the government hopes that Kenyan forces can help install a friendly administration in the neighboring area of southern Somalia, and strongly advocates moving toward the repatriation of refugees.

Policy responses

Having stressed the close relationship between environmental challenges and conflict, politics and governance processes in Somalia, this section considers how the environmental dimensions of crisis and migration might best be addressed in the future. Responses to movement in humanitarian crises often focus on dealing with already displaced populations, but it is equally if not more important to address what forces people to move. This requires working across what are often viewed as discrete policy fields and specialisms: bringing together elements of climate change adaptation, humanitarian relief, livelihoods development, disaster and post-conflict recovery, and human rights protection. This section explores three main areas of problems and policy making: the changing environment and people's resilience; the humanitarian crisis and recovery; and the protection of already displaced people. In each of these areas, local, national, and international actors have roles to play. In each area, there are strong cross-border elements to problems and potential solutions.

The changing environment and people's resilience

Rural livelihood systems in the Horn of Africa have adapted to a certain level of rainfall variability, but recent drought and environmental degradation have prompted major concerns. To the extent that recent rainfall variability may be linked to anthropogenic climate change, this implies global responsibility; the carbon footprint of the average Somali is minuscule by comparison with the nationals of the Organization for Economic Co-operation and Development (OECD) countries. In terms of adaptation, although party to the UN Framework Convention on Climate Change, the Somali government needs to extend control over rural areas before it can tackle key environmental issues. Facilitating water management systems, with an emphasis on access to water for the poor and flood defenses, and the regulation of charcoal production are important tasks, requiring the navigation of vested interests of powerful political actors. There are also more local possibilities for preventive practice mitigating the effects of natural hazards. In Somaliland, for example, certain conservation practices, such as soil bunds, forestation, runoff water harvesting, and gully control appear to have had some effect in mitigating land degradation, although thinly implemented (Vargas *et al.* 2009). In Puntland, some nongovernmental organizations (NGOs) are supporting the development of alternative livelihoods and fuel sources to combat charcoal production (Dini 2009).

Another problem is the deep vulnerability of many Somalis to any kind of shock, including drought (LeSage and Majid 2002). Many NGO and community interventions aim to mitigate the impact of drought by increasing people's resilience, broadly defined as the "ability to deal with adverse changes and shocks" (Béné *et al.* 2012, 11). For example, efforts are made to improve rural households' access to markets, and ensure the availability of effective financial services including savings, credit and insurance (Oxfam 2011). The task of increasing resilience is not just a technical challenge, but is deeply political and rights based: the most vulnerable people have been on the receiving end of systematic violence and marginalization for more than two decades. Stronger and higher-level political action is needed to address this. Finally, routine and coping mobility, including across borders, has long been a major source of resilience; militarized frontlines and closed borders have threatened this, and future policy should be sensitive to the value of mobility in sustaining rural livelihoods.

Drought response, humanitarian relief and rehabilitation

As drought unfolds into a crisis, the challenge becomes to mitigate the impact on lives and livelihoods and promote early recovery. This may also prevent people from needing to move. The Early Warning System functioned well in terms of providing alerts regarding severe food insecurity, signaling opportunities to take pre-emptive measures to support livelihoods. But early action was not taken, due to failures by domestic political actors and the international community already discussed. Two important recommendations have been made to enhance the link between early warning and action (see Lautze *et al.* 2012). First, the Early Warning System should be geared more toward meeting the information needs of domestic actors ("providers of first resort")—i.e. state agencies, civil society and community organizations, religious institutions, healthcare systems, local private-sector and diaspora associations—to help these actors to respond effectively to food insecurity and reinforcing elements of domestic political accountability. Second, there is a need to clarify the rights, resources and responsibilities of the international humanitarian system ("providers of last resort") in relation to early warning and action, to form a stronger compact against famine.

When drought is allowed to unfold into a severe crisis, we have seen how the humanitarian effort is hampered by the acute politicization of aid and the corrupt political economy surrounding it. But these problems highlight not so much a need to *depoliticize* humanitarian aid as a need for humanitarian agencies to redouble efforts to address the needs of the most vulnerable civilians—an inherently political act—while distancing themselves from particular state-building projects (Hammond and Vaughan-Lee 2012). Moreover, in this difficult context, the role of "alternative" actors (non-western donors, as well as Somali diaspora, religious and business groupings) and "alternative" approaches of providing emergency relief (such as cash-based interventions as opposed to food aid) are now widely recognized as a key part of the humanitarian landscape in Somalia.

Finally, there is the classic challenge of moving from relief to recovery, in the absence of durable political stabilization. Rehabilitation efforts have historically tended to be short-term in nature (food aid, some cash relief, emergency water provision and medical care, seed distribution, and irrigation rehabilitation), but need to be linked to—coming full circle—longer-term frameworks for building the long-term resilience of rural communities (Levine 2011). Early action, effective humanitarian relief and committed recovery strategies all have a role to play in stemming displacement and helping those already displaced by crisis.

Protecting the rights of displaced people

Once people are internally displaced, they are often in an extremely vulnerable situation, and securing their basic rights is a major challenge. Several ways forward have been suggested elsewhere (see Lindley 2013). The role of Somali sociocultural resources, including kinship, religion and diaspora support, have been increasingly prominent in the context of the international aid paralysis, and where possible international actors should work in harmony with these indigenous sociocultural protective capacities. However, the role of macro-political authorities is absolutely vital: the consolidating SFG—and whatever units it eventually recognizes as states within the federal model—as well as the secessionist government of Somaliland must be encouraged to protect *all* Somalis on their territory. Meanwhile, alongside targeted emergency assistance to destitute IDPs, they should be included in wider efforts to promote the livelihoods of the urban poor, to minimize long-term segregation and exploitation.

In the major refugee-hosting country, Kenya, as elsewhere, there are concerns about the shrinking of "asylum space" in response to the latest humanitarian crisis in Somalia (RCK 2012; HRW 2013b). In the context of increasing tension and abuse of refugees, and a Refugees Bill under consideration, which raises human rights concerns, strong leadership from UNHCR as well as NGO and civil society advocacy is essential (HRW 2013b). It is extremely unlikely that all refugees can or will return to Somalia, and organizations trying to protect refugees should continue to push for gradual pathways to more positive participation in society, taking account of their presence in urban and rural development planning, and supporting their mobility as key for livelihoods (Lindley 2011). Meanwhile, Kenya will continue to need major international aid to support one of the largest refugee populations in the world.

Although there is a common assumption that the "drought displaced" find it easy to return once the rains fall, the fear and impoverishment among many displaced Somali people, and the ongoing political uncertainty in their home areas, suggest otherwise. Despite improved rainfall, by June 2012, only 14 percent of refugees surveyed in Dadaab said they would consider returning and, by mid-2013, returns were still limited, despite the increased pressure on refugees in Kenya (RCK 2012). In light of these pressures, international and domestic actors must emphasize the need to uphold the principle of voluntary return. This further highlights the critical role of broader political processes in addressing displacement. Once again, policies cannot be developed in isolation. Any return movements of refugees and IDPs will need to connect with long-term efforts toward rehabilitation and building rural resilience if they are to be sustainable in the long run.

Conclusions

It is abundantly clear that drought poses a major and recurring challenge to the livelihoods of many people across the Horn of Africa. However, this chapter has shown that Somali mobility in 2011 cannot be boiled down to the simple epithet of "drought displacement." Such reductive terminology misrepresents the drivers of displacement and hides how drought interlocks with political processes, both historical and concurrent. Thus, problems arise with a single-sector approach to policy making in contexts of humanitarian crisis. Thinking across policy silos can be a professionally and politically uncomfortable business for those involved, but this chapter has highlighted several potentially fruitful avenues worth exploring.

Acknowledgments

With special thanks to Giulia Baldinelli for her assistance with background research, and Jennifer Hyndman for her comments on an earlier draft.

Notes

1 Coastal fishing and urban livelihoods are also affected by environmental issues (the tsunami, illegal fishing, toxic waste dumping, poor waste and water infrastructure, flooding, slum issues) but this is not the focus of this chapter.
2 The Integrated Phase Classification of food security defines famine as a situation where at least 20 percent of population in affected areas has extremely limited access to basic food; the global acute malnutrition rate exceeds 30 per cent; and death rates exceed 2 per 10,000 per day (FSNAU 2011).

References

Ali, D. and Gelsdorf, K. (2012) "Risk-averse to Risk-willing: Learning from the 2011 Somalia Cash Response," *Global Food Security*, 1(1): 57–63.
Amnesty International (2008) *Routinely Targeted: Attacks on Civilians in Somalia*, London: Amnesty International.
Béné, C., Godfrey Wood, R., Newsham, A. and Davies, M. (2012) "Resilience: New Utopia or New Tyranny? Reflection about the Potentials and Limits of the Concept of Resilience in Relation to Vulnerability Reduction Programmes," IDS Working Paper, No. 405, Brighton: Institute of Development Studies.
Bradbury, M. (2008) *Becoming Somaliland*, Oxford: James Currey.
Checchi, F. and Robinson, W.C. (2013) "Mortality among Populations of Southern and Central Somalia Affected by Severe Food Insecurity and Famine during 2010–2012," report for FAO/FSNAU and FEWSNET, May 2, Rome and Washington, DC.
De Waal, A. (1997) *Famine Crimes: Politics and the Disaster Relief Industry in Africa*, Oxford: James Currey.
Devereux, S. (2006) *Vulnerable Livelihoods in Somali Region, Ethiopia*, Brighton, UK: Institute of Development Studies at the University of Sussex, available online at: http://edoc.bibliothek.uni-halle.de/servlets/MCRFileNodeServlet/HALCoRe_derivate_00004563/IDS_Rr57.pdf
Dini, S. (2011) "Addressing Charcoal Production, Environmental Degradation and Communal Violence in Somalia: The Use of Solar Cookers in Bander Beyla," *Conflict Trends*, 2: 38–45.
Edwards, A. (2006) "Refugee Protection in Africa," *African Journal of International and Comparative Law*, 14: 204–233.
FSNAU (2010) "Quarterly Brief: Focus on Gu Season Early Warning," Nairobi, Kenya: Food Security and Nutrition Analysis Unit—Somalia, June 18.

FSNAU (2011) "Food Security and Nutrition Analysis Post Gu 2011," FSNAU Technical Series Report N. VI. 42, October 8, Nairobi, Kenya: Food Security and Nutrition Analysis Unit—Somalia.

Goth, B. (2011) "Col Iyo Abaar: War and Drought," Somalilandpress, available online at: http://somalilandpress.com/col-iyo-abaar-war-drought-22979.

Government of Kenya (2011) "Briefing on the Refugee and Drought Situation in the Country," July 21, available online at: http://reliefweb.int/node/435254.

Hammond, L. and Vaughan-Lee, H. (2012) "Humanitarian Space in Somalia: A Scarce Commodity," Humanitarian Policy Group Working Paper, London: Overseas Development Institute.

Hammond, L., Awad, M., Dagane, A., Hansen, P., Horst, C., Menkhaus, K. and Obare, L. (2011) "Cash and Compassion: The Somali Diaspora's Role in Relief, Development and Peace Building," report for the United Nations Development Program in Somalia.

Hansch, S., Lillibridge, S., Egeland, G., Teller, C. and Toole, M. (1994) *Lives Lost, Lives Saved: Excess Mortality and the Impact of Health Interventions in the Somalia Emergency*, Washington, DC: Refugee Policy Group.

Horn Relief (2009) "Rapid Assessment Report on the Current Drought Emergency in the Sanaag Region," available online at: http://reliefweb.int/sites/reliefweb.int/files/resources/52EBB809C2B320C2C1257602003097AC-Full_Report.pdf.

Horst, C. (2006) *Transnational Nomads: How Somalis Cope with Refugee Life in the Dadaab Camps of Kenya*, Oxford and New York: Berghahn Books.

HRW (2011) *You Don't Know Who to Blame: War Crimes in Somalia*, Washington, DC: Human Rights Watch.

HRW (2013a) *Hostages of the Gatekeepers: Abuses against Internally Displaced in Mogadishu, Somalia*, Washington, DC: Human Rights Watch.

HRW (2013b) *"You are All Terrorists": Kenyan Police Abuse of Refugees in Nairobi*, Washington, DC: Human Rights Watch.

Hyndman, J. (2010) *Dual Disasters: Humanitarian Aid After the 2004 Tsunami*, Sterling, VA: Kumarian Press.

Kapteijns, L. (2013) *Clan Cleansing in Somalia: The Ruinous Legacy of 1991*, Philadelphia: University of Pennsylvania Press.

Kolmannskog, V. (2010) "Climate Change, Human Mobility, and Protection: Initial Evidence from Africa," *Refugee Survey Quarterly*, 29 (3): 103–119.

Lautze, S., Bell, W., Alinovi, L. and Russo, L. (2012) "Early Warning, Late Response (Again): The 2011 Famine in Somalia," *Global Food Security*, 1: 43–49.

LeSage, A. and Majid, N. (2002) "The Livelihoods Gap: Responding to the Economic Dynamics of Vulnerability in Somalia," *Disasters*, 26 (1): 10–27.

Levine, S. (2011) "System Failure: Revisiting the Problems of Timely Response to Crises in the Horn of Africa," Humanitarian Policy Group Working Paper No. 71, London: Overseas Development Institute.

Lewis, I.M. (2003) *A Modern History of the Somali* (4th edn.), Athens, OH: Ohio University Press.

Lindley, A. (2010a) *The Early Morning Phone Call: Somali Refugees' Remittances*, Oxford and New York: Berghahn Books.

Lindley, A. (2010b) "Leaving Mogadishu: Towards a Sociology of Conflict-related Mobility," *Journal of Refugee Studies*, 23 (1): 2–22.

Lindley, A. (2011) "Between a Protracted and Crisis Situation: Policy Responses to Somali Refugees in Kenya," *Refugee Survey Quarterly*, 30 (4): 14–49.

Lindley, A. (2013) "Displacement in Contested Places: Governance, Movement and Settlement in the Somali Territories," *Journal of Eastern African Studies*, 7(2).

Lindley, A. and Hammond, L. (forthcoming) "Histories and Challenges of Crisis and Mobility in Somalia," in Lindley, A. (ed.), *Crisis and Migration: Critical Perspectives*, Abingdon: Routledge.

Little, P.D. (2003) *Somalia: Economy Without State*, Oxford: James Currey.

Lott, F.C., Christidis, N. and Stott, P.A. (2013) "Can the 2011 East African Drought be Attributed to Human-induced Climate Change?" *Geophysical Research Letter*, 40: 1–5.

Luling, V. (2002) *Somali Sultanate: The Geledi City-state over 150 Years*, London: HAAN.

Majid, N. and McDowell, S. (2012) "Hidden Dimensions of the Somalia Famine," *Global Food Security*, 1: 36–42.

Menkhaus, K. (2007) "The Crisis in Somalia: Tragedy in Five Acts," *African Affairs*, 106 (204): 357–390.

Oduori, S.M., Rembold, F., Abdulle, O.H. and Vargas, R. (2011) "Assessment of Charcoal Driven Deforestation Rates in a Fragile Rangeland Environment in North Eastern Somalia Using Very High Resolution Imagery," *Journal of Arid Environments*, 75 (117): 1173–1181.

Oxfam (2011) "Briefing on the Horn of Africa Drought: Climate Change and Future Impacts on Food Security," available online at: http://www.oxfam.org/sites/www.oxfam.org/files/briefing-hornofafrica-drought-climatechange-foodsecurity-020811.pdf.

RCK (2012) *Asylum Under Threat: Assessing the Protection of Somali Refugees in Dadaab Refugee Camps and Along the Migration Corridor*, Nairobi: Refugee Consortium Kenya.

Sen, A. (1981) *Poverty and Famines: An Essay on Entitlement and Deprivation*, Oxford: Clarendon Press.

Simons, A. (1995) *Networks of Dissolution: Somalia Undone*, Boulder, CO: Westview Press.

Tacoli, C. (2009) "Crisis or Adaptation? Migration and Climate Change in a Context of High Mobility," *Environment and Urbanization*, 21 (2): 213–225.

UNHCR (2011) "Donor Update: Somalia Situation Response," available online at: http://www.unhcr.org/4e157f499.pdf.

UNOCHA (2006) "Livelihoods and Food Security," Nairobi, Kenya: United Nations Office for the Coordination of Humanitarian Affairs in Somalia, available online at: http://www.somali-jna.org/downloads/Food%20Security.pdf.

UNOCHA (2011) "Mogadishu Update: IDP Drought Response," March 3, Nairobi: OCHA.

Van Hear, N. (1998) *New Diasporas*, London: UCL.

Vargas, R.R., Omuto, C.T., Alim, L.S., Ismail, A. and Njeru, L. (2009) "Land Degradation Assessment and Recommendation for a Monitoring Framework in Somaliland," FAO-SWALIM Technical Report, Nairobi, Kenya, available online at: http://www.faoswalim.org/ftp/Land_Reports/Cleared/L-10%20Land%20Degradation%20Assessment%20of%20a%20Selected%20Study%20Area%20in%20Somaliland.pdf.

Webersik, C. and Crawford, A. (2012) "Commerce in the Chaos: Bananas, Charcoal, Fisheries, and the Conflict in Somalia," *Environmental Law Reporter*, 42: 10534–10545.

9

ENVIRONMENTAL STRESS, DISPLACEMENT AND THE CHALLENGE OF RIGHTS PROTECTION

Roger Zetter and James Morrissey

Introduction

The actual and potential population displacement impacts of environmental stress and climate change provide a novel lens through which to examine how rights and, more specifically, the protection of those rights[1] are afforded to migrants, and how concepts of displacement are perceived in the context of protection and rights.

But a paradox lies at the heart of the current political discourse on the protection of rights and displacement in the context of climate and environmental drivers. In many countries, notably those likely to be most affected by these dynamics, there is increasing awareness of population displacement impacts—for example, the issue has high policy saliency in countries such as Bangladesh (Government of Bangladesh 2009) and in the national planning framework of "living with floods" in Vietnam (Government of Vietnam 2009). Yet, when it comes to considering how rights protection might be afforded to displaced populations impacted by these phenomena, then legal and normative frameworks are virtually silent. There is a "protection gap" (UNHCR 2010); and its existence is surprising given the scope of protection—concepts, norms and legal instruments—available to other groups of forcibly displaced populations and "involuntary migrants" in domestic and international law.

There are many plausible reasons why this silence might exist. There are the challenges of attributing causality, of defining migration as "forced" and, as such, of defining and identifying a specific category of displaced people whose rights may be threatened and in need of protection (Zetter, Boano and Morris 2008; Laczko and Aghazarm 2009; Zetter 2009, 2010). There is the difficulty of determining the duty bearers with the responsibility to protect: should protection be a moral imperative and a tool of restorative justice provided by developed countries, the major CO_2 emitters, or a short-term humanitarian response to life-threatening disasters? The fact that the most dramatic impacts of climate change will manifest in the future provides a powerful excuse to avoid complex and contentious international debate about attribution and thus protection needs. It can be argued that broadening the scope of protection for a new category of involuntary migrants runs the risk of further weakening or diminishing states' capacities or willingness, to fulfill their existing obligations under international law to protect the rights of particularly vulnerable groups such as refugees. A further argument is that it makes little sense to privilege individuals displaced by the impacts of climate change (or other forms of environmental stress) over other "involuntary migrants" moving for a variety of reasons, who are similarly outside already well-established categories, or conversely for whom there is established protection apparatus such as the 1998 Guiding Principles on Internal Displacement (or Guiding Principles) (McAdam 2011; UNHCR 2011; Kälin and Schrepfer 2012).

Notwithstanding the above arguments, this chapter proposes a different explanation for this lacuna in rights protection for those who are displaced, move anticipatorily or are trapped as a consequence of the impacts of climate change and environmental stress in the five countries of our research.[2]

We argue that latent conditions and factors explain how various forms of migration[3] and rights are shaped and instrumentalized by governments. This is because government policies related to migration and displaced populations, on the one hand, and legal and normative frameworks relating to rights protection, on the other, do not exist in a vacuum. Their conjuncture is mediated by politico-historical experiences and contemporary contextual factors which include: the past and present patterns and processes of migration and displacement; the socio-economic and developmental circumstances of the country which influence population mobility, especially internal migration; the political saliency of migration and displacement issues which conditions the extent to which migration policy features in government discourse and action; and the government regime and its disposition toward human rights.[4]

In other words, migration histories and politics shape the way in which migration policy regimes conceive, frame and articulate rights provisions for groups and individuals who could be displaced by environmental stress.

Our argument challenges the ahistorical and apolitical framing of the way in which environmental variables are claimed to shape mobility decisions in a context of environmental stress, and the rights that might obtain to this process—reasoning that dominates current research and policy analysis on this relationship.

In sum, it is through the analysis of the politics of migration and rights that we can better appreciate why it is that the governments in our study do not, as yet, accord rights to those who are displaced, or threatened by displacement, in the context of environmental stress such as climate change. The politics of migration and rights are played out at both the national level—the focus of this chapter—and at the micro or local level, which is elaborated elsewhere (Zetter and Morrissey 2014).

With the argument laid out, the chapter proceeds by first framing the issues of rights, protection and environmental stress, and then situating the five country cases before focusing on the main argument, which explores the reasons for the lacuna in rights protection and the significance of "context."

Framing the issues: protection, rights and environmental stress

Human rights protection in the context of involuntary migration is a long-standing, accepted and, indeed, an expanding concept and practice embedded in the responsibilities of states, and international and intergovernmental actors.

A 60-year-long history can be traced in which the concept and international legal and normative frameworks of rights protection have been developed to assist different categories of vulnerable migrants. This body of law addresses the problematic of migrants whether voluntary or forced, temporary or permanent, internal or cross-border.

In specific terms, it includes six main categories: refugees,[5] internally displaced persons (IDPs),[6] stateless people,[7] indigenous peoples,[8] trafficked people[9] and migrant workers.[10]

In addition, there is a portfolio of international human rights law that, while not migration specific, is relevant and increasingly invoked in the context of rights protection and advocacy for forced migrants; this body of law now has its own genus in subsidiary, complementary and temporary protection.[11]

Finally, the growing body of International Environmental Law implies some general, but not displacement-specific, protective provisions and remedies (McAdam and Saul 2008). For example, Principle 2 of the 1992 Rio Declaration[12] notes the responsibility of states to ensure that the sovereign right to resource exploitation does "not cause damage to the environment of other states or of areas beyond the limits of national jurisdiction," while Principle 3 indicates that the right to development "must be fulfilled so as to equitably meet developmental and environmental needs of future generations." These principles imply the need to protect the interests, if not the exactly the rights, of populations vulnerable to the negative environmental impacts of climate change. More specifically, in the context of this chapter, in 2011 the Cancun Adaptation Framework adopted "measures to enhance understanding, coordination and cooperation with regard to *climate change induced displacement*, migration and planned relocation" (emphasis added):[13] this was the first occasion on which mobility issues entered into the United Nations Framework Convention on Climate Change (UNFCCC) lexicon.

Thus rights exist for different categories of vulnerable migrants and forcibly displaced people, but what do we mean by protection? Although international law makes substantial reference to protection, it does not define it. In general terms, protection in relation to migration is concerned with safety, security, dignity and reducing vulnerability, as well as securing or safeguarding political, civil, social, economic and cultural rights for people on the move. More specifically, two polarities underpin the meaning of protection employed in this chapter: material and structural.

In *material* terms, protection may be conceived as a commodity—for example, physical assistance and shelter provision to people who may be temporarily displaced by disasters. And protection may also be seen as both responsive and remedial, as in humanitarian responses to life threatening disasters, property restitution, access to resettlement land (Giossi-Caverzasio 2001). It may also be conceived as preventative and environment building, for example, protecting from harm with disaster risk reduction (DRR) policies and livelihood safety nets. It is the material representation of rights protection that, we argue, dominates current thinking in the context of environmental displacement.

But the protection of rights may also be conceived in *structural* terms. For example, protection may occur at a high level of international actors—notably in some of the conventions referred to above—or at the local level of civil society organizations that advocate for, or empower, local communities to participate in decisions about their lives such as resettlement after disasters. Especially as a *structural* process, protection of migrants is highly politicized, not least in the context of environmental factors, as we shall see.

The role of humanitarian actors and especially governments speaks to a conception of protection as both commodity driven and responsive in the context of existential threats in disasters: this is the material representation of rights noted earlier. A conception of rights protection as a continuing process to tackle the structural and systemic inequalities and risks that underlie disaster vulnerabilities (Wisner *et al.* 2004) and the impacts of environmental stress, a conception that is inherently political, is far more problematic.

Environmental stress in general, and climate change in particular, potentially impinge upon the enjoyment of a wide range of internationally protected human rights—civil, political, social, cultural and economic, including freedom of movement. A case exists, therefore, for developing rights-based norms and instruments of protection to support the needs of groups and individuals who are forcibly displaced in a context of significant environmental stress, as well as communities (in both sending and receiving areas) impacted by this displacement. This situation presents new challenges and potentially significant obligations for national and international actors in relation to the likely numbers involved, their access to land and resources, participation in policy making, cross-border movement, and issues of security. In other words, ensuring rights and protection is part of the challenge of managing the consequences of environmental change, and particularly climate change.

However, with the exception of the 2009 African Union (AU) Convention for the Protection and Assistance of Internally Displaced Persons in Africa (Kampala Convention)[14] (see below), there are no international legal instruments or norms that deal specifically with the rights protection of those whose displacement could be attributed in some way to environmental or climatic factors. Moreover, despite having some advocates (Council of Europe 2008), it seems unlikely that the international community will agree on such a framework, a conclusion that several intergovernmental bodies have emphasized (UNHCR 2009).

Nevertheless, a rights protection "line of defense" is developing, which mainly rests on the contention that the majority of those displaced in situations where climate and environmental factors play a role will remain within their own countries. In these circumstances, the 1998 Guiding Principles on Internal Displacement might offer a potentially promising vehicle for extending rights protection to this "new" category. In this regard, Article 5.4 of the Kampala Convention makes specific mention of the obligation on states parties to protect people who have been internally displaced by climate change.

One major problem with this argument, however, is the growing appreciation of the manner in which environmental stress operates on migration decisions in part through its impact on the economic conditions of the household—in other words triggering a labor migration strategy rather than an involuntary displacement response. As Koser (2011) points out, the instruments for rights protection would thus require a framework which ensures the rights of labor migration, an omission that was essential to get the Guiding Principles accepted in the first place.

Situating the case study countries

The five case study countries—Kenya, Bangladesh, Vietnam, Ethiopia and Ghana—offer a sample of "climate-stressed" countries, rendering them suitable for the study of climate change impacts, through the use of analogy (Morrissey 2009). In addition they all display high levels of vulnerability to the impacts predicted likely to accompany climate change, and thereby represent contexts in which protection concerns at the nexus of environmental change and displacement are relevant. Notably, however, in terms of social context, dominant environmental stresses and climate change vulnerabilities, the five countries vary considerably in relation to:

- environmental conditions—highlands, lowlands, deltas, riparian environments, drylands, tropical regions and wetlands;
- environmental stresses and climate change impacts—desiccation, rising sea levels, changes in rainfall regimes, erosion and salination;
- internal and regional migratory processes and patterns;
- slow-onset change and extreme weather scenarios;
- legal and normative rights protection apparatus;
- governance and civil society structures;
- capacities to protect rights.

Equally, with respect to the awareness of, and response to, environmental stress and climate change, the five countries have developed varied policy frameworks:

- environmental and climate change policies are strongly developed and mainstreamed in Ethiopia, Vietnam and Bangladesh, less so in Kenya and Ghana;
- Vietnam, Kenya and Bangladesh have formally adopted a comprehensive suite of responses—national climate change plans or strategies, and more advanced

disaster preparedness, DRR and mitigation plans; Ethiopia has, in preparation, a climate resilience plan, a policy for adaptation to climate change and a disaster risk management strategy program and investment policy framework; likewise, Ghana has a draft National Climate Change Adaptation Strategy;

- policies focus on rapid-onset natural disasters and humanitarian responses (especially Bangladesh); only Vietnam locates climate change responses in a wider developmental discourse, although the approach in Ethiopia (and to a lesser extent Ghana) strongly inclines to a developmental perspective (particularly agricultural development), mainly in terms of vulnerability and poverty reduction;
- the emphasis on DRR and rapid-onset disasters limits the development of institutional and policy responses to the growing specter of slow-onset climate change;
- the governmental and institutional infrastructure and capacity is well developed in Vietnam and Bangladesh, less so in Ethiopia, Kenya and Ghana; nonetheless, in all the countries there are significant implementation and coordination gaps;
- the engagement of civil society actors varies considerably between the countries in terms of both the scope of activity and the roles adopted;
- similarly, public awareness and debate varies in intensity—very active in Bangladesh, and more circumscribed in Vietnam and Ethiopia where the media are controlled;
- technical expertise especially in the area of environmental law is limited in all the countries.

These characteristics condition, to a significant degree, the context within which the discourse on displacement in general and human rights in particular is framed in the five countries. In all of them, we find a significant "rights protection gap" for both groups and individuals displaced, or susceptible to displacement in the context of environmental stress, including climate change. This reflects the contested conjuncture of these migration-related issues and legal and normative rights protection to which the chapter now turns.

The displacement and rights protection nexus

The core of the argument, sketched in the introduction, is that we need to turn to latent conditions and factors to explain how migration and rights are shaped and instrumentalized by governments. This is because legal and normative rights protection frameworks, on the one hand, and government policies related to migration and displacement, on the other, do not exist in a vacuum. Rather, migration histories and politics shape the way in which migration policy regimes are conceived and framed, and how rights are articulated for those groups and individuals displaced in a context of environmental stress and climate change. With the exception of Ghana, both migration and rights are sensitive issues in the political discourse in the case study countries—the conjuncture of the two is especially sensitive—and this creates tensions, which are played out in unusual and different ways.

Kenya

For the vast majority of Kenyans, land is the principal basis of livelihoods, which are rendered increasingly vulnerable by the combination of high population growth and declining environmental conditions, together driving rural out-migration. Moreover, traditional pastoral communities—an important minority group in Kenya—have been increasingly susceptible to ecosystem vulnerability, weakening coping capacity and thus government sedentarization strategies. Large-scale spontaneous rural-to-urban migration has produced extensive irregular settlement and illegal land appropriation.

Against this background, questions of migration and population displacement in Kenya remain highly politicized as a result of their close relationship with a matrix of issues around land, political power, unequal access and social grievances. These issues and their relationships with one another can be traced back to the colonial period and its practices of eviction (i.e. forced migration) and unequal development, as well as the country's organization into a number of ethno-territorial blocks (a condition which colonial policies served to exacerbate). Such a context has rendered ongoing practices of eviction and displacement as powerful means for shaping political exclusion, claiming power, and accessing resources. It is these conditions that underlie the violence and conflict-induced displacement following elections in 1992, 1997 and 2007 (with 660,000 displaced). And it is a legacy that conditions how displacement, in the specific context of climate change and environmental stress, is addressed.

The Kenyan Constitution provides some level of rights protection for displaced persons. However, like the other four countries, Kenya has struggled to incorporate the rights-based norms of the Guiding Principles and the more recently proposed national guidelines on IDPs—the 2009 Draft National Policy on the Prevention of Internal Displacement and the Protection and Assistance to Internally Displaced Persons in Kenya—into its national legal or normative frameworks. Regional frameworks, such as the Great Lakes Pact and the Kampala Convention have some relevance for the protection of IDPs. Kenya has yet to sign and ratify the convention and is therefore not legally bound by its provisions.

The 2007 post-election violence and the unprecedented scale of displacement and widespread distribution of displaced people across the country, disclosed crucial gaps in Kenyan jurisdiction regarding specific legal or normative frameworks to deal with internal displacement and the protection of the rights of people who are internally displaced, whether by environmental and climatic or other factors. The government and all political actors (including civil society and human rights non-governmental organizations (NGOs)) have confronted and initiated wide-ranging reform on three interlinked matters: human rights, the legacy of the country's internal displacement, and restructuring the institutional and operational framework for DRR, disaster management and response.

The response is not simply the result of the post-election violence, but rather the realization that population displacement was already a widespread problem, among which a principal reason was "natural disasters such as drought in large parts of Northern Kenya that often lead to conflicts over natural resources like water leading to displacement" (Government of Kenya 2008). The adoption of a comprehensive framework on IDPs, as proposed in the National Policy initiative and underscored by the Kampala Convention, could be a milestone in rights protection in Kenya and could provide a platform for extending the scope of the reform process to environmentally displaced people. The domestic proposals, however, concentrate on addressing the displacement impacts of political violence, the peaceful reconstruction and rehabilitation of the country and natural disasters. Protecting the rights of those displaced by other environmental stresses and slow-onset climate change-related phenomena (such as slow processes of desiccation, worsening rainfall regimes, reduced grazing and decreases in the return period for flood events) is not part of this agenda at present.

Given the political complexity of rights-based issues and the relative fragility of the Kenyan state, political resistance and the scarcity of resources and capacities continue to hinder progress of protection initiatives. For example, neither the Disaster Management Act of 2008, nor the 2010 Disaster Management Plan address the question of displacement, not even short-term displacement.

Thus the rights of those susceptible to the displacement effects of climate change and environmental stress remain in a vacuum in Kenya's legal and normative frameworks since the underlying dynamics of power and historical grievances in relation to land and migration are themselves yet unresolved.

Bangladesh

Bangladesh's sensitivity to issues of displacement and migration results from the fact that such phenomena manifest as significant outcomes of the country's formative moments. Massive population displacements were first associated with the 1947 Partition of India. The exodus left a legacy of political, social and cultural trauma in the region, which was reignited by the war leading to Bangladesh's independence in 1972. Once again, these events produced huge population upheavals—perhaps up to ten million people temporarily displaced—followed by a severe famine in 1974, which led to substantial population migration, mostly toward India. Since independence, population movement has been complicated by a number of factors that augment this legacy. The cross-border family, cultural and linguistic links that remain from these three major migratory episodes underpin the continuing flux of population movement from Bangladesh to India. These informal transnational communities are reinforced by the presence of millions of Bangladeshis in India who have migrated, mainly from the environmentally fragile coastal areas in the southwest of the country. This largely undocumented population remains unacknowledged both sides of the border. Population displacement in the Chittagong Hill Tracts and the continuing marginalization of the Rohingya—again suppressed from political debate—constitute another dimension of the population displacement history of Bangladesh.

Thus, the terms "displacement" and "displaced people" are yet to gain explicit recognition in legal and normative frameworks, as if systematically excised from the national consciousness because of the episodic traumas of past forced migration. Within this context, although large-scale population displacement is an acknowledged outcome of climate change (IOM 2010), the plans and policies dealing with the impacts only contain extensive provision for mitigation and post-disaster relief and recovery measures. They are, however, silent on the likely large-scale displacement in future, or on planned resettlement as a strategy. Moreover, to the extent that population displacement is acknowledged, this is more as a future challenge, while the current preference lies with mitigation and adaptation policies to "contain" the challenge.

Turning to the challenge of rights protection, Bangladesh has a very active civil society and well developed constitutional provisions for civil and political rights. In addition, terms such as "environmental refugees" or even "climate victims" appear in official Bangladeshi documents. Importantly, however, these terms are not formally defined, nor is there any indication of how such people would be identified and their needs met or their rights enacted. Illustrative of these dynamics, despite such wide-ranging discourse on "environmental refugees," Bangladesh has not acceded to the 1951 Convention, there is also no legal definition of IDPs and the Guiding Principles have not been incorporated into domestic laws (Kälin *et al.* 2010). Two exceptions exist, but neither applies to the current topic.[15] As such the rights of people displaced or susceptible to displacement in the context of environmental stress and climate change, are yet to gain explicit recognition in this legal and constitutional framework.

Two examples of environmental stress—one concerning extreme weather, the other slow onset—highlight the consequences of this deficit.

In 2009, Cyclone Aila left at least 500,000 people temporarily landless and homeless. The fortunate ones reinstalled themselves following the cyclone. Those who permanently lost their land simply joined the broad category of poor and landless displaced. There were, and are, no longer-term policies for rehabilitation or relocation and there is no machinery to define what rights those who are permanently displaced might expect and how these might be protected.

River bank erosion, perhaps displacing a million a year (Abrar and Azad 2004), has been increasing in the last few years. The displaced are part of a process of silent and incremental forced migration. Compensation measures exist. But, most displaced persons have to manage by themselves because the land redistribution and compensation process inadequately defines their rights. Among many procedural limitations, the redistribution system lacks transparency and it is the previously larger and politically more powerful landowners who benefit, whereas the majority of the displaced become progressively more marginalized and impoverished, either as landless laborers in nearby villages or by moving to towns and cities.

Vietnam

The dominant contextual feature shaping Vietnam's national policy making on the rights of groups displaced by environmental stress pertains to the centrally planned economy and a socialist government established in the 1970s. Not only did this shift in political and economic organization entail the forcible relocation of approximately 6.7 million people (likely a significant underestimate) between 1976 and 1985 (Guest 1998; Dang 2005); in its ongoing operation the strict regulation of migration has been a core component of the centrally planned economy, which to a large extent was concerned with limiting migration to the cities. Although regulations controlling migration through the household registration system (*ho khau*) and employment policies were relaxed in the late 1980s as the Vietnamese economy took on an increasingly market orientation, unregulated migration is still viewed unfavorably: registered migrants and people who do not move having greater access to essential services than migrants moving outside of the government's managed system.

Under such conditions, there is no awareness of individualized rights-based approaches, or discourse on protection, in the apparatus of the one-party state. The government has come to interpret "displacement" as a reactive and uncontrolled process, in contrast to its proactive planned relocation strategies and regulated migration policies which relocated about 6.6 million people (about 8 percent of the population) between 2004 and 2009. There is no mention of displacement or resettlement in government policy documents, with the term "relocation" being preferred and, accordingly, there is no scope to apply the Guiding Principles. The state's view on un-managed migration is reflected in the invisibility of unregistered migrants in the state system and therefore the question of rights does not arise.

Nonetheless, certain basic rights, including freedom of movement (paradoxically) and residence, and equality of access of all citizens, including migrants, to these rights are enshrined in Vietnam's 1992 Constitution. However, the concept and observance of human rights and their protection are still nascent. The one-party state, which has been concerned with rising pressure to reform its political structures, has little interest in such issues, while civil society offers little compensation for this void.

None of the above is meant to suggest that there are no people for whom explicit protections would be of value. Nor is it to suggest that there are no relevant policies for protecting the well-being of people exposed to environmental stress and for whom mobility might prove a useful response. Regarding the former, the increasingly relaxed approach taken to migration, as part of the liberalizing of the Vietnamese economy, means that spontaneous migrants have outnumbered registered migrants since the end of the 1990s. Indeed, the number of such migrants is expected to grow significantly as environmental and economic pressures increase.

Regarding the latter, recent government policies and programs specifically aim to relocate people living in disaster-prone areas. By 2015, at least 135,000 households living in high-risk areas will be relocated from the most flood-prone parts of the Mekong delta to safe areas. Thus, in contrast to Bangladesh for example, in Vietnam, climate change is mainstreamed as a developmental, not a humanitarian, policy concern. In this context the 2007 National Strategy for Natural Disaster Prevention, Response and Mitigation to 2020 (NS–NDPAM), advances the key strategy of "living with floods" (i.e. adaptation and planned relocation) in areas where flooding is a regular part of life. In addition, for communities that are designated for compulsory government relocation or disaster mitigation programs, the schemes provide compensation and loans for house construction. Inevitably implementation is complex, and access and eligibility criteria do not necessarily work equitably. Evidence suggests that these projects may be pushing poor households into poverty—replicating the experience of development induced relocation schemes elsewhere in the world (Dun 2011).

In Vietnam, both the historico-political experience of migration and the ideological attitude to rights and rights protection contrast sharply with the other countries discussed so far in this chapter. While the outcomes, in relation to rights protection for those susceptible to displacement related to environmental stress or climate change, are little different from the other cases, there are, as we have seen, significant contrasts in relation to policies dealing with the displacement impacts.

Ethiopia

Understanding the manner in which issues around climate change and population displacement are being framed and addressed in Ethiopia requires an appreciation of two important historical processes. The first is the legacy of the Derg, the socialist government that ruled the country from 1975 to 1991, and its overthrow by an insurgent group, which has gone on to rule the country as the Ethiopian People's Revolutionary Democratic Front (EPRDF). The second concerns the extent to which the EPRDF has reverted to a pattern of authoritarianism as a means to ensure its longevity as the ruling party (Tronvoll 2009).

In terms of the former of these events, the most notable is the extent to which the Derg used a major drought in the 1980s to justify large-scale, violent (in effect forced) resettlement strategies. Since such strategies were principally aimed at countering the efforts of insurgent forces rather than securing livelihoods for those individuals experiencing drought, the lasting impact of the process has been to arouse popular suspicion of relocation programs as a means to address environmental problems. As a result the current government's approach is to focus on the provision of relief to environmentally stressed areas and on transforming livelihoods so as to minimize the imperative to move in the first place.

While such policies have been effective to some extent as a means for addressing concerns about mobility, pressure on land, worsening ecological conditions and liberalization of the Ethiopian economy have all contributed to a growing trend of rural–urban migration. Thus, in addition to the efforts at building an agricultural safety net and transforming rural livelihoods, the government has begun to develop a significant focus on social protection in urban areas. Efforts in this regard include:

the formulation of a draft labor policy that (currently)[16] makes reference to protecting migrants moving for reasons of environmental stress; the implementation of a broad-ranging social protection framework that, while not explicit about migration, includes groups who could also be migrants (such as the elderly, child-headed households, and widows); and a program for assisted voluntary resettlement. The government does not use the term "IDPs" and has not implemented the Guiding Principles. However, although it has yet to ratify its contents, Ethiopia is a signatory to the Kampala Convention and this may initiate some form of institutional mandate.

Understanding the formulation of these policies and their likely (in)effectiveness in ensuring the rights of migrants responding to environmental stress, however, requires more than an account of the technical capacities of the Ethiopian state to implement such policies. Rather it requires an appreciation of the EPRDF's shift toward authoritarianism since 2005, and its ongoing efforts at reasserting its dominance. To this end, the government uses access to vital resources (land, work, social protection), all of which it controls through state apparatus, as a means to consolidate political power. Dissent is suppressed and human rights and civil society formation are resisted.

Under such conditions the government has shied away from international agreements on human rights, which could be used as a benchmark of its failure to meet its obligations to its citizens and thereby undermine its claims to legitimacy. In addition it has actively undermined political rights and democratic institutions. Much of this process has been facilitated by international funding of Ethiopia, whose geo-strategic importance (particularly in the "War on Terror") outweighs international concern at rights abuses.

In this context, the positive elements of expanded social protection for, and efforts at ensuring the material well-being of, all migrants in Ethiopia—including those responding to environmental stress—should be viewed with caution. As much as these may appear as positive initiatives, the institution of such material rights may well come at the expense of political rights.

In sum, given the governmental regime of Ethiopia has an ideological resistance to rights, rights protection and empowerment, the implications for those whose livelihoods are susceptible to environmental stress are somewhat similar to those in Vietnam. However, the historico-political experience of controlled migration through forced resettlement has produced rather more circumspect policy responses to the actual and potential displacement impacts of environmental stress: these are development led and top down, as in the case of Vietnam, but more concerned with adaptation and livelihood protection.

Ghana

Internal and international migration, as well as hosting refugees from the region, have long been features of Ghanaian society. Yet Ghana stands in contrast to the other four countries, because the historical determinants of the current patterns and policies on migration and displacement have rendered migration less contentious and, to the extent that policies have been developed in this arena, they are less politicized than in the other cases.

During colonial times, land expropriation and the resultant forced displacement and relocation of population were significant. Colonial settlement patterns (which were focused on the coast) and development strategies (land taxation, underdevelopment of the north generating a surplus labor pool for the south) entrenched regional labor migration strategies initially through forced labor practices. In the post-colonial era, these patterns have been incorporated into contemporary livelihoods, which now rely on well-formed cultural norms around migration. Young people migrate to raise the capital needed for later in life: to acquire land and housing, pay for a wedding, and fund their children's education in the sending region. These patterns of mobility rely on other institutional resources, such as established networks across the country.

A second colonial legacy is that, today, about 70 percent of the land is under customary administration. This is a flexible administrative structure allowing customary actors to enforce local norms and taboos, while at the same time granting access to land, including to migrants, thereby facilitating migration.

Thus, while present-day internal movement is strongly linked to colonial and post-colonial politico-historical determinants, it has been absorbed into Ghana's social and economic fabric, thus rendering it far less politically sensitive than the comparative cases of Kenya and Ethiopia. Similarly, the ethnic interests that feature strongly in shaping the history of migration discourse in Bangladesh and Kenya are not replicated in Ghana.

Third, there has been episodic and small-scale refugee and "forced" internal displacement in Ghana, but not to the degree experienced in Kenya or Ethiopia. Accordingly it has limited political saliency.

The outcome of these conditions is that neither migration nor displacement is as politically charged as in the other four countries. Indeed, the politico-historical legacy has further entrenched migration in the region, where mobility is a long-standing process for dealing with environmental variability. The economy of the north is reliant upon the economy of the south, as is the economy of the Sahel reliant on the economy of the Savannah.

However, this is not to deny that the issue of migration itself is a pressing one. Rural to urban migration, farmer–herder conflicts in the transition zones, the displacement impacts of ecological degradation in the Sub-Saharan north and tropical coastal south, and growing acknowledgement of the potential impacts of environmental stress and climate change on population mobility, are all present.

Recognition of these pressures remains limited as yet. In the case of people displaced by natural hazards and disasters, there are provisions under the 1996 National Disaster Management Act, which define natural disasters where assistance can be invoked. In addition, the definition of a disaster provided by this Act could, in principle, be extended in order to designate people displaced by slow-onset events such as climate change as people affected by environmental disaster, thus invoking provisions of the National Disaster Management Organization (NADMO).

Nevertheless, displacement "induced" by environmental degradation in the three northern regions of the country and the coastal belt are already symptomatic of the emerging problem of whether, and if so how, the rights of those currently affected will be protected. Climate change awareness is gathering momentum, but the government has yet to respond, in a comprehensive manner, to the call by some NGOs and civil society organizations to mainstream climate change issues into all its policies, projects and programs as a national priority and political commitment. This requires, among other initiatives, the strengthening of domestic legislation and national institutions and enhancing governance structures, in particular those responsible for promoting and protecting the rights of people who will be most impacted by climate change.

In the same vein, migration and displacement issues are not yet linked to rights concerns and there is no rights protection architecture of norms and instruments dealing with population migration. Accordingly, there is little evidence that Ghana seeks to implement norms set out in the Guiding Principles, although there are provisions to protect the rights of people affected by development or project induced displacement, including compensation and relocation.

In terms of international migration and displacement, Ghana has signed but not ratified the Kampala Convention. On the other hand, as a member of the Economic Community of West African States (ECOWAS), it does accede to the regional migration initiatives, which support relatively free population movement, a potentially significant mechanism as Sub-Saharan environmental stress intensifies across the region.[17]

All this falls far short of a comprehensive rights-based response to migration and development pressures. Again, this reflects a lack of political saliency of migrant rights compared to the other countries.

In sum, the discourse in Ghana focuses on reconciling environmental pressures with socio-economic priorities in order to achieve sustainable goals of national development, rather than on population displacement as in Bangladesh and Vietnam. This parallels the situation of Ethiopia, although the outcome of a very different political regime and latent politico-historical accounts of migration and displacement. In this regard, Ghana could be said to be adopting a developmental rather than a humanitarian default response, such as that of Bangladesh, for example.

As regards a rights protection framework for the displaced, even compared with the uneven efforts of Bangladesh and Kenya, Ghana's progress is very limited.

That migration and rights are less politically sensitive than in the other countries is an important message, and one which may place Ghana in a better position to address the rising pressures of environmental stress and population displacement than the other case study countries.

Conclusions

In challenging the apolitical and ahistorical framing of the environmental stress–displacement–rights nexus, this chapter has argued that the legacy of migration histories shapes how migration is understood, which in turn shapes how rights protection in relation to migration processes is manifest (or not) in the contemporary political discourses of the case study countries. Moreover, the chapter has argued that there is resistance to institutionalizing rights based norms and legal instruments, which threaten existing power structures. This conjuncture of forces, in turn, has a significant bearing on how the case study countries are responding to the emerging pressures of displacement induced by environmental factors such as climate change.

Episodic migration histories and the complex political milieu within which migration sits—the latter often intimately tied up with access to land and land rights—are highly sensitive phenomena in four of the five countries. A reluctance to acknowledge the presence of internal migration and to develop policies to manage the social and economic consequences is the outcome of this politico-historical legacy. Failure to advance the Guiding Principles is symptomatic of the reluctance to develop policies and strategies to tackle internal migration.

These characteristics underpin the countries' disinclination to respond to internal migration and, more specifically, population displacement induced by environmental factors, as matters of policy concern. The unwillingness to engage with migration as an arena of public policy constitutes both the backcloth to, and an explanation of, the reluctance of all the governments to develop policy frameworks, which will effectively tackle the current and future population displacement impacts associated with climate change and environmental stress.

It could be argued that the lack of political will and commitment also reflects competing national priorities in which environmentally related population displacement is predominantly a future challenge and thus of a lower order than more immediate developmental and poverty reduction goals. While not denying these competing agendas, our contention is that resistance to engage with the politics and polices for migration is underpinned by the reluctance of these countries—all of which have somewhat insecure governments—to address human rights issues. This is revealed in their failure to develop legal and normative frameworks to protect the rights of migrants in general and, more specifically, in relation to environmentally displaced people.

In this regard, whereas in Ethiopia and Vietnam—and less clearly in Ghana—the policy response to these displacement impacts is predicated on "development led" strategies, Kenya and Bangladesh incline to frame these outcomes in terms of a humanitarian "default" position, not a developmental challenge. Yet, irrespective of the political regime and the alternative framings of policy responses, the protection of rights in the context of environmental stress is appropriated essentially in terms of *material rights*—restoration of livelihoods, livelihood safety nets, and resettlement to safer ground. This enables governments to acknowledge material needs while subverting the structural challenge of affording *political rights*—empowerment, decision making and full participation in, for example, resettlement schemes. In the contest for political power, the provision of material rights reduces calls for empowerment that threaten the power of the political élites. Paradoxically, but not perhaps surprisingly, the more "development led" strategies of Ethiopia and Vietnam do not open up political space for the protection of rights and civil society.

Despite their diverse and contrasting histories and political regimes, and the differing political saliency of migration and displacement issues, there are only modest differences in their disposition to human rights protection which, in general, is weakly embedded in the political discourse. The governments are, on the whole, unsympathetic to promoting human rights policies and practices. Ghana stands out as an exception, but only to the extent of benign neglect rather than explicit resistance.

In the contested ground of development and rights, it is not that rights are of a lower order than development; rather that retaining political control is more important than rights. Consequently, rights are afforded only in material terms, with little or no possibility to affect the political power structures and elites (although in Ghana, as we have seen, this looks rather different). Thus rights are only incorporated into development in a fashion that prevents them from challenging the contemporary structures of power.

The implications of this analysis for the implementation of some form of "guiding principles" for protecting the rights of people moving under conditions of environmental stress are significant. Given the enduring political denial of migration and displacement as a policy and social challenge, and the fragility of the governments which mediates their disinclination to develop systematic and structural responses to the protection of rights, the prognosis for protecting the rights of those displaced by changing environmental or climate conditions is, accordingly, poor.

The analysis suggests that limited effort is likely to be put into adopting such principles and, if adopted, similarly little energy will be expended in implementing them. This problem is not easily resolvable in the climate change context. This is because both high emissions-producing states and states likely to be worst affected by climate change have an incentive to maintain the notion of adaptation as a technical challenge (of material provision) rather than a social challenge (of political empowerment). On the part of emitting countries, such a framing ameliorates the imperative to get emissions rapidly under control by implying that the impacts of such emissions could be offset in the future through technical or material transfers. More problematic, it becomes politically difficult to tie reparations for climate change (such as adaptation funding) to conditionalities, given the likelihood that those setting the conditions have generated the problem of climate change.

For countries that are, or will be, affected by climate change, as this chapter has argued, pragmatically providing material resources and responses to people who are, or are likely to be, displaced reduces the need to diversify and share political power with other groups, organizations and interests. More fundamentally, from the evidence of the case studies, it seems that the resources likely to be made available by these measures could well be appropriated to consolidate political power: this is one of the major concerns around relocation as a means to address environmental stress.

The analytical framework highlights the need to appreciate the conflicting interests apparent in providing rights protection not just for those susceptible to environmental stress, but across many other socio-economic policy areas. This, in turn, emphasizes the need to disaggregate the framework of the various actors involved in the provision of those protections and the different contexts within which they operate. In other words, who should do what to ensure implementation of norms and principles for rights protection? As we have seen in the case of civil society structures—notably the contrast between Bangladesh, on the one hand, and Ethiopia and Vietnam, on the other—they have variable leverage and impact in terms of social protection and play different roles in different contexts.

On the whole, civil society structures remain weak in their capacity, and the resources they have to hand, to advocate, promote and coordinate an active defense of human rights. On the other hand, civil society and local community organizations in all the countries (most notably Bangladesh) have experience in coping with disaster relief—including temporary population displacement—and thus potentially provide a platform for developing a capacity to respond to the rights of populations increasingly displaced, or susceptible to be so, by the impacts of environmental change.

Nevertheless, connecting these displacement and rights-based capacities to the broader responsibilities, which the case study governments have for rights protection in the context of environmental stress and displacement, remains both highly contested and deeply problematic.

Notes

1 We use the concept of rights as principles articulating how people should be treated— embodied in the Universal Declaration of Human Rights and other international and national instruments dealing with civil social and political rights—and the concept of protection as the means by which these principle become manifest. The terminologies "rights protection" and "protection of rights" are used synonymously. Our chapter deals with both "rights protection" and "rights *and* protection"—that is to say the degree to which rights are, or are not, enshrined in different contexts, but also the extent to which these rights are effectively manifest in the different societies.

2 Bangladesh, Vietnam, Kenya and Ghana were studied in the first research project on this subject (Zetter 2011). Ethiopia has been added in the current research funded by the John D. and Catherine T. MacArthur Foundation Grant.

3 On the distinction between "migration," "displacement" and "involuntary migration," we use the terms "displacement" and 'involuntary migration" as a sub-set of "migration." Migration is characterized by complex migratory motivations, whereas displacement implies more of an unplanned, involuntary phenomenon. However, since the thesis of our chapter is the way issues of migration in general shape attitudes toward displacement in particular, we use both "migration" and "displacement," depending on the context.

4 It is possible that the argument may extend to other groups of "crisis migrants" who do not fit within already established policy and legal instruments for protection, although that is beyond the reach of the current chapter.

5 1951 Geneva Convention relating to the Status of Refugees and the 1967 Protocol; 1969 OAU Convention Governing the Specific Aspects of the Refugee Problem in Africa, 1984 the Organization of American States (OAS) Cartagena Declaration on Refugees.

6 1998 Guiding Principles on Internal Displacement, 2009 African Union Convention on the Protection and Assistance of Internally Displaced Persons in Africa.
7 1954 Convention relating to the Status of Stateless Persons and the 1991 Convention on the Reduction of Statelessness.
8 1991 International Labour Organization (ILO) Convention 169 on the Rights of Indigenous People and the 2007 UN Declaration on the Rights of Indigenous Peoples.
9 2000 UN Convention against Transnational Organized Crime (protocols on trafficking); 2008 Council of Europe Convention on Action against Trafficking in Human Beings.
10 1990 International Convention on the Protection of the Rights of All Migrant Workers and Members of their Families.
11 1948 Universal Declaration of Human Rights, 1966 International Covenant on Civil and Political Rights, 1966 International Covenant on Economic, Social and Cultural Rights (Articles 6 and 7), 1981 Convention on the Elimination of All Forms of Discrimination against Women, 1984 Convention against Torture and Other Cruel, Inhuman or Degrading Treatment or Punishment, 1989 Convention on the Rights of the Child.
12 Declaration of the United Nations Conference on Environment and Development (Rio Declaration), UN Doc. A/CONF/151/26/Rev.1 (1992).
13 Decision 1/CP.16, The Cancun Agreements: Outcome of the work of the Ad Hoc Working Group on Long-Term Cooperative Action under the Convention, in Report of the Conference of the Parties on its sixteenth session, Addendum, Part Two: Action taken by the Conference of the Parties, UNFCCC/CP/2010/7/Add.1, 15 March 2011, para. 14 (f), available online at: http://unfccc.int/resource/docs/2010/cop16/eng/07a01.pdf.
14 The African Union Convention on the Protection and Assistance of Internally Displaced Persons in Africa 2009, available online at: http://www.internal-displacement.org/8025708F004BE3B1/%28httpInfoFiles%29/0541BB5F1E5A133BC12576B900547976/$file/Convention%28En%29.pdf.
15 The exceptions are: those displaced by development projects, yet thousands of applicants still await compensation as much as fifty years later. Even here, only landowners are entitled to compensation: small landowners and the landless—poor and vulnerable people—are not accommodated in the process. The second exception relates to those displaced by the 1947 Partition of India.
16 This draft policy was not finalized at the time of writing and still had to go before parliament for approval.
17 Space prevents development of an argument that the ECOWAS agreement could also be conceived as a component of the politico-historical processes that accommodate migration.

Acknowledgments

This chapter is based on research funded by the John D. and Catherine T. MacArthur Foundation entitled "Environmentally Displaced People: Rights, Policies and Labels," Grant No 10-94408-000-INP. The authors are grateful for the support of the Foundation. An earlier version of this chapter was presented as a keynote paper at the "ClimMig" Conference on Human Rights, Environmental Change, Migration and Displacement held at the Ludwig Boltzmann Institute of Human Rights, University of Vienna, September 20–21, 2012. We record our thanks to our field research team: Jane Chun, Marie Pécoud, Malika Peyraut and Augustine Yelfaanibe.

References

Abrar, C.R. and Azad, S.N. (2004) *Coping with Displacement: Riverbank Erosion in North-West Bangladesh*, Dhaka, Bangladesh: RDRS/North Bengal Institute/RMMRU.

Council of Europe (2008) "Environmentally Induced Migration and Displacement: A 21st Century Challenge," report of the Committee on Migration, Refugees and Population Parliamentary Assembly Council of Europe, December 23, Doc. 11785:2, available online at: http://assembly.coe.int/Documents/Adopted Text/ta09/EREC1862.htm.

Dang, N.A. (2005) "Vietnam Internal Migration: Opportunities and Challenges," Regional Conference on Migration and Development in Asia, unpublished.

Dun, O. (2011) 'Migration and displacement triggered by floods in the Mekong Delta, Vietnam' International Migration 49 (Supplement s1): e200-e223, DOI: 10.1111/j.1468-2435.2010.00646.x.

Giossi-Caverzasio, S. (ed.) (2001) *Strengthening Protection in War: A Search for Professional Standards*, Geneva: ICRC, available online at: www.icrc.org/eng/assets/files/other/icrc_002_0956.pdf.

GoB (Government of Bangladesh) (2009) "Climate Change Strategy and Action Plan," Ministry of Environment and Forests, Dhaka, Government of the People's Republic of Bangladesh.

GoK (Government of Kenya) (2008) "Report of the Findings of the Commission of Inquiry into the Post-Election Violence (CIPEV) in Kenya, ('Waki Commission')," Nairobi, Government of Kenya.

GoV (Government of Vietnam) (2009) *Climate Change, Sea Level Rise Scenarios for Vietnam*, Hanoi: Ministry of Natural Resources, Government of Vietnam, June.

Guest, P. (1998) "The Dynamics of Internal Migration in Vietnam," UNDP, Discussion Paper 1, Hanoi, Vietnam.

IOM (International Organization for Migration) (2010) "Policy Dialogue on Environment, Climate Change and Migration in Bangladesh," Dhaka, International Organization for Migration, May.

Kälin, W. and Schrepfer, N. (2012) "Protecting People Crossing Borders in the Context of Climate Change Normative Gaps and Possible Approaches," UNHCR, Division of International Protection, PPLA/2012/01 Legal and Protection Policy Research Series.

Kälin, W., Williams, R.C., Koser, K. and Solomon, A. (eds.) (2010) "Incorporating the 'Guiding Principles on Internal Displacement' into Domestic Law: Issues and Challenges," *Studies in Transnational Legal Policy No. 41*, Washington: Brookings-Bern and the American Society of International Law.

Koser, K. (2011) "Climate Change and Internal Displacement: Challenges to the Normative Framework," in Piguet, E., Pécoud, A. and de Guchteneire, P. (eds.), *Migration and Climate Change*, Paris: UNESCO/Cambridge: CUP: 289–305.

Laczko, F. and Aghazarm, C. (2009) *Migration, Environment and Climate Change: Assessing the Evidence*, Geneva: IOM.

McAdam, J. (2011) "Swimming against the Tide: Why a Climate Change Displacement Treaty is Not the Answer," *International Journal of Refugee Law*, 23(1): 2–27.

McAdam, J. and Saul, B. (2008) "An Insecure Climate for Human Security? Climate-induced Displacement and International Law," Sydney Centre for International Law, Working Paper No. 4, Faculty of Law, University of Sydney, available online at: http://sydney.edu.au/law/scil/documents/2009/SCILWP4_Final.pdf.

Morrissey, J. (2009) "Environmental Change and Forced Migration: A State of the Art Review," *Workshop on Environmental Change and Forced Migration*, Oxford: University of Oxford: 1–49.

Tronvoll, K. (2009) "Ambiguous Elections: The Influence of Non-electoral Politics in Ethiopian Democratisation," *Journal of Modern African Studies*, 47(3): 449–474.

UNHCR (2009) "Forced Displacement in the Context of Climate Change: Challenges for States under International Law," paper submitted by the Office of the UNHCR in cooperation with the NRC and the UNSGSR on IDP to the 6th session of the Ad Hoc Working Group on Long-Term Cooperative Action under the Convention (AWG-LCA 6), May 19.

UNHCR (2010) "Closing Remarks," 2010 High Commissioner's Dialogue on Protection Gaps and Responses, Palais des Nations, Geneva, December 9, available online at: www.unhcr.org/4d0732389.html.

UNHCR (2011) "Summary of Deliberations on Climate Change and Displacement," Geneva, UNHCR, April, available online at: www.unhcr.org/cgi-bin/texis/vtx/home/opendocPDFViewer.html?docid=4da2b5e19&query=Summary%20of%20Deliberations%20on%20Climate%20Change%20and%20Displacement%20April%202011.

Wisner, B., Blaikie, P., Cannon, T. and Davis, I. (2004) *At Risk: Natural Hazards, People's Vulnerability and Disasters* (2nd edn.), London, Routledge.

Zetter, R. (2009) "Protection and the Role of Legal and Normative Frameworks," in Laczko, F. and Aghazarm, C. (eds.), *Migration, Environment and Climate Change: Assessing the Evidence*, Geneva: International Organization for Migration: 285–441.

Zetter, R. (2010) "Protecting People Displaced by Climate Change: Some Conceptual Challenges," in McAdam, J. (ed.), *Climate Change and Displacement in the Pacific: Multidisciplinary Perspectives*, Oxford, Hart Publishing: Ch. 10.

Zetter, R. (2011) "Protecting Environmentally Displaced People: Developing the Capacity of Legal and Normative Frameworks," Refugee Studies Centre, 2010, report commissioned by UNHCR and Governments of Switzerland and Norway, available online at: http://www.unhcr.org/4da2b6189.pdf.

Zetter, R., Boano, C. and Morris, T. (2008) "Environmentally Displaced People: Understanding Linkages between Environmental Change, Livelihoods and Forced Migration," Forced Migration Policy Brief No.1, Refugee Studies Centre, University of Oxford, available online at: http://www.google.com/url?sa=t&rct=j&q=&esrc=s&source=web&cd=3&cad=rja&uact=8&ved=0CEIQFjAC&url=http%3A%2F%2Fwww.rsc.ox.ac.uk%2Ffiles%2Fpublications%2Fpolicy-briefing-series%2Fpb1-environmentally-displaced-people-2008.pdf&ei=ePczU9HVArLd7QaPx4CYCg&usg=AFQjCNGIIZXUY1ET-EzxawFCCOg42Z-8tA&sig2=qclUkWm1atU4F-bL2WHVSg&bvm=bv.63808443,d.ZGU.

Zetter, R. and Morrissey, J. (2014) "The Environment–Mobility Nexus: Reconceptualizing the Links between Environmental Stress, Mobility and Power," in Fiddian-Qasmiyeh, E., Sigona, N., Loescher, G. and Long, K. (eds.), *Oxford Handbook of Refugee and Forced Migration Studies*, Oxford: OUP: 342–54.

10

ENHANCING ADAPTATION OPTIONS AND MANAGING HUMAN MOBILITY IN THE CONTEXT OF CLIMATE CHANGE

Koko Warner and Tamer Afifi

Introduction

Climate change is likely to worsen the situation in parts of the world that already experience high levels of stressors to livelihood and food insecurity. The consequences of greater variability of climatic factors like rainfall conditions affect the livelihoods and safety of vulnerable people. Less predictable seasons, more erratic rainfall, unseasonable events or the loss of transitional seasons can decrease food security, the livelihoods of millions of people, and the migration decisions of vulnerable households in areas that are already under climate stress (e.g. from drought, flooding, and temperature variation in areas like the Horn of Africa and parts of Asia). In order to make informed decisions about adaptation planning, development, and a transition to a more climate-resilient future, policy makers and development actors need a better understanding of the linkages among changes in the climate, household livelihood and food security profiles, and migration decisions—particularly in crisis situations.

Since at least the mid-1980s, scientists have linked environmental change to human mobility (El-Hinnawi 1985).[1] Early debates emerged around future projections and predictions of the number of "environmental migrants" (Myers 2005; Brown 2008). More recently, conceptual and empirical studies have examined broad relationships between environmental factors and human mobility in different situations (Jäger *et al.* 2009; Warner *et al.* 2009).[2] These studies have identified broad patterns as a point of departure for further, more nuanced work on the interactions of climatic and socio-economic factors (Brown 2008; Hugo 2008; Laczko and Aghazarm 2009; Morrissey 2009; Tacoli 2009; Jónsson 2010; P. Martin 2010; S. Martin 2010; Afifi 2011).

Research since that time has determined that environmental factors do play a role in human mobility (Jäger *et al.* 2009; Warner *et al.* 2009; Government Office for Science 2011), and emphasizes that some people who are more exposed to environmental stressors—particularly farmers, herders, pastoralists, fishermen and others who rely on natural resources and the weather for their livelihoods—may be the least able to move very far away, if at all (Betts 2010; Black *et al.* 2011). In the decades ahead, these potentially "limited mobility" populations could face deteriorating habitability of their traditional homelands, with fewer options for moving to more favorable places in safety and dignity. A subset of literature on environmental change and human mobility is emerging that examines climate change and climatic risks. This chapter fits into that rapidly growing sub-set of literature. The implications of climate change—ranging from extreme weather impacts like flooding, storms and drought to slower, incremental changes in regional climate patterns, as well as glacial melt, sea level rise and ocean acidification—for a wider scope of issues related to population movement in the medium and longer term have driven a quest for better understanding the circumstances under which climatic factors affect human decisions about whether to leave, where to go, when to leave, and whether (and when) to return.

Up to the present, research relating environmental change to human mobility confirms that environmental factors play a role in migration. Yet in the context of climate change, scholarly literature on migration ranges across a host of climatic stressors and geographies, making it difficult to date to solve the debate whether migration is a form of adaptation or an indicator that people are moving under "crisis" conditions not normally considered part of "positive" adaptation (e.g. limits to adaptation as currently understood; Kates, Travis and Wilbanks 2012).

This chapter presents new evidence from a study on the relationship between rainfall variability, food and livelihood insecurity, and human migration in eight countries. The findings illustrate the importance of understanding how households use migration as a risk management strategy when faced with climatic stressors like changes in rainfall. The findings further demonstrate four household profiles that help address the question of the role that migration does or does not play in adaptation to climate change. The main findings show that while all communities surveyed used migration when faced with climatic stressors, those with "resilient" characteristics were able to benefit from the strategy. In contrast, those households that lacked such characteristics (such as access to land, education, social networks, formal and informal institutions) had migration outcomes that slowly reduced their development base and can be considered as "erosive coping strategies." These insights raise issues about anticipatory movements (some beneficial and others problematic) as well as trapped populations. The chapter concludes with a few reflections for adaptation and development policy.

Climate change and migration: a framework for understanding migration as adaptation or crisis

The "Where the Rain Falls" (Rainfalls) research[3] was undertaken to answer the question "under what circumstances do households use migration as a risk management strategy when facing rainfall variability and food insecurity?" (Warner *et al.*

2012, 2). The Rainfalls research expands insights into how human mobility may develop in the context of a changing climate where rainfall patterns are expected to shift notably in timing (seasonality), quality (extreme events, intensity of rainfall), and distribution (geographically) in coming decades. This research administered a household survey (n = 1300) and participatory research (n = 2000 respondents) in eight rural districts: Northern Thailand (Lamphun Province); Peru, Central Andes (Huancayo Province); Vietnam, Mekong Delta (Dong Thap Province); Central India (Janjgir District, Chhattisgarh); Northern Bangladesh (Kurigram District); Guatemala, Western Highlands (Cabricán Municipality); Northern Ghana (Nadowli District, Upper West Region); Northern Tanzania (Same District, Kilimanjaro Region).

Rainfall variability affects migration via livelihood systems

Rainfall variability affects migration via livelihood systems, illustrated in the research framework employed in the case studies (Figure 10.1). This framework facilitated an examination of the interrelationships and pathways that affect household risk management and migration decisions as they relate to rainfall, food and livelihood (Warner and Afifi 2013).

In this framework, the livelihood security of the studied households is influenced by rainfall variability (an independent variable influencing livestock and crop production). These factors, plus the factor of land ownership, help shape the food security situation of the household, which is also structured by external processes. In the framework, a notion of "degree of vulnerability" (taking into account the degree of economic diversification, number of household members of working age, financial situation, and other factors) is used to indicate the range of available coping and adaptation strategies for households. Research findings are based primarily on fieldwork-generated qualitative and quantitative data (Rademacher-Schultz *et al.* 2012). Where secondary data have been used, this is indicated by colored boxes at the edge of the research area boxes. The framework illustrates the feedback loops within the community systems (e.g. initial conditions change dynamically due to the interlinkages and interactions of household decisions).

The more household economy and food consumption depend on rainfall, the more sensitive they may be to changes in rainfall variability. Other factors, such as the availability of a range of coping strategies, further affect how households use migration to manage climatic risks.

Study findings about household migration decisions

Household characteristics in districts surveyed

Table 10.1 summarizes the households surveyed in eight rural districts in eight countries. The last column of the table shows the total number of households surveyed in the Rainfalls research sites. At each site, between 130 and 206 house-

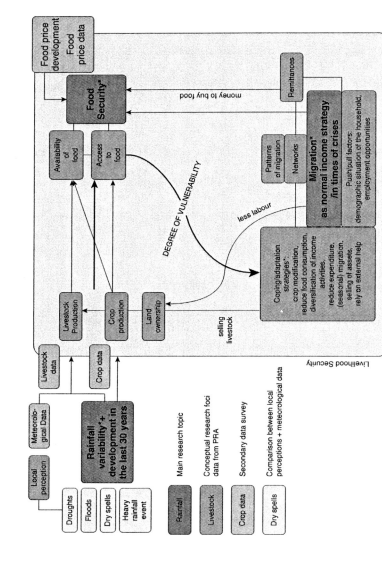

FIGURE 10.1 Research foci, methods and data sources

Source: Christina Rademacher-Schulz and Verena Rossow (2012, in Warner *et al.* 2012)

Note: This figure provides an overview of the three major research foci of the Rainfalls project and the sources of information, as well as how the data were triangulated using qualitative and quantitative methods (survey, participatory community research, interviews, etc.).

TABLE 10.1 Households surveyed in eight case study research sites

	Lamphun, Thailand	Huancayo, Peru	Chhattisgarh, India	Dong Thap, Vietnam	Kurigram, Bangladesh	Nadowli, Ghana	Same, Tanzania	Cabrián, Guatemala	
Households interviewed (n)	206	150	180	150	150	158	180	136	total = 1310
Approximate % of local population	31,7	29,9	12,8	8,6	2,3	27,2	11,9	18,5	mean = 17,8
Female-headed households interviewed (%)	14,6	20,6	7,7	6,6	2,7	12	23	15	mean = 12,8
Female interviewees (%)	14,6	75,3	18,3	44,7	19	20	58,1	63	mean = 39,1
Average age of the interviewees	49,62	42,14	43,58	44,4	45	47,75	47,39	37,04	mean = 44,7
Household size (average)	4,31	5,03	6,64	4,3	5,1	7,03	6,08	6,79	mean = 5,6
Household dependency ratio	0,49	0,88	0,70	0,46	0,80	0,93	1,29	1,10	
Average years of schooling of HH-head	4,16	7,56	5,93	5,2	3,3	2,78	5,16	3,12	mean = 4,7
Average years of schooling of HH-members aged 14+	5,82	8,42	7,48	6,7	4,6	4,02	6,06	3,57	mean = 5,9
Poor (1.25–2.5 US$/cap/day) %	78	82	55	68,6	66	n/a	n/a	61,6	mean = 51,4
Households facing food shortages in last year (%)	29,1	82,6	43,9	43	75,3	52,5	84	52,9	mean = 52,9
Landless households (%)	2,4	43,3	24,4	31	36	6	6,7	2,9	mean = 19,1
Land-scarce HH—small land holding (%)	44,6	39,3	36,1	26	48	3,8	24,8	65	mean = 35,9
Medium land holdings (%)	22,3	8,6	12,8	36,6	13	33	49	24	mean = 24,8
Above average land holdings (%)	30,6	8,8	26,1	6,6	3	43,6	19,3	6	mean = 17,9
Average land holding size (ha)	2,856	0,54	1,18	2,4	0,5	7,02	1,815	0,54	mean = 2,1 ha
Households with migrants (%)	67	63,3	41,7	60	43,3	76,6	53,9	23,5	mean = 57
Migrants seeking livelihood diversification (%)	76,00	75,6	87,7	69,6	90	82,8	78,4	97,1	mean = 80,6

Source: Warner et al. (2012)

holds were surveyed. In six of the eight districts, this number represented at least 10 percent of the local (district) population. Of the households surveyed, a median value of 13.3 percent were headed by females. In most districts surveyed, respondents were poor and often land was scarce, and most used migration in one form or another.

- **Definition of dependency ratio:** Ratio of household members typically not in the labor force (the dependent part—age ranges 0–14 and >64) and those typically in the labor force (the productive part—age range 15–64). It is used to measure the pressure on productive household members.
- **Definition of land scarce varies by country:** Thailand <=10 Rai or 1.6 ha; Peru 0.1–5.0 ha; India <= 1 acres; Vietnam 0.1–1.0 ha; Ghana 0.1–1.0 ha; Bangladesh 0,1–0,7 ha; Tanzania 0.01–1.75 acres; Guatemala <0,44 ha.
- **Definition of medium-sized farm varies by country:** Thailand 10.01 to 20 Rai; India 1.01–2 acres; Ghana <5ha; Tanzania 1.76–4 acres; Guatemala >0.44 and <1 ha.
- **Definition of above average-sized farm varies by country:** India >=2 acres; Ghana >5.01 ha; Tanzania >=4.01 acres; Guatemala >1 ha.

Livelihood and food security related to rainfall variability

Rural people in the eight research locations perceive climatic changes in the form of rainfall variability, and these perceptions shape household risk management decisions. The most common changes reported relate to the timing, quality, quantity, and overall predictability of rainfall, including: delayed onset and shorter rainy seasons; reduced number of rainy days per year; increased frequency of heavy rainfall events, and more frequent prolonged dry spells during rainy seasons. These perceived changes correlate with an analysis of local meteorological data over the last three decades.

Access to land of sufficient quality to support household food consumption and income needs was an important issue in the research areas. Landlessness and land scarcity was manifest in median values of 15.5 and 37.7 percent of households surveyed respectively, with these households in each site manifesting distinct characteristics relevant to their mobility decisions (discussed below). The average land holding for households across all sites was 1.5 hectares of productive land (excluding grazing land for livestock).

Household migration experience managing rainfall stress on household income and consumption

Migration—seasonal, temporal and permanent—plays an important part in many families' struggle to deal with rainfall variability and food and livelihood insecurity. Migration for this type of household can be seasonal (less than six months), temporal (more than six months) or permanent, with the nearest places with more favorable livelihood opportunities as areas of destination.

Households manage climatic risks such as changes in rainfall variability by migration. Migration—seasonal, temporal and permanent—plays an important part in many families' struggle to deal with rainfall variability and food and livelihood insecurity, and was reported to have increased in recent decades in a number of the research sites. Rainfall was observed to have a more direct relationship with household migration decisions in research sites where the dependence on rainfed agriculture, often with a single harvest per year, was high and local livelihood diversification options were low. Pressure on rainfall-dependent livelihoods is likely to grow as a driver of long-term mobility in the coming decades if vulnerable households are not assisted in building more climate-resilient livelihoods in situ.

Table 10.2 summarizes migration experience in the households sampled in the respective case studies.

- In the household survey, sometimes respondents did not give a clear answer, which made the interviewer drop the respective question. In other cases, respondents gave two answers where the question required only one answer. Therefore, in some exceptional cases in this table (particularly the cases of Thailand, Peru and India), adding up percentages gives a sum of slightly less or more than 100 percent.
- Seasonal migration is defined as yearly recurring migration over periods of less than six months per year.
- Temporal migration is defined as a move from the household of origin during at least six months per year to a place within the country or abroad with the purpose of working, studying, or family reunification, over a distance that forces the concerned person to settle at the destination and stay overnight.
- Current migration means that a person is currently away for the purpose of migration.
- Returned migration is defined as the return of a once-migrated household member who has not migrated again in more than one year

Migration was found to have increased in recent decades in a number of the research sites. Households with more diverse assets and access to a variety of adaptation, livelihood diversification, or risk management options can use migration in ways that enhance resilience. Migration related to rainfall variability and stress on livelihoods and food production systems was found to be almost entirely within national borders. It was predominantly male, but with growing participation by women in a number of countries. Seasonal, temporal or permanent migration patterns were observed. Migration was undertaken largely by individual household members (with India as the exception where entire nuclear families moved together). Migration in the studies is largely driven by livelihood-related needs (household income) in most countries, but with a growing number of migrants seeking improved skill sets (e.g. through education) in countries like Thailand, Vietnam and Peru. The research across the studies shows a mix of rural–rural and rural–urban migration, with more productive agricultural areas (Ghana, Bangladesh, Tanzania), nearby

TABLE 10.2 Demographics and types of migration in the research areas

Indicators	Lamphun, Thailand	Huancayo, Peru	Chhattisgarh, India	Dong Thap, Vietnam	Kurigram, Bangladesh	Nadowli, Ghana	Same, Tanzania	Cabrican, Guatemala	
Total number of households	206	150	180	150	150	158	180	136	total = 1310
HH with migration experience %	67	63	42	60	43	77	49	19	mean =59,13
Migrant demographic information									
Total number of migrants	224	160	212	168	89	257	204	35	total = 1349
Male %	61	64	62	63	97	69	68	77	mean = 70,06
Female %	39	36	38	37	3	31	32	23	mean = 29,94
Average age of migrants	23,18	24,43	21,1	27,6	37	22,68	24,95	22,8	mean = 25,47
Education level of migrants (average years of schooling)	8,48	8,88	6,1	7,6	3,5	4,06	5,7	4,83	mean = 6,14
Marital status of migrants									
Single %	43	33	19	58	11	40	45	20	mean = 33,68
Married %	50	46	70	39	89	53	47	46	mean = 54,85
Other %	7	21	11	3	0	7	8	34	mean = 11,47
Purpose and temporal aspects of migration choices									
Migration motivated by need to earn livelihood %	76	76	88	70	90	83	40	97	mean = 77,37
Migration motivated to improve skills, education %	18	14	2	18	10	9	20	3	mean = 11,74
Other %	6	10	10	13	0	8	41	0	mean = 10,89

Type of migration

Seasonal %	66	67	66	36	80	58	50	17	mean = 54,85
Temporal %	6	33	28	64	20	37	43	80	mean = 38,91
Permanent %	28	0	6	0	0	5	7	3	mean = 6,24

Migration status

Current %	42	46	58	50	84	68	47	NA	mean = 56,25
Returned %	60	53	42	50	16	32	53	NA	mean = 43,77

Source: Warner et al. (2012)

urban centers (Peru, India), mining areas (Ghana), and industrial estates (Thailand, Vietnam) as the most common destinations.

Migration experiences varied across the research sites. Table 10.3 summarizes the livelihood and food security and migration across the research sites.

Implications of household migration decisions: household characteristics and sensitivity to rainfall variability and food/livelihood security

The findings on rainfall variability, food and livelihood security, and migration help build understanding about how households use migration to manage risk or to survive when faced with changing rainfall patterns that affect food and

TABLE 10.3 Livelihood and food security, and migration, across the research sites

Research site	Geography	Findings
Northern Thailand (Lamphun Province)	Upland and riverine	Diverse livelihoods and access to assets and services make migration a matter of choice in Lamphun Province
Northern Thailand Peru Central Andes (Huancayo Province)	Highland	Livelihood options and migration strategies in Huancayo Province vary by elevation and proximity to urban centers
Northern Thailand Vietnam Mekong Delta (Dong Thap Province)	Delta lowland	Landless, low-skilled poor of Hung Thanh Commune have few options, despite a rising economic tide
Northern Thailand Central India (Janjgir District, Chhattisgarh)	Irrigated lowland	Poor households in Janjgir–Champa still must rely on seasonal migration for food security, despite irrigation, industrialization and safety net
Northern Thailand Northern Bangladesh (Kurigram District)	Riverine lowland	Migration is a key coping strategy for poor households in Kurigram, but one with high social costs
Northern Thailand Guatemala Western Highlands (Cabricán Municipality)	Highland	Little livelihood diversification and limited migration opportunities leave people of Cabricán with few good options
Northern Thailand Northern Ghana (Nadowli District, Upper West Region)	Savannah woodland	High dependence on rain-fed agriculture in Nadowli District contributes to continued reliance on seasonal migration as a coping strategy
Northern Thailand Northern Tanzania (Same District, Kilimanjaro Region)	Upland and riverine lowland	Migration is a common coping strategy for smallholder farmers and livestock keepers struggling for food security in Same district

Source: Warner *et al.* (2012)

livelihood security. Four broad household profiles emerged from the research, showing how the circumstances in which households use migration as a risk management strategy lead to different "adaptation" or "crisis-like" outcomes. The first pattern has to do with "resilient" households, the remaining three fall along a spectrum of households on different levels of "vulnerability." This section refers to findings from individual case studies from the Rainfalls study. More research is needed along these lines in the future.

Households that use migration to improve their resilience (successful migration)

Across all case studies, these households use migration as one successful risk management or livelihood strategy amongst a wider range of options. The profile of such households was low income or poor, but with adequate access to a variety of livelihood options and assets (social, political, financial) to enable the household to be less sensitive to rainfall stressors. Children in these households typically had three to five years more education than parents, with migrants usually in their early twenties, single, aspiring to better livelihood opportunities, and able to send remittances back home. Migration, first and foremost, is an accessible option for these households to enhance livelihood security and resilience for the entire household, including members left behind. Second, migration is an active, positive choice associated with capturing an opportunity that benefits the household. For instance, in these households, migrant remittances facilitate investments in education, health, and assets that enhance the welfare of the household in ways that make it less susceptible to rainfall stressors.

For example, Sakdapolrak, Promburom and Reif (2013) showed that 51 percent of households in Lamphun Province, Thailand, considered the impact of rainfall-related environmental stress on their livelihoods to be significant. Three-quarters of households suffer from lower income due to declining crop yields and deceasing income from agriculture as a result of exposure to environmental stress. However, diversified on- and off-farm (less sensitive to rainfall variability) income generation activities, access to financial resources through community funds, and assistance from the local government reduce household vulnerability to rainfall-related stress and food insecurity. In Lamphun, diverse livelihoods and access to assets and services make migration a matter of choice, not necessity.

This pattern was found to some degree in the research in Peru. In their research on Huancayo Province, Peru, Milan and Ruano (2013) found that livelihood options and migration strategies vary by elevation and proximity to urban centers. The impact of changing rainfall on food production is severe for 53 percent of households and two thirds of surveyed households reported crop damage and lower crop yields from changing rainfall patterns. Rainfall changes affect the ability of households to feed themselves and earn livelihoods and 42 percent experience substantial negative impacts on household income. Migration facilitates lowered dependence on agriculture-based livelihoods and expanded employment opportunities in non-farming activities in nearby urban areas.

For the next two groups, impacts of migration on households facing rainfall stressors depend on the degree of success migrating members have in securing food or resources to obtain food.

Households that use migration to survive, but not flourish

For this group, migration is a way to avoid the worst consequences of rainfall variability and food insecurity, but few or inadequate livelihood diversification or in situ adaptation options available mean that households may be "just getting by." These families are usually land-poor, and while they may have access to livelihood diversification strategies, these options are often insufficient to ensure food security for the household. Migrants are usually heads of household in their mid-forties. Children in these households have—within a four-month average—the same level of education and skill sets as their parents. These families have less access to social institutions and less access than the previous group to other forms of livelihood diversification or measures to cope with rainfall-related stressors on livelihoods and food security.

While migration for these households is somewhat accessible—they have the assets necessary to migrate—the migration choice is more risky than for contented households. The households in this group can easily slide into vulnerability if migration proves to be erosive or if rainfall stressors overwhelm their capacity to cope. For these households, migration may perpetuate cycles of debt (migration requires an initial outlay), and periodic hunger (if migration is unsuccessful). Migration might not have been the first choice if more viable in situ options were available or accessible. Migration for such households is often seasonal or temporary, to obtain food directly, or to obtain resources to access food. Migration, therefore, serves as a stop-gap measure, allowing temporary relief from rainfall variability and the impacts of crop failure or decline on the household economy, but it does not transform households or release them from the poverty cycle.

This pattern of migration related to managing the risks of changing rainfall patterns for livelihood and food security is illustrated in cases from Bangladesh, Vietnam and India: Etzold *et al.* (2013) showed that in Kurigram District, Bangladesh, migration is a major "coping strategy" to address unfavorable economic and unexpected environmental conditions, including the local implications of rainfall variability: 89 percent of households are affected economically by prevailing weather patterns and rainfall variability. Longer dry spells and frequent droughts are a "very important" migration reason for 39 percent and 36 percent of households, respectively. In the district surveyed, both of these climatic variations have severe impacts on local agricultural production and thus on people's livelihoods. Landless, low-skilled and poor households (depending on rain-fed agriculture for both their livelihoods and food security) are most sensitive to rainfall variability and utilize rural-rural migration to manage climatic risks. Although migration allows households to survive the impacts of rainfall variability on household consumption and economy, there are negative social consequences associated with this risk management choice.

Khoa, Thao and van der Geest (2013) did research in the Thap Muoi District in Vietnam (in the Mekong Delta). They found that of the households surveyed, the majority noted the adverse effects of heavy rainfall, shifting seasonality of rainfall and a higher frequency of rainy days on crop yields and non-farm income sources: 89.5 percent of these households reported negative effects of changing rainfall patterns on household economy. Migration in Thap Muoi District is a short-term risk management strategy, if households face difficulties attaining livelihood security locally. However, the impacts on longer-term resilience can be very negative. For landless and low-skilled households, rural-rural migration can help fill household income gaps if successful, but can also interrupt skill building and education.

Murali and Afifi (2013a, 2013b) found in Jangir District in Chhattisgarh State in India that migration is one of the most important strategies employed by the residents of the research villages to cope with rainfall variations/climatic changes and food insecurity. There, migration is often the last resort for resource-poor and landless households, especially when they are unable to access or benefit from livelihood options in situ. The authors found that migration does not increase resilience or provide better long-term opportunities. Family migration keeps households intact but amplifies negative longer-term effects on livelihood diversification, for example by interrupting children's schooling and household skill building.

Households that use migration as a last resort and erosive coping strategy

Another profile of households included those for whom migration is an erosive coping strategy (i.e. one that makes them more vulnerable or prevents them from escaping poverty). These households are similar to the previous group: they are landless or land scarce, poor, and have few or no options to diversify livelihoods away from crop and livestock production. Children from these households have the same (low) level of education as their parents. Migrants from these households compete for unskilled labor in the agricultural sector (and sometimes in urban settings). The profile of such migrants in the Rainfalls research was head of household, mid-forties, married with dependents. These households are also "just getting by," and do not have access to or are unable to capture in situ adaptation or livelihood diversification options. Typical coping measures when faced with rainfall stressors on livelihoods and food availability include reducing food consumption, the quality of food consumed, selling assets or seeking help from others in the village. As these households may already have limited mobility, focus group discussions indicated that entire villages may face similar challenges and be in a poor position to help one another in times of need (co-variation of risks).

When such migrants leave during the hunger season to find food or resources to access food, household members left behind can be more vulnerable to a variety of environmental as well as social stressors. Migration both is a last resort to avoid the worst consequences of food insecurity and may require actions—such as attempting to access credit to pay for migration expenses—that leave the household deeper in poverty. Furthermore, for these populations, repeated environmental shocks and stressors—and repeated migration—erode their livelihoods, food security and asset base enough to make migration inaccessible. This pattern can be seen in small numbers in all the cases but is more pronounced in countries that generally face greater challenges with poverty and food insecurity and low livelihood diversification options for climate-sensitive sectors.

Rademacher, Schraven and Mahama (2013) found in Nadowli District in Ghana that migration at new points in time (e.g. during the main food production periods) is mainly due to livelihood and food insecurity linked to climatic and environmental factors affecting rain-fed agriculture. They found that the most important triggers of migration among households are crop production decline; rainy season shifts; unemployment; longer drought periods causing unreliable harvest; increased drought frequency. Households use migration to bridge income gaps but are often unable to improve overall well-being for those household members left behind. In Nadowli, female-headed households are more vulnerable to shifting rainfall variability, face a higher degree of food insecurity, have fewer household members of working age, possess less land and engage slightly less in migration than male-headed households.

The research by Afifi, Liwenga and Kwezi (2013) in Same District in Tanzania showed that changes in rainfall variability translate directly into impacts on food security. Surveyed households identified drought as the major hazard to livelihoods. In Same, rainfall changes affect the food production of more than 80 percent of households "a lot" and there are strong linkages between unpredictable and changing weather patterns and the decision to migrate. The top three factors affecting household migration decisions in Same are: (1) increased drought frequency; (2) longer drought periods; and (3) water shortage. While the majority of migrants are male and young, women now represent one-third of the total and out-migration from Same District is a mix of rural–rural and rural–urban migration.

Households that cannot use migration and are struggling to survive in their areas of origin

The final profile of households includes those that have been described as "trapped populations" in the literature: households that do not possess the assets necessary to migrate, even to cope with food insecurity, or who cannot access migration options. These are often landless or land scarce households in very poor areas. Characteristics of these households (or individual members within the household) include: female-headed households who may have multiple burdens of needing to care for agricultural land and care for young children or elderly; households where—often—a main breadwinner has already left the household in search of other livelihood options; households with few able-bodied workers in relation to dependents such as children, elderly or disabled persons.

These households face acute food production and consumption shortfalls when rainfall varies, and they report having too little to eat at multiple times in a given year. These households tend to have few or no diversification options, and limited migration options. For trapped households or populations, repeated environmental shocks and stressors can continue to erode their asset base and increase their food and livelihood insecurity. In Guatemala, remote, food-insecure communities face a situation where they have few good options—high sensitivity to rainfall, few local options to diversify risks or livelihoods, and migration options that are too expensive (to a major city or another country), too risky, or to places with similar challenges.

Examples of such households were found in all research areas, but the research in Guatemala provided more examples of communities with few good options either to stay or to leave. Milan and Ho (2013) shared results from Cabricán, Guatemala, showing that 97 percent of migration is motivated by attempts to secure stable household consumption and income generated by rain-fed agriculture. Rainfall affects food production of 68 percent of households surveyed. Migration opportunities (seasonal within Guatemala and long term to the US) are reducing due to decreased demand for labor and difficulties in reaching destinations. Households reported concerns about the long-term viability of their farming systems and food availability, but were also not in a position to use migration to ameliorate the risks which changes in rainfall patterns pose for household economy and household consumption. Some very vulnerable households in Bangladesh, especially women-headed households, were also found to have characteristics of "trapped populations" unable to improve their situation in situ or via migration to improve livelihood and food security (Etzold *et al.* 2013).

Modeling results and future scenarios

This section relates the four household profiles to an agent-based modeling approach applied in the Tanzanian case to explore under what scenarios rainfall variability and food security have the potential to become significant drivers of human mobility in particular regions of the world in the next two to three decades.

In order to understand the potential for rainfall to become a significant driver of human mobility in the future, it is important to identify the range of impacts that likely scenarios may have upon migration flows. By investigating the impact of rainfall variability on household- and community-level factors such as food and livelihood security, the influence of such variability on the decisions made by individual migrants can be further understood. Using the Rainfalls case study sites as examples of locations where changes in rainfall might contribute to increased food insecurity and human mobility, a process of future-oriented simulation and analysis provides a valuable opportunity to understand the circumstances under which rainfall variability might become a significant driver of migration.

The Rainfalls Agent-Based Migration Model (RABMM) represents vulnerability and migration decision making at two levels of agent analysis: the household and the individual, both of which can be generated from the household survey data collected in each case study location. The RABMM is designed to represent the degree of vulnerability of households to rainfall variability-induced changes in livelihood and food security, and the subsequent impact of these upon the migration of household members. The research identified a range of impacts that likely scenarios may have upon migration flows and showed that rainfall changes have the potential to become a significant driver of human mobility in the future.

Tanzanian results: migration from 2014–2040 under drier, wetter and extremely drier/wetter rainfall scenarios

Using the conceptual framework described above, the Tanzanian RABMM outputs the number of migrants originating from contented and vulnerable households across the case study villages (Smith 2013).

Figure 10.2 shows modeling results under the various scenarios for migrants from "resilient" households.

The results of the modeling for contented migration seen in Figure 10.2 show a much lower level of sensitivity to changes in rainfall than is the case for vulnerable migration. The mean annual normalized rate of contented migration under the scenario is 0.05, only 5 percent greater than that seen under the "average" scenario.

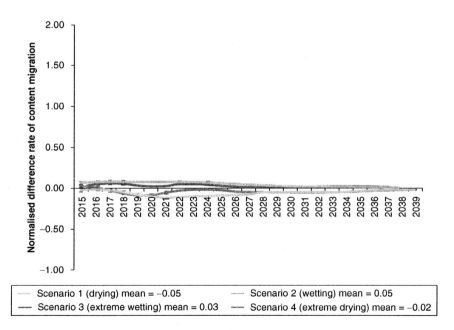

FIGURE 10.2   Normalized difference rate of content migration

Source: Smith (2013)

Note: Five-year moving averaged normalized difference in the rate of RABMM modeled contented migration. Error bars indicate the envelope of changes modeled under five member ensembles.

In contrast, Figure 10.3 shows the modeling results for "vulnerable" households in Tanzania's Same district.

The agent-based modeling results from the Rainfalls project are pertinent to discussions of crisis migration in the context of climate change: households using migration to build resilience ("contented migration") show a much lower level of sensitivity to changes in climatic patterns than is the case for vulnerable migration. Vulnerable households have a higher sensitivity to different rainfall scenarios and feel imminent needs to change their situation through migration (maybe akin to crisis migration). Changes in rainfall patterns can impact food and livelihood security in the future and have the potential to increase the vulnerability of many households worldwide.

The two figures illustrate the key finding of the modeling exercise in Tanzania and the main message from the Rainfalls study: resilient households use migration in ways that appear to reduce their sensitivity to climate stressors over time (Figure 10.2), while vulnerable households use migration in ways that either do not affect their climate sensitivity over time, or may exacerbate it through related actions such as selling land or productive assets, migration interrupting skills building and

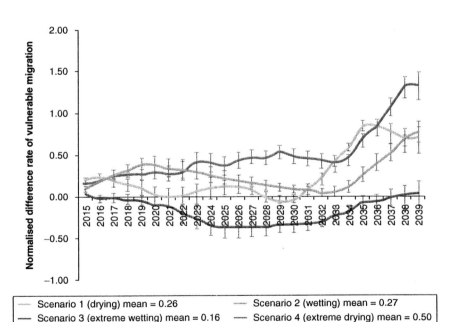

Scenario 1 (drying) mean = 0.26   Scenario 2 (wetting) mean = 0.27
Scenario 3 (extreme wetting) mean = 0.16   Scenario 4 (extreme drying) mean = 0.50

FIGURE 10.3   Normalized difference rate of vulnerable migration

Source: Smith (2013)

Note: Five-year moving averaged normalized difference in the rate of RABMM modeled vulnerable migration. Error bars indicate the envelope of changes modeled under five member ensembles.

children's education (as seen in the Indian case). For these vulnerable households, when facing scenarios of changing rainfall variability, particularly of extreme drying, migration rises notably over time.

Case study and modeling results illustrate the circumstances under which migration decisions occur—showing that both "content" and "vulnerable" households use migration, but in markedly different ways that either enhance resilience or reinforce a downward spiral of vulnerability to climatic and other stressors.

Longer-term thinking about human mobility, adaptation and development

The data reveal for the first time in a comparable global study distinct household profiles of "resilience"—those households in which migration is one of a variety of adaptation measures that progressively reduce their climate sensitivity—and "vulnerability"—those households in which migration—or an inability to migrate ("trapped populations," see Kniveton, Smith and Black 2012)—is part of a spectrum of erosive coping strategies in which sensitivity to climatic stressors is maintained or exacerbated.

The findings of this research call for a far more nuanced approach to the discussion of human mobility in the context of climate change than is currently the case. The research indicates that migration has beneficial outcomes for those affected by climate change only in households that have diverse livelihood possibilities. Other households may survive through migration but are unlikely to flourish or may even experience erosion in their living standards. Still others are trapped in place, unable to migrate to avoid the worst effects of climate change.

These findings point toward the key importance of the types and quality of adaptation measures chosen by countries. Many of these activities happen under the umbrella of economic and social development. Increasingly, a newer set of (ideally) complementary activities are emerging under the umbrella of climate adaptation efforts at the national level—spurred by international climate policy discussions, for example under the UN Framework Convention on Climate Change (UNFCCC).

Current risk management approaches (many of which include mobility) may be insufficient or inappropriate in a changed climate situation in the future. It will be important to incorporate long-term time horizons (or "climate foresight") as opposed to simple "impact/vulnerability" mapping (that results in providing short-term "coping" strategies) in adaptation planning. There is a likelihood that—at least in the medium and longer term—the humanitarian response could be overwhelmed by growing disaster-related human mobility. It is possible that disaster risk reduction and measures to avoid loss and damage may not be aligned with the rising and potentially permanent changes associated with desertification, sea level rise, ocean acidification, loss of geologic and other freshwater sources, etc., which can add pressure to human mobility. This underscores the need for new thinking about managing and planning for the impacts of climate change upon human mobility, ranging from migration to displacement to relocation. It will be increasingly important to develop approaches that consider shifts in the baseline situation of many regions. The Adaptation Committee—established under the UNFCCC in 2010 as part of the Cancún Adaptation Framework—and managers of climate finance could benefit from such longer-term perspectives when recommending or funding adaptation related activities (in this case, those that relate to mobility).

There is a need for longer-term planning mechanisms related to human mobility, which may be difficult to attain in the context of voluntary, non-binding international cooperation. It would be useful to include "transformational" adaptation strategies (see Kates *et al.* 2012) when current incremental adaptation strategies for current climatic conditions in specific locations may not suffice. For example, typical community based adaptive activities in coastal Bangladesh include providing assistance to vulnerable communities in low-lying areas with house-plinth raising to keep the houses above flood level, and rainwater harvesting to give them clean drinking water against salinity intrusion in surface and ground water supplies (Huq *et al.* 2011). A longer-term strategy would include empowering, training, and building the skills of rising generations (e.g. children) in those communities to be in a position to adjust to change (not only variability). This may include development-related resilience building such as enabling people to get climate-appropriate and/or better-paying jobs in nearby towns (over the next decade) so that they could then take the rest of their families to join them, as survival in their present location may become gradually more and more difficult. This kind of longer-term strategy—an "empowered mobility" strategy (as opposed to "forced migration" or even "planned relocation")—could center decisions on participatory processes at the community and household level around when, how, where and who moves with the household/community.

Conclusions

In summary, household vulnerability to rainfall variability affects food and livelihood security outcomes and migration choices and patterns. Households with more diverse assets and access to a variety of adaptation, livelihood diversification, or risk management options—through social networks, community or government support programs, and education—can use migration in ways that enhance resilience. Those households which have the least access to such options—few or no livelihood diversification opportunities, no land, little education—usually use internal migration during the hunger season as a survival strategy in an overall setting of erosive coping measures which leave or trap such households at the margins of decent existence.

In coming decades, the way countries manage adaptation will drive patterns of population distribution in areas of the world that are very sensitive to climate change. Such areas include mountain regions, densely populated deltas, and arid and semi-arid locations where rain-fed crop and livestock production are already under pressure. New research presented in this article suggests that the question to be asked regarding the interactions between global (and local) climatic change and human migration is not *whether* environmental drivers are the *sole* driving factors of mobility, but rather *how factors interact* to shape migration choices. A more nuanced understanding of how climatic and other variables interact to affect migration choices will help shape adaptation investments to ensure that whatever strategies households do use—including migration—*contribute to increased resilience to climate change*. Governance of human mobility must evolve to manage these changes.

Notes

1 El-Hinnawi (1985) introduced the first definition for "environmental migrants" in a United Nations Environmental Program (UNEP) report. His definition has been refined and made more comprehensive by other authors and institutions, such as the International Organization for Migration (IOM) in 2007.
2 Jäger *et al.* (2009) synthesized the results of the "Environmental Change and Forced Migration Scenarios" project (EACH-FOR, www.each-for.eu)—the first global survey of its kind employing fieldwork to investigate environmental change and migration in 23 case studies; Warner *et al.* (2009, "In Search of Shelter") brought EACH-FOR results to policy makers, particularly in the UNFCCC process.
3 This section draws on new findings from the project "Where the Rain Falls: Understanding Relationships between Changing Rainfall Variability, Food and Livelihood Security, and Human Mobility" (Rainfalls), undertaken by the UN University Institute for Environment and Human Security and CARE. The Rainfalls work is supported by the AXA Group and the John D. and Catherine T. MacArthur Foundation. It explores the interrelationships among rainfall variability, food and livelihood security, and human mobility in a diverse set of research sites in eight countries. While climate change affects nearly all aspects of food security—from production and availability, to the stability of food supplies, access to food and food utilization (Schmidhuber and Tubiello 2007)—the Rainfalls research focused on linkages between shifting rainfall patterns and food production and the stability of food supplies (Jennings and McGrath 2009). The central focus of the initiative was to explore the circumstances under which households in eight case study sites in Latin America, Africa, and Asia use migration as a risk management strategy when faced with rainfall variability and food and livelihood insecurity (see www.wheretherainfalls.org and www.ehs.unu.edu).

References

Afifi, T. (2011) "Economic or Environmental Migration? The Push Factors in Niger," *International Migration*, 49(1): e95–e124 (International Organization for Migration, Special Issue, Oxford, UK: Wiley Online Library).

Afifi, T., Liwenga, E. and Kwezi, L. (2013) "Rainfall Induced Crop Failure, Food Insecurity and Outmigration in Same-Kilimanjaro, Tanzania," *Climate and Development*. DOI:10.1080/17565529.2013.826128.

Betts, A. (2010) "Survival Migration: A New Protection Framework," *Global Governance*, 16(3): 361–382.

Black, R., Bennett, S.R.G., Thomas, S.M. and Beddington, J.R. (2011) "Climate Change: Migration as Adaptation," *Nature*, 478(7370): 447–449.

Brown, O. (2008) *Migration and Climate Change*, International Organization for Migration (IOM): Research Series No. 31. Geneva: IOM.

El-Hinnawi, E. (1985) "Environmental Refugees," United Nations Environment Programme, Nairobi.

Etzold, B., Ahmed, A., Hasan, S. and Neelormi, S. (2013) "Clouds Gather in the Sky, but No Rain Falls: Vulnerability to Rainfall Variability and Food Insecurity in Northern Bangladesh and its Effects on Migration," *Climate and Development*. DOI: 10.1080/17565529.2013.833078.

Government Office for Science (2011) *Foresight: Migration and Global Environmental Change: Future Challenges and Opportunities*, Final Project Report, London: Government Office for Science.

Hugo, G. (2008) *Migration, Development and Environment*, Geneva: International Organization for Migration.

Huq, S., Warner, K., Loster, T., Rhyner, J. and Harmeling, S. (2011) Personal communication about community based adaptation, short-term adaptation vs. longer-term transformational adaptation, July 21, Dhaka.

International Organization for Migration (IOM) (2008) "Discussion note: Migration and environment" Ninety-fourth session, MC/INF/288, 1 November 2007. IOM, Geneva.

Jäger, J., Frühmann, J., Grünberger S. and Vag, A. (2009) "Environmental Change and Forced Migration Scenarios Project Synthesis Report," Deliverable D.3.4 for the European Commission.

Jennings, S. and McGrath, J. (2009) "What Happened to the Seasons?" Oxfam GB Research Report, UK.

Jónsson, G. (2010) "The Environmental Factor in Migration Dynamics: A Review of African Case Studies," Working Paper 21, Oxford: International Migration Institute.

Kates, R.W., Travis, W.R. and Wilbanks, T.J. (2012) "Transformational Adaptation When Incremental Adaptations to Climate Change are Insufficient," *Proceedings of the National Academy of Sciences*, 109: 7156–7161.

Khoa, N.V., Thao, N.C. and van der Geest, K. (2012) "Where the Rain Falls" project. Case study: Viet Nam. Results from Dong Thap Province, Thap Muoi District. Report No. 8. Bonn: United Nations University Institute for Environment and Human Security (UNU-EHS).

Kniveton, D.R., Smith, C.D. and Black, R. (2012) "Emerging Migration Flows in a Changing climate in Dryland Africa," *Nature Climate Change*, 2: 444–447.

Laczko, F. and Aghazarm, C. (eds) (2009) *Migration, Environment and Climate Change: Assessing the Evidence*, Geneva: International Organization for Migration.

Martin, P. (2010) "Climate Change, Agricultural Development, and Migration," Background Paper for the Transatlantic Study Team on Climate Change and Migration, German Marshall Fund.

Martin, S. (2010) "Rethinking the International Refugee Regime in Light of Human Rights and the Global Common Good," in Hollenbach, D. (ed.), *Driven from Home: Protecting the Rights of Forced Migrant*, Washington, DC: Georgetown University Press.

Milan, A. and Ruano, S. (2013) "Rainfall Variability, Food Insecurity, Migration and Trapped Populations in Cabricán, Guatemala," *Climate and Development* (forthcoming).

Milan, A. and Ho, R. (2013) "Livelihood and Migration Patterns at Different Altitudes in the Central Highlands of Peru," *Climate and Development*. DOI: 10.1080/17565529.2013.857589.

Morrissey, J. (2009) "Environmental Change and Forced Migration: A State of the Art Review," Refugee Studies Centre Background Paper, Oxford: Refugee Studies Centre.

Murali, J. and Afifi, T. (2013a) "Rainfall Variability, Food Security and Human Mobility in the Janjgir-Champa District of Chhattisgarh State, India," *Climate and Development*. DOI: 10.1080/17565529.2013.867248.

Murali, J. and Afifi, T. (2013b) "Where the Rain Falls" Project. Case Study: India. Results from Janjgir-Champa District, Chhattisgarh State. Report No. 4. Bonn: United Nations University Institute for Environment and Human Security (UNU-EHS).

Myers, N. (2005) "Environmental Refugees: An Emergent Security Issue," 13th Economic Forum, Prague, May 23–27.

Rademacher-Schulz, C., Afifi, T., Warner, K., Rosenfeld, T., Milan, A., Etzold, B. and Sakdapolrak, P. (2012) "Rainfall Variability, Food Security and Human Mobility. An Approach for Generating Empirical Evidence," *Intersections*, No. 10, Bonn: UNU-EHS.

Rademacher-Schulz, C., Schraven, B. and Mahama, E.S. (2013) "Time Matters: Shifting Seasonal Migration in Northern Ghana in Response to Rainfall Variability and Food Insecurity," *Climate and Development*. DOI: 10.1080/17565529.2013.830955.

Sakdapolrak, P., Promburom, P. and Reif, A. (2013) "Why Successful In-situ Adaptation with Environmental Stress does Not Prevent People from Migrating? Empirical Evidence from Northern Thailand," *Climate and Development*. DOI: 10.1080/17565529.2013.830955.

Schmidhuber, J. and Tubiello, F.N. (2007) "Global Food Security Under Climate Change," *Proceeding of the National Academy of Science*, 104: 19703–19708 (doi:10.1073/pnas.0701976104).

Smith, C. (2013) "Modeling Migration Futures: Development and Testing of the Rainfalls Agent-based Migration Model–Tanzania," *Climate and Development*. DOI: 10.1080/17565529.2013.872593.

Tacoli, C. (2009) "Crisis or Adaptation? Migration and Climate Change in a Context of High Mobility," in Guzmán, J.M., Martine, G., McGranahan, G., Schensul, D. and Tacoli, C. (eds.), *Population Dynamics and Climate Change*, New York: UNFPA/London: IIED.

Warner, K. and Afifi, T. (2013) "Where the Rain Falls: Evidence from Eight Countries on the Circumstances under which Households Use Migration to Manage the Risk of Rainfall Variability and Food Insecurity," *Climate and Development*. DOI: 10.1080/17565529.2013.835707.

Warner, K., Afifi, T., Henry, K., Rawe, T., Smith, C. and de Sherbinin, A. (2012) "Where the Rain Falls: Climate Change, Food and Livelihood Security, and Migration," Global Policy Report of the Where the Rain Falls Project, Bonn: CARE France and UNU-EHS.

Warner, K., Ehrhart, C., de Sherbinin, A., Adamo, S. and Chai-Onn, T. (2009) *In Search of Shelter: Mapping the Effects of Climate Change on Human Migration and Displacement*, CARE/CIESIN/UNHCR/UNU-EHS/World Bank, Geneva: United Nations University, Institute for Environment and Human Security.

11

COMMUNITY RELOCATIONS

The Arctic and South Pacific

Robin Bronen

Introduction

The Intergovernmental Panel on Climate Change (IPCC) predicts that climate change will displace 150 million people by 2050. Erosion, flooding and sea level rise will be the primary causes. In the Arctic, increased temperatures are melting sea ice rapidly. Record minimum levels have been recorded since 2007. These climate-induced biophysical changes threaten coastal communities. Connected by the Pacific Ocean, the Newtok Traditional Council (NTC) in Alaska and Tulele Peisa, a Carteret Islands non-governmental organization (NGO) in Papua New Guinea, are mobilizing their communities to relocate because the ocean that provides them food now eats their shorelines. In Newtok, thawing permafrost and increased storm surges accelerate erosion, causing homes and community infrastructure to fall into the water. In the Carterets, king tides inundate the land, creating swamps where malarial mosquitoes breed. Areas that previously held food gardens are now underwater. Saline intrusion destroys the drinking water supply in both communities.

This chapter examines permanent community relocation through the case study analysis of two communities: Newtok, Alaska, and the Carteret Islands in the Autonomous Region of Bougainville in Papua New Guinea. In the Newtok and Carteret Islands, climate change adaptation means the permanent relocation of the entire community. In both communities, the local governing tribal councils chose to relocate because the traditional disaster relief reduction strategies of sea wall construction have not been able to prevent storm surges, flooding and accelerated rates of erosion from endangering residents and destroying the habitability of each community. The policy and practical challenges have been enormous. The people of the Newtok and Carteret Islands are facing a humanitarian crisis because of the lack of an institutional framework that can guide each community's relocation. Despite these challenges, each is creating community-based relocation strategies to respond to the climate-induced biophysical changes threatening people's lives. Their experience demonstrates the need to create a relocation governance framework, based in human rights doctrine, which can guide both local communities and tribal, local, regional and national government action when climate-induced biophysical change threatens community habitability and the lives of residents.

Climate change and population displacement

Climate-induced biophysical change threatens the lives, livelihood, homes, health, and basic subsistence of human populations that have inhabited the Arctic and small islands in the tropical and subtropical ocean for millennia. Relocation may be the best adaptation response for communities whose current location is uninhabitable or vulnerable to future climate-induced biophysical threats. Relocation is a process whereby a community's residents, housing and public infrastructure are reconstructed in another location (Abhas 2010).

In Alaska, dozens of communities located along the navigable waters of the coasts and rivers are threatened by accelerated rates of erosion and flooding. Some indigenous communities have determined that relocation is the only solution that will protect them from these climate-induced environmental changes. In 2003, the US Government Accountability Office (GAO) investigated the community relocations occurring in Alaska, finding that three communities wanted to relocate and that erosion affected 184 communities (GAO 2003). Six years later, in 2009, the GAO supplemented its 2003 report and found that the number of communities seeking to relocate quadrupled to twelve (GAO 2009). Despite the recognition that these Alaskan communities are imminently threatened by climate change and relocation is the only adaptation strategy that can protect them, no institutional community relocation framework exists in the United States. The institutional and statutory barriers to relocation have challenged traditional governance responses to extreme weather events. Despite these challenges, one community, Newtok, located in western Alaskan on the Bering Sea, is actively relocating. The NTC is leading the relocation effort for the 360 residents of the community.

Papua New Guinea is a country located in the southern Pacific Ocean, comprising low-lying islands or coral atolls surrounded by extensive coral reefs. Atolls are long, thin landmasses, which average approximately two meters above sea level, with little change in elevation (Holthus *et al.* 1992). The combination of extreme weather events that create storm surges and flooding, with sea level rise, threatens these islands. Due to the small size and low elevation of many inhabited islands, whole atolls have become unsafe or unsuitable to permanent habitation because migration away from the coast is either not feasible or provides limited protection (Holthus *et al.* 1992). As in the Arctic, climate change is aggravating coastal erosion, which damages people's homes, exacerbates pressure on scarce land resources, and increases the vulnerability of island populations to extreme weather events (UNDP 2009, 5).

Several atoll communities, including those located on Carteret, Mortlock, Tasman and Fead islands, need to relocate because of these changes. These island groups are located in eastern Papua New Guinea, to the east of Buka Island, with

Tasman and Mortlock directly east, the Carteret Islands to the northeast and Fead Islands to the north of the Carterets (O'Collins 1988). The Carteret Islands are located 115 kilometers from Buka Island. The NGO, Tulele Peisa, is working with the clan elders of the Carteret Islands, to facilitate the islanders' relocation (Tulele Peisa 2009). As in Alaska, there is no institutional relocation framework in Papua New Guinea or the Autonomous Region of Bougainville (the regional government for these island communities), to guide the relocation effort and prioritize government efforts amongst these communities.

Climate-induced environmental changes impacting habitability in Alaska and the Carteret Islands

Article 1(2) of the UN Framework Convention on Climate Change (UNFCCC) defines "climate change" as an alteration in "the composition of the global atmosphere . . . which is in addition to natural climate variability observed over comparable time periods." Climate change is most often associated with temperature changes in the earth's atmosphere. According to the IPCC, the global average of land and sea surface temperature has increased since 1861 (IPCC 2001).

Nine of the ten warmest years in the 132-year temperature record occurred during the twenty-first century. Only one year during the twentieth century, 1998, was warmer than 2011 (NOAA 2011). The warmest year on record in the United States is 2012, with 2010 and 2005, which were 1.15 degrees Fahrenheit above average, the second warmest years (NOAA 2013). These temperature changes impact the hydrosphere, cryosphere, atmosphere and biosphere. As a consequence, numerous and diverse climate-induced environmental changes will occur and affect the totality of the environment where humans exist. Current and projected changes include: contraction of snow-covered areas and shrinking of sea ice; sea level rise and higher water temperatures; heavy precipitation events; and increased intensity of typhoons and hurricanes (IPCC 2007a, 8, 14).

Rising sea-surface temperatures will cause an increased intensity of storms and cyclones, with both higher peak wind speeds and heavier levels of precipitation (IPCC 2007a, 30). Storms can create atmospheric pressure changes, which can lead to sudden increases in sea level height, creating large waves. Combined with high tide levels, these waves may reach up to 2.3 meters and can inundate coastal communities (UNDP 2006, 39–51).

Sea level rise and thawing permafrost are two climate-induced environmental changes that, combined with the loss of natural coastal barriers, such as coral reefs, severely impact the habitability of communities in the Arctic and Papua New Guinea. Sea level rise is accelerating and expected to worsen over the next century due to increased rates of ice sheet mass loss from Antarctica and Greenland (IPCC 2007a, 47; Nicholls and Cazenave 2010). Increased temperatures, causing ocean water to expand and glaciers to melt, will contribute to sea level rise (Nicholls and Cazenave 2010). Sea level rise will contribute to flooding, sea surges, erosion and salination of land and water (IPCC 2007b, 689; IPCC 2012).

Permafrost—permanently frozen soil which is the glue that makes land habitable in the Arctic—is thawing due to warming temperatures. The 2007 IPCC physical science report has documented that the temperature of the top layer of permafrost has increased by up to 37 degrees Fahrenheit since the 1980s (IPCC 2007c, 339).

Warming temperatures also affect ocean ecosystems and cause a loss of the natural barriers that protect coastal communities from sea surges, erosion and floods. In the Arctic, sea ice is decreasing in thickness and extent with record minimum levels of Arctic sea ice have been recorded since 2002 (Serreze 2008/2009, 3–4). Scientific observations of the Arctic sea ice extent during the summer of 2007 documented a new record low, with 23 percent less ice coverage measured than the previous record of September 2005, a loss equivalent to the size of California and Texas combined (Serreze 2008/2009, 3–4). This trend continues, with 2012 marking the lowest Arctic sea ice extent on record (NSIDC 2012).

The decrease in extent of Arctic sea ice coupled with warming temperatures has caused a delay in freezing of the Bering and Chukchi Seas (Shulski and Wendler 2007, 122, 124). Near the shore, pack ice has historically provided a protective barrier to coastal communities (Shulski and Wendler 2007, 122, 124). Since the 1980s the Arctic seas have been remaining ice-free approximately three weeks longer in the autumn (Hufford and Partain 2005, 1). The delay in freezing of the Arctic seas has left many communities exposed to the autumnal storms that originate in the Pacific, and occur primarily between August and early December (Hufford and Partain 2005; Shulski and Wendler 2007, 122, 124). The loss of Arctic sea ice combined with thawing permafrost is causing severe erosion and storm surges.

In the tropical and subtropical oceans, coral reefs and mangroves protect coastal communities from extreme weather events and storm surges. Depending on the health and physical and ecological characteristics, such as fragmentation, coral reefs can absorb approximately 70–90 percent of the energy of wind-generated waves (United Nations Environment Programme (UNEP) World Conservation Monitoring Centre 2006, 5). The roots and stems of mangroves also dissipate the energy and size of waves (UNEP World Conservation Monitoring Centre 2006, 15).

Coral reef mortality and reef degradation have increased dramatically in the past 20–50 years (IPCC 2007b). Temperature increases exacerbate the decrease in healthy coral reefs worldwide (IPCC 2007b). Ocean acidification also impacts the health of coral reefs, which are the foundation for all atolls. The ocean is becoming more acidic due to the sequestration of the carbon dioxide in the atmosphere. As the ocean acidifies, coral reefs will stop growing and may start to dissolve, destroying a critical natural barrier for coastal communities. Mangroves can also lose their ability to protect coastal communities because of extreme weather events, which can damage their root system (UNEP World Conservation Monitoring Centre 2006, 6).

Climate-induced environmental change and human mobility

Because of these disparate climate-induced environmental changes, individuals and communities will be forced to migrate. The climate change drivers of migration can be segregated into three distinct categories: extreme weather events, such as hurricanes and cyclones; the depletion of ecosystem services by ongoing environmental change, such as drought, thawing permafrost and sea level rise; and the combination of repeated extreme weather events and ongoing environmental changes that accelerate and are exacerbated by these extreme weather events (Bronen 2010).

Each of these climate change drivers will cause distinct patterns of human migration, based on a combination of the temporal nature of the migration and the demographics of the population movement. These patterns can also be separated into three categories. First, climate-induced ongoing environmental change, such as drought, sea level rise, and thawing permafrost, will initially most often affect people at the individual or household levels rather than create a mass population exodus. With this type of migration, socio-economic factors may play a more dominant role in the decision to move, or remain in place, because households that have more resources will be able to move, if desired, to escape the impact of the environmental change (Bronen 2010). Extreme weather events will most often cause mass displacement where entire communities are forced to evacuate, generally for a temporary period. Due to direct and imminent threats, socio-economic factors at the individual or household levels will be less influential to the decision to evacuate although they may influence the ability to do so, resulting in some populations requiring assistance to move out of harm's way. Finally, mass displacement may occur, where entire communities are forced to relocate permanently because of the combination of extreme weather events coupled with ongoing and accelerating environmental change that causes community locations to become entirely uninhabitable. In this situation, it is also unlikely that individual and household socio-economic factors will play a critical role in the decision to relocate. This situation is similar to the displacement caused by the construction of dams, where an entire location is inundated with water and the financial resources available to an individual or household will not prevent the displacement.

With respect to each type of migration, strategies are required to ensure that institutional responses are appropriate and human rights are protected. Extreme weather events often result in humanitarian crises and displace large numbers of people. The Hyogo Framework for Action, endorsed by the UN General Assembly, encouraged countries to build disaster resilience and develop and implement a systematic disaster risk management approach, and created a ten-year plan to reduce loss of lives and social, economic and environmental assets due to disasters by 2015. National governments play a key role in implementing the Hyogo Framework through the established financial and organizational mechanisms of governing and managing disaster risks (IPCC 2012). The ability of governments to implement disaster risk management responsibilities differs significantly across countries, depending on their capacity and resource constraints, but in the majority of countries, national systems have been strengthened by applying the principles of the Hyogo Framework to mainstream risk considerations across society and sectors (IPCC 2012).

Extreme weather events can cause populations to be trapped, perhaps because of inadequate or discriminatory implementation of disaster risk reduction (DRR) and response activities, or because the population has no higher ground to which they can evacuate. This is the situation in both the Carteret Islands and Newtok, where the community locations are minimally above sea level and no higher ground is within close proximity to allow for temporary evacuation.

Slow ongoing environmental change is not traditionally included within the scope of the humanitarian institutional response to disasters because these environmental changes do not normally create the emergency mass population displacement typically associated with humanitarian crises. Drought is the sole exception. Erosion, which is the primary cause for population displacement in Alaska and the Carteret Islands, is an example of a slow ongoing environmental change that does not typically cause mass population displacement and is not included in the category of environmental events that warrant a disaster response from the humanitarian community. In the United States, erosion is not included in the list of environmental events defined by statute in the Stafford Act, and consequently communities like Newtok are not eligible for funding for disaster relief unless a storm causes severe flooding and inundates the community (Bronen 2011). DRR and development-related institutional responses may provide appropriate mechanisms to respond, through erosion control, for example, and either prevent displacement or facilitate migration on an individual or household level. Populations in this situation could often be trapped if households or individuals do not have the resources to adapt to these environmental changes and the institutional response does not alleviate the impacts.

This chapter focuses on the third type of displacement, permanent community relocations, caused by extreme weather events combined with gradual ongoing environmental changes, such as erosion and thawing permafrost. The extreme weather events create an immediate humanitarian crisis and can cause the displacement of entire communities. In this situation, the traditional humanitarian response is to repair and rebuild in the community's original location and implement DRR strategies so that community residents can be protected in place. However, socioeconomic, demographic, and health-related differences and differences in governance, access to livelihoods, entitlements, and other factors influence local coping and adaptive capacity, and pose disaster risk management and adaptation challenges from the local to national levels.

As stated above, slow, ongoing environmental changes, such as erosion, may be a catalyst for DRR activities and sustainable development policies. However, when extreme weather events combine with ongoing environmental change, traditional preventive DRR strategies, such as erosion control, may not be able to protect populations in place. As a consequence, entire communities become uninhabitable and community relocation may be the only adaptation strategy that can protect these populations.

The case studies of Newtok, Alaska, and the Carteret Islands, Papua New Guinea, demonstrate the ways these changes can affect the habitability of a community, threaten people's lives and cause them to choose community relocation as the only adaptation strategy that can protect them from these climate change impacts.

Communities impacted by climate-induced ecological change

Alaska

Newtok, a Yup'ik Eskimo village, is located near the Bering Sea in western Alaska. Approximately 320 residents reside in about sixty houses (Cox 2007). The Ninglick River borders Newtok to the south; to the east is the Newtok River (USACE 2006). A combination of increased temperatures, thawing permafrost and decreased Arctic sea ice is causing accelerating erosion, moving the Ninglick River closer to the village (Cox 2007). The State of Alaska spent about $1.5 million to control the erosion between 1983 and 1989 (USACE 2008a). Despite these efforts, erosion associated with the movement of the Ninglick River is projected to reach the school, the largest structure in the community, by about 2017 (USACE 2008a).

Six extreme weather events between 1989 and 2006 exacerbated these gradual ecological changes. Five of these events precipitated Federal Emergency Management Agency (FEMA) disaster declarations (ASCG 2008). FEMA declared three disasters between October 2004 and May 2006 alone (ASCG 2008). These three storms accelerated the erosion and repeatedly "flooded the village water supply, caused raw sewage to be spread throughout the community, displaced residents from homes, destroyed subsistence food storage, and shut down essential utilities" (USACE 2008a, 5). Public infrastructure that was significantly damaged or destroyed included the village landfill, barge ramp, sewage treatment facility and fuel storage facilities (Bronen 2011). The only access to the community is by barge during the summer or by airplane. The barge landing, which allows for most delivery of supplies and heating fuel, no longer exists, creating a fuel crisis. Salt water is affecting the potable water (Cox 2007).

In 1994, the NTC analyzed six potential relocation sites to start a relocation planning process. In September 1996, Newtok inhabitants voted for the first of three times, to relocate to Nelson Island, nine miles south of Newtok (Cox 2007). Subsequent votes occurred in May 2001, and most recently in August 2003. Newtok obtained title to their preferred relocation site, which they named Mertarvik, through a land-exchange agreement negotiated with the US Fish and Wildlife Service. Construction of pioneer infrastructure, including a multi-purpose evacuation center and barge landing, began at the relocation site in 2009 (Bronen 2011).

State, federal, and tribal government and non-governmental agencies have issued numerous reports documenting the social-ecological crisis faced by Newtok residents and the habitability of the relocation site (Bronen 2011). These reports are a model for the type of documentation needed to demonstrate that relocation is the only feasible solution to protect community residents from climate-induced environmental change.

The NTC commissioned the oldest report, in 1984, which documented the impact of erosion on their community (Bronen 2011). Twenty years later, in 2004, the NTC commissioned a second report, the Newtok Background for Relocation Report, which summarized the previous erosion studies, mapped the advancing Ninglick River to show the scope of erosion, documented the social-ecological impacts of erosion on the village, and developed a tentative timeline for the short-term and long-term relocation of residences (ASCG 2004) The report also described the NTC's evaluation of each potential village relocation site, including "collocation" to one of four existing communities or relocation to one of six potential new sites in the region (ASCG 2004). In addition, it contained the results of the 2003 resident survey, which asked Newtok residents to vote on relocation alternatives (ASCG 2004). This second report was instrumental to the community's relocation effort because it provided background documentation to government agencies and officials to justify the village's relocation efforts and to support the NTC's requests for government assistance in this process (Cox 2007).

In addition, the US Army Corps of Engineers (USACE or the Corps of Engineers) funded a report in 2008 that analyzed five alternative responses to the social and ecological crisis facing Newtok village residents (USACE 2008b). These alternatives included: taking no action; staying in place with erosion and flood control; collocation; relocation funded and orchestrated solely by the Corps of Engineers; and a collaborative relocation effort (USACE 2008b). The report found that a coordinated relocation effort was in the best interests of Newtok residents, explaining:

> With no Federal and state action, relocation efforts will be piecemeal and uncoordinated and will increase ultimate costs many times over a coordinated, efficient relocation plan. Local efforts will take many years and the existing significant risk to health, life, and property will continue in Newtok. The disintegration of these people as a distinct tribe may result from splitting the community in two or more locations for many years as they relocate under their own efforts.
>
> *(USACE 2008b, 15)*

The Corps of Engineers also specifically rejected the collocation alternative, finding that "[c]ollocation would destroy the Newtok community identity" (USACE 2008b, 16).

It also issued several reports that evaluated the habitability of Newtok's relocation site and confirmed the NTC's conclusion that Mertarvik is a suitable relocation site (USACE 2008a, 2008b).

The Newtok Planning Group (NPG) emerged in May 2006 from an ad hoc series of meetings. It is unique in Alaska in its multi-disciplinary and multi-jurisdictional structure, and consists of about twenty-five state, federal, and tribal governmental and non-governmental agencies that all voluntarily collaborate to facilitate Newtok's relocation (Bronen 2011). From the NPG's inception, the NTC has led the relocation effort. On June 9, 2011, the NTC unanimously approved a set of guiding principles ("Maligtaquyarat") for the community's relocation to Mertarvik (AgnewBeck 2012).

These guiding principles are based on the Yup'ik way of life and are as follows.

- Remain a distinct, unique community—our own community.
- Stay focused on our vision by taking small steps forward each day.
- Make decisions openly and as a community and look to elders for guidance.
- Build a healthy future for our youth.
- Our voice comes first—we have first and final say in making decisions and defining priorities.
- Share with and learn from our partners.
- No matter how long it takes, we will work together to provide support to our people in both Mertarvik and Newtok.
- Development should:

  - Reflect our cultural traditions.
  - Nurture our spiritual and physical wellbeing.
  - Respect and enhance the environment.
  - Be designed with local input from start to finish.
  - Be affordable for our people.
  - Hire community members first.
  - Use what we have first and use available funds wisely.

- Look for projects that build on our talents and strengthen our economy.

(AgnewBeck 2012, no page)

These guiding principles govern every aspect of the relocation process and have been integrated into the strategic relocation master plan, which guides federal and state government participation in the relocation effort (AgnewBeck 2012).

The institutional and statutory barriers to relocation are enormous. With no designated relocation funding, each government agency must follow its own budgetary and funding prioritization criteria to allocate funding for Newtok's relocation effort. The lack of a population base at the relocation site and the uncertainty of when and if people will be able to relocate because of the lack of any infrastructure has impacted Newtok's ability to receive state funding for capital expenditures at their relocation site. For example, Newtok's request for state funding for an airstrip, the only means of transportation to the relocation site, was initially placed at a lower prioritization than other communities seeking this funding where people are already living and need state funding to maintain and rebuild this infrastructure.

Despite these challenges, the NPG has been exceptionally creative in finding existing funding sources to support the relocation effort and in coordinating different funding streams to meet cost sharing eligibility requirements as well as increasing the revenue available for the relocation effort. The NPG has also been hugely successful in designing and implementing a strategic relocation plan that upholds the individual human rights of Newtok residents as well as the collective human rights of the indigenous community. However, due to the lack of an institutional framework, the relocation process remains painfully slow, at great risk to Newtok residents as erosion continues to accelerate and threaten them.

The Carteret Islands

The Carteret Islands are within the jurisdiction of the Autonomous Bougainville Government (ABG)—the government of the Autonomous Region of Bougainville, Papua New Guinea, a collection of dozens of islands in eastern Papua New Guinea. The government was established in 2000 following a peace agreement between the Government of Papua New Guinea and the Bougainville Revolutionary Army (Boege 2011).

Originally the Carteret Islands comprised six atolls, but the sea has divided one, so that seven atolls, which are 1.2 meters above sea level, now encircle the lagoon (Boege 2011). Five of these atolls—Han, Piul, Iangain, Ialassa and Iesila—are inhabited (O'Collins 1988).

Similar to Newtok, erosion has plagued the Carteret Islands for decades. Despite the construction of sea walls and planting of mangroves to protect against the sea, Carteret Islanders state that as well as one of the atolls dividing, more than 50 percent of their land has eroded into the sea since 1994 (Boege 2011). Salt water has flooded the gardens on the atolls for many years, which is causing a food shortage (Boege 2011). As early as 1964, government officials recognized the inadequacies of sea walls to protect the Carteret Islanders from the effects of flooding and erosion (O'Collins 1988).

During the 1970s, the Carteret Islanders and the provincial government began planning for relocation to Bougainville Island. However, finding culturally appropriate land for the relocation, which would allow for the growing of subsistence foods and access to fishing, proved difficult and delayed the implementation of the relocation scheme (O'Collins 1988). There was also concern over conflicts with local fishing communities that would be impacted by resettled islanders' access to their fishing grounds, and inadequate land for subsistence gardens (O'Collins 1988).

The relocation scheme began to be implemented in the 1980s when ten Carteret families moved to the relocation site on Bougainville Island. These families had to adjust to dwellings constructed with non-culturally appropriate materials and consumption of non-traditional foods, while the children had to walk a long distance to attend schools along roads with motor vehicle traffic to which they were unaccustomed (O'Collins 1988). In addition, the relocation scheme did not include economic development activities to assist the relocated families' transition to a cash economy, causing them to depend on government assistance to survive (O'Collins 1988). As a result of these challenges, at least three of the initial ten families returned to the Carteret Islands within a few years of relocation. The relocation scheme was eventually abandoned in the late 1980s when civil war broke out on Bougainville Island, and many families returned to the Carteret Islands or dispersed to other parts of Bougainville Island.

A new effort to relocate the Carteret Islanders began in 2006. At the same time that the NTC began leading the NPG to address the accelerated rates of erosion in Newtok, the Council of Elders of the Carteret Islands held a series of meetings to discuss the worsening impact of sea surges on their islands. They decided that the Carteret people needed their own indigenous organization to plan and implement a voluntary relocation program. In September 2007, they formed a local the NGO, called Tulele Peisa, "Sailing the Waves on Our Own," under the Investment Promotion Authority in Papua New Guinea. The Tulele Peisa vision is "To maintain our cultural identity and live sustainably wherever we are" (Tulele Peisa 2009, 5).

Tulele Peisa developed the Carteret Integrated Relocation Project (CIRP), a community-led relocation model, to coordinate the voluntary relocation of 1,700 Carteret Islanders to Bougainville (Displacement Solutions 2008). Tulele Peisa is dedicated to ensuring that the relocation process is culturally appropriate for the Carteret Islanders who are relocating and for the host communities of the relocation sites (Displacement Solutions 2008). Since 2007, Tulele Peisa has conducted landholder negotiations with the Catholic Church, which resulted in the church allocating eighty-one hectares of land to the Carteret Islanders (Boege 2011). The location of the relocation site is critical because Tulele Peisa seeks to ensure there is sufficient land with secure tenure available for the Carteret families to be economically self-sufficient and have secure food resources (Displacement Solutions 2008).

Working with the host communities has been a critical component of the relocation program because of the complexities involved in relocating people from their traditional homelands to integrate into existing communities that are geographically, culturally, politically and socially different (Displacement Solutions 2008). Work with the host communities began with exchange programs involving chiefs, women and youth to raise awareness about the rising sea levels on the Carteret atolls and the impact on the islanders. From these exchanges, Tulele Peisa, in association with the ABG, the Catholic Church and Government Health and Education Agencies, worked with the host communities on Bougainville to ensure that there is adequate land, infrastructure and economic opportunities to include the Carteret peoples within the host communities (Tulele Peisa 2009). CIRP seeks to ensure that the host communities will benefit from the relocation through upgrading of health facilities and schools and through training and resources for new income generation opportunities (Tulele Peisa 2009).

Similar to the NTC, Tulele Peisa is working with and coordinating the efforts of a range of local, national and international organizations, which are providing technical support for the relocation program. Tulele Peisa is making the decision about the types of technical assistance they need to move their relocation effort forward so that the relocation process is culturally appropriate and supports the islanders' long-term self-sufficiency at the new site. Technical assistance programs include training on food security, organic gardening, financial management training, community health programs, home management skills and land use management. Maintaining access to their traditional fishing grounds is a critical component of the relocation effort so that people can still rely on this food source even though they no longer reside on the Carteret Islands (Tulele Peisa 2009).

In 2007, the ABG also developed a relocation plan, separate from the efforts of Tulele Peisa, for the Carteret Islands. Since developing this plan the ABG has been in protracted negotiations with landowners to secure land for the relocation effort. Unlike the efforts of the NPG in Alaska, the work of the ABG and Tulele Peisa is not coordinated because the Bougainville authorities have not decided how to work with Tulele Peisa.

In both the United States and Papua New Guinea, the lack of an institutional relocation framework has significantly delayed the relocation efforts of both the NTC and Tulele Peisa. Both have had to navigate enormous barriers and challenges to move the relocation effort forward. However, despite the lack of a framework in the United States and Alaska, the state and federal governments have figured out ways to support the relocation efforts of the NTC. As a consequence, the NTC can rely on the state and federal government agencies to assist them with funding requests to the state and federal legislature and also for the technical support to construct the infrastructure needed at their relocation site. In comparison, Tulele Peisa has not received similar support from national, regional and local governmental agencies and has had to rely on the international community for funding and technical assistance.

Climate-induced community relocations require an adaptive governance framework based in human rights

Climate-induced community relocations present one of the most significant challenges for governments responsible for protecting their citizens and for those displaced. This type of permanent population displacement, when community relocation is required to protect residents from climate-induced environmental changes, which alter ecosystem services permanently, cause extensive damage to public and private infrastructure, and repeatedly place people in danger, is described as *climigration* (Bronen 2010; Bronen and Chapin 2013).

The community relocations of Newtok and the Carteret Islands demonstrate the complexity of this process. Population relocation can cause immense economic, social and cultural impacts. As Hodgkinson *et al.* (2008) state, "displacement is a form of adaptation that creates particular vulnerabilities requiring protection." In order to protect the collective human rights of communities, as well as the individual human rights of community residents forced to relocate because of climate-induced environmental change, an adaptive governance framework based in human rights doctrine needs to be created.

Enormous protection gaps exist for the communities and community residents forced to leave their homes permanently. The protection gaps include legal protection for residents of small island states, such as the Maldives, where no higher ground exists within the country (Maldives 2008). Protection gaps also include the ability to exercise collective and individual rights; the physical safety and security of community residents; long-term durable solutions which require permanent community relocation; and humanitarian and development assistance and services.

In Newtok, Alaska, and the Carteret Islands, Papua New Guinea, community residents made the decision that the permanent relocation of their entire community was the only adaptation strategy that could protect their lives. In addition, both the NTC and Tulele Peisa are determined to create self-sufficient communities that do not need government handouts. Workforce development is a key component of each community's relocation plan. Despite government acknowledgement and concurrence that community relocation is the only adaptation strategy that can protect residents' lives, both communities have experienced enormous difficulty accessing the funding and resources they need to implement the relocation.

Each community's relocation experience demonstrates that four factors compel the creation of a specific adaptive governance framework, based in human rights doctrine, to protect the rights of those living in communities that are no longer habitable due to climate-induced environmental change. These factors are: (1) nation-state governments have a duty to protect populations that reside within their jurisdiction; (2) no human rights instruments provide protections for planned community relocations; (3) relocation human rights guidelines must ensure the protection of collective rights, because climate change impacts the habitability of entire communities whose residents will be forced to permanently relocate; and (4) the human rights of host communities must also be protected.

Nation-state governments have a duty to protect their citizens

Nation-state governments have an obligation to protect vulnerable populations from climate-induced environmental change, which threatens the civil, economic, social and cultural rights fundamental to the inherent dignity of individuals as well as collective society. A nation-state government's protection of human rights is a critical threshold for that nation-state to claim sovereignty over its citizens and also is a minimum test for international legitimacy (Hathaway 1991; CICISS 2001, ¶2.18, ¶2.20). International law defines sovereignty as the legal identity of a state and signifies a nation-state government's capacity to make authoritative decisions about the people and resources within its territory (Montevideo Convention on the Rights and Duties of States 1934).

The duty to protect is inherent in the concept of sovereignty and implies that the nation-state government has the primary responsibility for the protection of populations within its jurisdiction (CICISS 2001, ¶2.15). The duty to protect is also considered a seminal principle for United Nations membership and for the attainment of international peace and security (CICISS 2001). Nation-state governments have a primary duty to protect the human rights, recognized in international human rights conventions, of their citizens if they are a Party to these conventions (CICISS 2001).

The duty to protect has three core principles (CICISS 2001, 2.14–2.15). Prevention is the most important and is defined as the responsibility to address the primary causes of crises that threaten populations. The responsibility to react means that a nation-state government must respond appropriately to situations requiring humanitarian assistance. The third principle requires a nation-state government to provide resources to reconstruct after a humanitarian crisis occurs (CICISS 2001).

International legal doctrine also specifically outlines the responsibilities of a nation-state government to protect internally displaced populations. The Inter-Agency Standing Committee (IASC) has provided operational standards that incorporate human rights protections to guide nation-state government response to natural disaster victims (IASC 2006, 33). In this context, protection means securing the physical safety of natural disaster victims, and also securing all of the human rights guaranteed in international human rights law (UN Economic and Social Council 2006, ¶4–8). The obligation also includes the responsibility to minimize damage caused by natural hazards (UN Economic and Social Council 2006, ¶4–8).

Emphasizing the duty of a nation-state government to protect its citizens, the IASC stated:

> Those affected by natural disasters, including those displaced by such events, remain, as residents and most often citizens of the country in which they are living, entitled to the protection of international human rights law as well as, if applicable, of international humanitarian law subscribed to by the State concerned or applicable as customary international law. They do not lose, as a consequence of their being displaced or otherwise affected by the disaster, the rights of the population at large. At the same time, they have particular needs which call for specific protection and assistance measures that are distinct from those required by individuals who were not adversely affected by the disaster.
> *(IASC 2008, 7)*

Similarly, the Guiding Principles on Internal Displacement (Guiding Principles), which are non-binding soft law principles, incorporate this sovereign responsibility to protect internally displaced populations. The responsibility includes the duty to provide safe access to housing "at the minimum, regardless of the circumstances, and without discrimination" (UNHCR 1998). The Pinheiro Principles on Housing and Property Restitution echo the principle that nation-states have an obligation to guarantee human rights protections to persons affected by internal displacement and emphasize the obligation to protect human rights related to housing and property restitution (COHRE 2007 ¶8.2). States are required to comply with obligations that they have accepted under human rights treaties through ratification.

The failure to protect is a human rights violation. For example, the European Court of Human Rights found that government officials in Kabardino-Balkariya (Russia) violated the right to life of community residents in 2000 when they failed to implement land-planning and emergency relief policies even though they were aware of an increasing risk of a large-scale mudslide. The Court also noted that the population had not been adequately informed about the risk (Budayeva 2008).

The duty to protect means that nation-state governments are responsible for implementing adaptation strategies. Communities will need a continuum of responses, from protection in place to community relocation, to adapt to climate-induced environmental change. Disaster and hazard mitigation are critical components of this continuum in order to assess vulnerabilities and develop disaster mitigation strategies so that protection in place is possible (Bronen 2011).

Community relocation challenges governmental and humanitarian institutions mandated to respond to emergencies. The inability of government agencies to change their approach from protection in place to relocation imperils communities. For communities whose current location is no longer habitable or lies entirely within vulnerable coastal or riverine areas, this means that they may be unable to receive government or humanitarian funding to repair and rebuild damaged infrastructure due to government regulations or institutional policies which prevent expenditures on infrastructure built within these high-risk areas. As a consequence, communities such as Newtok are unable to receive funding to repair damaged and deteriorating sanitation and sewage systems that endanger their health.

Second, no comprehensive governance framework exists that can evaluate when communities and government agencies need to shift their work from protection in place to community relocation. No method exists to determine whether and when a community can no longer be protected in place with traditional flood control and erosion protection devices. This determination requires a sophisticated integrated social and environmental assessment in order to evaluate whether a community needs to relocate in its entirety to a new location, can gradually move some of the infrastructure and residents to a location close to the original community or can be protected in place.

In order to use resources efficiently and prevent or exacerbate humanitarian crises, social ecological indicators can assess vulnerability and guide the design of adaptation strategies for communities and government agencies in order to transition from protection in place to community relocation. In Alaska, government agencies have proposed the following social-economic indicators to prioritize the climate change risk to the community and assess the appropriate adaptation response: (1) risk to life or safety during storm or flood events; (2) loss of critical infrastructure; (3) threats to public health; and (4) loss of 10 percent or more of residential dwellings (Bronen 2011; Bronen and Chapin 2013).

Finally, in some instances, the traditional humanitarian response to repair and rebuild so residents can return to their community may not be the most appropriate solution. In high-risk coastal or riverine areas, populations may be endangered if they return to their original location because of the inability of the traditional DRR activities to provide protection.

Funding is also a significant issue for communities that have decided that relocation is the only adaptation strategy that provides protection. Without a governance framework to authorize the expenditure of funds specifically for relocation, communities may be without the financial resources to implement the adaptation strategy.

These factors support the creation of an adaptive governance framework, which can dynamically respond to communities' needs as climate change impacts habitability and residents' safety. Multi-level institutional capacity-building is essential so that governmental and non-governmental agencies can respond to disasters, promote DRR strategies and expand their ability to respond to environmental changes that are ongoing and not the result of an extreme weather event.

A human rights framework is critical to the design and implementation of this adaptive governance framework to ensure that nation-state governments focus on the protection of freedoms that are fundamental to collective society, and that relocation only occurs when there are no other feasible solutions to protect vulnerable populations. Lack of capacity and resources can limit a government's ability to protect the economic, social and cultural rights of populations within its jurisdiction (Kolmannskog 2008, 23, 35; Zetter 2010, 144). If human rights protections cannot be realized because of inadequate resources, then working for institutional capacity building through expansion or reform can be a part of the international obligations generated by the recognition of these rights (Sen 2004, 4).

International cooperation is an essential component of the successful design and implementation of adaptation strategies. The UNFCCC clearly articulates the need for international cooperation in the development and implementation of adaptation strategies, including planned relocations, and specifically states that developed country Parties shall: "[A]ssist the developing country Parties that are particularly vulnerable to the adverse effects of climate change in meeting costs of adaptation to those adverse effects" (UN 1992, art. 4(1)(b)).

Human rights instruments are needed to protect relocating communities

Existing human rights instruments fail to protect communities needing to relocate because of climate-induced environmental change. The 1951 Convention relating to the Status of Refugees (1951 Refugee Convention) is the only treaty that creates an international structure to respond to human migration. Refugee law is based on the fundamental premise that the ordinary bonds between citizen and state have been broken and a person is outside of their country of origin because the nation-state government is unable or unwilling to protect them (Hathaway 1991). As a consequence, the 1951 Refugee Convention actually relieves a nation-state government of its obligation to protect the human rights of its citizens (Hathaway 1991). This underlying premise of refugee law conflicts with the nation-state obligation to guarantee human rights protections for all populations within its jurisdiction.

In the situation of climate-induced population displacement, communities should still be able to rely on national protection to respond to their humanitarian crisis. In addition, while refugee status is conferred on an individual basis, climate-induced environmental change will cause entire communities to relocate. The 1951 Refugee Convention provides no mechanism to confer refugee status in this situation.

Two international human rights documents that concern displacement are the Guiding Principles and the IASC Operational Guidelines on the Protection of Persons in Situations of Natural Disasters, but these are not adequate to address the complex issues and human rights implications of climate-induced community relocations for several reasons.

First, emergency mass population displacement caused by sudden extreme weather events is clearly different from planned relocations. Both the IASC guidelines and the Guiding Principles do not provide for the prospective needs of populations planning their permanent relocation and do not provide any guidance on how communities can sustain themselves and create the necessary infrastructure to provide for basic necessities without the assistance of humanitarian aid (Brookings-Bern Project on Internal Displacement 2011). Most importantly, these documents do not clearly define a mechanism for communities themselves to make the decisions regarding the process of relocation. The IASC guidelines state the need for informed consent and participation in decisions regarding the relocation process, but as discussed below, these principles are different from the ability to make decisions about the relocation (Brookings-Bern Project on Internal Displacement 2011).

The IASC guidelines to respond to natural disasters were developed to respond to situations when pre-planning is not possible (Brookings-Bern Project on Internal Displacement 2011). The guidelines outline minimum core human rights obligations under the International Covenant on Economic, Social and Cultural Rights, such as the duty to provide food, shelter and health services, which a nation-state government must provide after the occurrence of a natural disaster (Brookings-Bern Project on Internal Displacement 2011). However, the guidelines assume that humanitarian aid organizations will provide these basic necessities to populations displaced by natural disasters and do not describe how displaced populations can provide these basic necessities for themselves.

The fact that these guidelines do not incorporate mechanisms for community self-sufficiency is a significant protection gap for communities facing permanent relocation. Both the NTC and Tulele Peisa are determined to incorporate workforce development opportunities into the relocation process. In Newtok, the NTC sent a group of community residents to vocational training so that they could learn how to build some of the infrastructure at the relocation site. Tulele Peisa has focused on agro-forestry as a means to provide income to community residents and as well as grow food for personal consumption.

The Guiding Principles also do not provide sufficient human rights protections for those facing climate-induced community relocation. This document is not a binding international treaty or convention, but the UN General Assembly has recognized the Guiding Principles as "an important international framework for the protection of internally displaced persons" (UN General Assembly Resolution 60/1 2005, para. 22). Although the Guiding Principles include persons displaced by natural disasters, the primary focus of these guidelines is displacement caused by the state's inability or unwillingness to protect populations from political, religious, ethnic or otherwise discriminatory persecution or violence. In comparison, as stated above, those displaced by climate-induced ecological change should be able to continue to rely on state protection.

Second, both documents are based on the premise that populations may be able to return to their original home. Climate-induced environmental change will cause permanent population displacement. Enormous differences exist between policy and human rights protections for temporary and permanent population displacement.

The World Bank has also developed guidelines and institutional requirements for population resettlements caused by government or government-supported actors constructing infrastructure projects, such as dams, and for preventive resettlements as part of a DRR strategy (World Bank 2004; Correa 2011). The guidelines, however, do not incorporate human rights protections and only provide for informed consent and participation of affected communities (World Bank 2004).

Relocation human rights guidelines must ensure the protection of collective rights

Climate-induced displacement will affect entire communities whose residents will collectively need protection from the threats caused by climate change. International human rights conventions, such as the UN Declaration on the Rights of Indigenous Peoples, recognize the rights of peoples collectively, and that indigenous peoples have the collective right to the fundamental freedoms articulated in the Universal Declaration of Human Rights and international law. Like these documents, a human rights instrument that addresses climate-induced population displacement must ensure the protection of collective rights because climate change impacts the habitability of entire communities whose residents will be forced to permanently relocate. These rights include the collective right to relocate as a community, as well as the collective right to make decisions regarding where and how a community will relocate. No human rights protocol currently contains a community right to make these decisions.

For the residents of both Newtok, Alaska, and the Carteret Islands, Papua New Guinea, the collective right to decide to relocate as a community is the most important right to protect. The residents of each community are making all of the decisions related to the relocation effort to ensure that, despite the enormous loss of connection to the land on which they have each dwelled, they will be able to recreate their community at the relocation site to preserve their cultural heritage and ensure the long-term sustainability of their community.

The human rights of host communities must also be protected

A human rights instrument developed to respond to climate-induced displacement must also ensure that human rights protections are extended to those living in communities that provide sanctuary for those displaced by climate change. Host populations may experience shortages of water, sanitation, shelter and essential health services as a result of the increase in population (IASC 2006, 10). Schools may also struggle to provide educational services if there is an influx of displaced students.

Human rights protections for host populations will ensure that host communities benefit from the relocation, preserve or improve their standard of living and also prevent conflicts and competition with the displaced populations (Abhas 2010). In Papua New Guinea, Tulele Peisa has developed several programs to ensure that host communities are involved in the relocation process, including providing funding for host community infrastructure so that the host community is not burdened by the increase in population (Tulele Peisa 2009).

Conclusion

As climate change causes increasing frequency and intensity of extreme weather events, and ongoing environmental change that renders entire communities uninhabitable and requires community relocation, an adaptive governance framework based in human rights protections needs to infuse the relocation process. Guiding Principles of Climigration need to be created to specifically protect the collective and individual rights of these communities so that human rights protections are embedded within the relocation process (Bronen 2011). A multi-level relocation governance framework also needs to be designed so that communities have the ability to relocate when the traditional erosion and flood control devices can no longer protect residents in place. In this way, a model adaptation strategy can be created that facilitates an effective transition from protection in place, to community relocation that can serve as a model for governments throughout the world.

Acknowledgments

I am grateful to Sally Russell Cox and the members of the NPG and the NTC whose working relationship has been an inspiration. I am also grateful to Ursula Rakova for inviting Stanley and George Tom from the NTC to the Carteret Islands to begin a collaboration between the Arctic and Small Island Developing States. I also thank Many Strong Voices and the National Science Foundation for providing funding to make the collaboration possible.

References

Abhas, K.J. (2010) *Safer Homes, Stronger Communities: A Handbook for Reconstructing after Natural Disasters Global Facility for Disaster Reduction and Recovery*, Washington, DC: World Bank.

AgnewBeck (2012) *Strategic Management Plan Newtok to Mertarvik*, Anchorage: AgnewBeck.

Arctic Slope Consulting Group (ASCG) (2004) *Newtok: Background For Relocation Report*, Anchorage: Arctic Slope Consulting Group.

ASCG (2008) "Village of Newtok, Local Hazards Mitigation Plan," ASCG Inc. of Alaska Bechtol Planning and Development, Newtok, available online at: http://commerce.alaska.gov/dnn/dcra/PlanningLandManagement/NewtokPlanningGroup.aspx.

Boege, V. (2011) "Challenges and Pitfalls of Resettlement Measures: Experiences in the Pacific Region," Center on Migration, Citizenship and Development Working Paper Series, Bielefeld.

Bronen, R. (2010) "Forced Migration of Alaskan Indigenous Communities Due to Climate Change," in Afifi, T. and Jäger, J. (eds.), *Environment, Forced Migration and Social Vulnerability*, London and New York: Springer-Verlag.

Bronen, R. (2011) "Climate-induced Community Relocations: Creating an Adaptive Governance Framework Based in Human Rights Doctrine," *NYU Review of Law and Social Change*, 35(2): 101–148.

Bronen, R. and Chapin, F.S. (2013) "Adaptive Governance and Institutional Strategies for Climate-induced Community Relocations in Alaska," *Proceedings of the National Academy of Sciences*, Washington, DC.

Brookings-Bern (Brookings-Bern Project on Internal Displacement) (2011) *IASC Operational Guidelines on the Protection of Persons In Situations of Natural Disasters*, Washington, DC: Brookings-Bern Project on Internal Displacement.

Budayeva and others v. Russia [2008] Applications nos. 15339/02, 21166/02, 20058/02, 11673/02 and 1534/02, European Court of Human Rights.

COHRE (Centre on Housing Rights and Evictions) (2007) *The Pinheiro Principles: United Nations Principles on Housing and Property Restitution for Refugees and Displaced Persons*, Centre on Housing Rights and Evictions.

CICISS (Canadian International Commission on Intervention and State Sovereignty) (2001) *Responsibility to Protect*, Canada: International Development Research Centre.

Correa, E. with Fernando, R. and Sanahuja, H. (2011) *Populations at Risk of Disaster: A Resettlement Guide*, Washington, DC: World Bank.

Cox, S. (2007) *An Overview of Erosion, Flooding, and Relocation Efforts in the Native Village of Newtok*, Anchorage, Alaska: Alaska Department of Commerce, Community and Economic Development.

Displacement Solutions (2008) *The Bougainville Resettlement Initiative Meeting Report*, Canberra, Australia: Displacement Solutions.

GAO (Government Accountability Office) (2003) *Alaska Native Villages: Most Are Affected by Flooding and Erosion, But Few Qualify for Federal Assistance*, Washington, DC: GAO.

GAO (2009) *Alaska Native Villages: Limited Progress Has Been Made on Relocating Villages Threatened By Flooding and Erosion*, Washington, DC: GAO.

Hathaway, J.C. (1991) "Reconceiving Refugee Law as Human Rights Protection," *Journal of Refugee Studies*, 23: 113–131.

Hodgkinson, D., Burton, T., Dawkins, S., Young, L. and Coram, A. (2008) "Towards a Convention for Persons Displaced by Climate Change: Key Issues and Preliminary Responses," *New Critic*, 18, available online at: http://www.ias.uwa.edu.au/new-critic/eight/hodgkinson.

Holthus, P., Crawford, M., Makroro, C. and Sullivan, S. (1992) *Vulnerability Assessment for Accelerated Sea Level Rise—A Case Study: Majuro Atoll, Republic of the Marshall Islands*, South Pacific Environment Program Reports and Study Series 60.

Hufford, G. and Partain, J. (2005) *Climate Change and Short-term Forecasting for Alaskan Northern Coasts*, Anchorage: National Weather Service.

Inter-Agency Standing Committee (IASC) (2006) *Protecting Persons Affected by Natural Disasters: IASC Operational Guidelines on Human Rights and Natural Disasters*, Washington, DC: Brookings-Bern Project on Internal Displacement.

IASC (2008) *Human Rights and Natural Disasters: Operational Guidelines and Field Manual on Human Rights Protections in Situations of Natural Disasters*, Washington, DC: Brookings-Bern Project on Internal Displacement.

IPCC (2001) "Summary for Policy Makers," in Houghton, J.T., Ding, Y., Griggs, D.J., Noguer, M., van der Linden, P.J., Dai, X., Maskell, K. and Johnson, C.A. (eds.), *Climate Change 2001: The Scientific Basis. Contribution of Working Group I to the Third Assessment*

Report of the Intergovernmental Panel on Climate Change, Cambridge and New York: Cambridge University Press.

IPCC (2007a) "Summary for Policymakers," in Solomon, S., Qin, D., Manning, M., Chen, Z., Marquis, M., Averyt, K.B., Tignor, M. and Miller, H.L. (eds.), *Climate Change 2007: The Physical Science Basis*, Contribution of Working Group I to the Fourth Assessment Report of the Intergovernmental Panel on Climate Change, Cambridge and New York: Cambridge University Press.

IPCC (2007b) "Climate Change 2007: Impacts, Adaptation and Vulnerability," Contribution of Working Group II to the Fourth Assessment Report of the IPCC, in Parry, M.L., Canziani, O.F., Palutikof, J.P., van der Linden, P.J. and Hanson, C.E. (eds.), Cambridge and New York: Cambridge University Press, available online at: http://www.ipcc.ch/ipccreports/ar4-wg2.htm.

IPCC (2007c) "Observations: Changes in Snow, Ice and Frozen Ground," in Solomon, S., Qin, D., Manning, M., Chen, Z., Marquis, M., Averyt, K.B., Tignor, M. and Miller, H.L. (eds.), *Climate Change 2007: The Physical Science Basis*, Cambridge and New York: Cambridge University Press.

IPCC (2012) "Summary for Policymakers," in Field, C.B., Barros, V., Stocker, T.F., Qin, D., Dokken, D.J., Ebi, K.L., Mastrandrea, M.D., Mach, K.J., Plattner, G.-K., Allen, S.K., Tignor, M. and Midgley, P.M. (eds.), *Managing the Risks of Extreme Events and Disasters to Advance Climate Change Adaptation*, A Special Report of Working Groups I and II of the Intergovernmental Panel on Climate Change, Cambridge and New York: Cambridge University Press.

Kolmannskog, V.O. (ed.) (2008) *Future Floods of Refugees: A Comment on Climate Change, Conflict and Forced Migration*, Norway: Norwegian Refugee Council, available online at: http://www.nrc.no/arch/_img/9268480.pdf.

Maldives (2008) Submission of the Maldives to OHCHR Study, Human Rights Council Resolution 7/23: "Human Rights and Climate Change".

Nicholls, R.J. and Cazenave, A. (2010) "Sea-level Rise and its Impact on Coastal Zones," *Science*, 328 (1517).

NOAA National Climatic Data Center (2011) "State of the Climate: Global Snow and Ice for September 2011," October, available online at: http://www.ncdc.noaa.gov/sotc/global-snow/2011/9.

NOAA National Climatic Data Center (2013) "State of the Climate: Global Snow and Ice for September 2013," October, available online at: http://www.ncdc.noaa.gov/sotc/national/2012/13/supplemental/page-4

NSIDC (2012) "Arctic Sea Ice Extent Settles at Record Seasonal Minimum," *Arctic Sea Ice News and Analysis*, September 19, available online at: http://nsidc.org/arcticseaicenews/.

O'Collins, M. (1988) "Carteret Islanders at the Atoll Resettlement Scheme: A Response to Land Loss and Population Growth," in Pernetta, J.C. (ed.), *Potential Impacts of Greenhouse Gas Generated Climate Change and Projected Sea Level Rise on Pacific Island States of the SPREP Region*, University of Papua New Guinea.

Sen, A. (2004) "Elements of a Theory of Human Rights," *Philosophy and Public Affairs*, 32.

Serreze, M.C. (2008/2009) "Arctic Climate Change: Where Reality Exceeds Expectations," *Witness Arctic*, 3–4, available online at: http://www.arcus.org/witness-the-arctic/2009/1.

Shulski, M. and Wendler, G. (2007) *The Climate of Alaska*, Fairbanks: University of Alaska Press.

Tulele Peisa (2009) "Carteret Islands Integrated Relocation Program Project Proposal," Bougainville, Papua New Guinea, available online at: http://ourworld.unu.edu/en/wp-content/uploads/2009/06/carterets-integrated-relocation-program-proposal.pdf.

UN (1951) "UN Convention Relating to the Status of Refugees," United Nations General Assembly, New York.

UN (1992) "Framework Convention on Climate Change," United Nations General Assembly, New York.

UN (2005) "General Assembly Resolution, 2005," 60/1, UN Doc A/Res/60/1, October 24.

UNDP (United Nations Development Program) (2006) *Developing a Disaster Risk Profile of the Maldives*, available online at: http://preventionweb.net/go/11145.

UNDP (United Nations Development Program) (2009) *Maldives: Integrating of Climate Change Risks Into Resilient Island Strategy Global Environment Facility*, Maldives: UNDP.

UN Economic and Social Council (2006) "Specific Groups and Individuals: Mass Exoduses and Displaced Persons," Report of the Representative of the Secretary-General on the human rights of internally displaced persons, Walter Kälin, Commission on Human Rights, 62nd session, E/CN.4/2006/71.

UN Environment Program World Conservation Monitoring Centre (2006) *In the Front Line: Shoreline Protection and Other Ecosystem Services from Mangroves and Reefs*, Cambridge: UNEP.

UNHCR (United Nations High Commissioner on Refugees) (1998) "Guiding Principles on Internal Displacement," E/CN.4/1998/53/Add.2.

USACE (2006) *Alaska Village Erosion Technical Assistance Program: An examination of erosion issues in the communities of Bethel, Dillingham, Kaktovik, Kivalina, Newtok, Shishmaref, and Unalakleet*, Anchorage: US Army Corps of Engineers.

USACE (2008a) *Revised Environmental Assessment: Finding of No Significant Impact: Newtok Evacuation Center: Mertarvik, Nelson Island, Alaska*, Anchorage: US Army Corps of Engineers, available online at: http://commerce.alaska.gov/dnn/dcra/PlanningLand-Management/NewtokPlanningGroup.aspx.

USACE (2008b) "Section 117 Project Fact Sheet," Anchorage: US Army Corps of Engineers, available online at: http://commerce.alaska.gov/dnn/dcra/PlanningLand-Management/NewtokPlanningGroup.aspx.

World Bank (2004) *Involuntary Resettlement Sourcebook: Planning and Implementation in Development Projects*, Washington, DC: World Bank.

Zetter, R. (2010) "Protecting People Displaced by Climate Change: Some Conceptual Challenges," in McAdam, J. (ed.), *Climate Change and Displacement: Multi-Disciplinary Perspectives*, Oxford: Hart Publishing.

12

SOMETHING OLD AND SOMETHING NEW

Resettlement in the twenty-first century

Anthony Oliver-Smith and Alex de Sherbinin

Introduction

As climate change impacts increase in severity, complex disasters affect ever-grow-ing numbers of people, land grabs and resource-related conflicts become more common, and political instability and conflict spreads in Africa and the Muslim world, there is little doubt that there will be an increase in involuntary population displacements in this century (Walker, Glasser and Kambli 2012). Even were these other factors to remain unchanged, population growth in affected areas would dic-tate greater numbers of displaced persons. Yet with more evidence that the Earth's climate is changing more rapidly than previously thought (Bryssea *et al.* 2013) and the significant environmental and economic disruptions that would be inherent in a likely 4°C rise in global temperatures (New *et al.* 2011), there is a need to anticipate how changes in the volumes of displacement will affect various populations, and how humanitarian and development actors will respond.

The mechanisms of climate-related displacement will be multiple. The first will be an increase in natural hazard impacts. Although climate scientists have been hesitant to draw conclusions from any single event (Field *et al.* 2012), the widespread occur-rence of extremes exceeding any precedents in recent years—widespread floods in Mozambique (2000, 2001, 2007), Kenya (2006) and Pakistan (2011); devastating hurricanes in the US (2005 and 2012); and widespread heat waves, drought and forest fires in Russia (2010), the US (2011–12), and Australia (2012–13)—have led many to conclude that an age of increasing extremes is upon us (Hansen, Sato and Ruedy 2012; Trenberth 2012). With such disasters come increases in displacement, and humanitarian groups have begun to track climate change's incremental impact (UNOCHA *et al.* 2009; Yenotani 2011). Beyond natural disasters, there is the increasing likelihood of gradual but irrevocable productive land loss to sea level rise and drying trends in areas previously suitable for rain-fed agriculture (Schellnhuber *et al.* 2006; New *et al.* 2011), which will trigger more gradual population displacements and outmigration (de Sherbinin, Warner and Ehrhart 2011a; Foresight 2011; de Sherbinin *et al.* 2012).

Conflict is the largest contributor to long-term displacement. Evidence suggests that climate variability is associated with an increased incidence of certain kinds of conflict (Hsiang, Meng and Cane 2011; Fjelde and von Uexkull 2012). Higher temperatures, or greater drought or flood incidence, will be "threat multipliers" for conflicts (CNA 2011). Finally, large-scale climate change mitigation and adaptation projects such as biofuel plantations, large dams, and coastal defenses may also displace people (de Sherbinin, Castro *et al.* 2011). Beyond these climate-related factors, displacements will likely increase as a result of land grabs for agricultural production (Borras and Franco 2012), and complex "natural" and technological disasters (e.g. the 2011 Tohoku earthquake and tsunami off the coast of northeastern Japan, and associated meltdown at the Fukushima Daiichi Nuclear Power Plant complex) with catastrophic impacts.

Anticipating this increase, some have called for greater attention to organized resettlement or planned relocation as possible responses (de Sherbinin, Castro *et al.* 2011; Ferris 2012). Resettlement has both positive and negative aspects. On the positive side, it can represent an important protection for vulnerable communities that would otherwise be left to their own devices. The UK government's Foresight Project (Black *et al.* 2011) recently concluded that there will be many who are "trapped" by environmental change and unable to move on their own, implying a need for government assistance. On the negative side, the track record of resettlement associated with large infrastructure and development projects—so called development-forced displacement and resettlement (DFDR)—has been poor. The fact that results for disaster-induced displacement and resettlement (DIDR) have been marginally better suggests that there is hope, however, for better results (Correa 2011a, 2011b).

Resettlement is a term that has different meanings for different actors. For development actors, DFDR is a population movement planned directly by the government or private developers, which generally entails an allocation of new homes and/or lands to replace lost homes or in compensation for lost lands, and possibly financial transfers and development activities to restore or establish new livelihoods.[1] In DFDR individuals and/or groups (communities) are moved within a country to another location that most often had been previously unsettled or thinly settled. For humanitarian and refugee policy actors, resettlement refers to a process in which refugees from conflict or political oppression are resettled on an individual and/or family basis in existing communities, often overseas. Although demands for international resettlement are likely to increase, we restrict ourselves in this chapter to a focus on DFDR and displacement by disasters.

For purposes of clarity, the terms displacement and resettlement must be both separated and defined. There is no necessary linkage between displacement and resettlement. Many populations have been displaced, but little formal resettlement has taken place. Many populations have been displaced by a variety of causes and have simply migrated without assistance, often joining migrant streams to available rural or urban destinations. The term *displacement* therefore refers to the movement of population from their place of usual residence to another area. This movement is forced in the sense that in the absence of a project or a disaster residents would not have chosen to leave the area (de Sherbinin, Castro and Gemenne 2010). The movement can be either internal or international, and can be temporary or permanent. Displacement has generally taken place because of the occurrence of a disaster, a conflict or a development project that renders the abode of a population uninhabitable, either temporarily or permanently.

Resettlement generally is a process planned directly by the government or private developers for a displaced population, and generally entails an allocation of new homes and/or lands to replace lost homes or in compensation for lost lands, and possibly financial transfers and development activities. Resettlement is a process that is planned and administered most frequently by the state or its agents, and increasingly by private interests in which individuals and/or groups (communities) are moved to another location. Resettlement, though not inevitable after displacement, may be undertaken individually or in small groups, but in larger contexts of communities, formal resettlement following a disaster or driven by a development project, involves the planned reestablishment of displaced peoples in a new location with appropriate settlement design, housing, services, and an economic base to enable the community to reconstitute itself and achieve adequate levels of resilience to normal social, economic, political and environmental variation. That such satisfactory outcomes have not been common constitutes a major humanitarian crisis.

This chapter examines something old—lessons from resettlement praxis and existing guidelines—and something new—the emerging guidelines and potential future trajectories of resettlement in the context of climate change and its anticipated impacts.

Lessons learned from resettlement praxis

Formal DFDR projects may involve a number of the following elements: settlement design, the provision of housing assistance, compensation for lost resources, and public services such as schools, medical facilities and livelihood options. Therefore, theories of resettlement must be theories about reconstruction and the recovery of losses. Communities that have been displaced and resettled, whether by development projects or disasters, are communities that must be reconstructed, either by themselves or with assistance (Oliver-Smith 2005; Birkmann *et al.* 2012). In either case, an infrastructure has to be built to replace the one that has been lost and a community, as a social body, has to reconstitute itself, whatever the cause of displacement. People affected by a resettlement project are confronted with a complex, cascading sequence of events and processes most often involving: dislocation, homelessness, unemployment, the dismantling of families and communities, adaptive stresses, loss of privacy, political marginalization, a decrease in mental and physical health status, and the daunting challenge of reconstructing one's ontological status, family and community (Colson 1971; Scudder 1981; Cernea 1990, 1997; Oliver-Smith 2005; Birkmann et al. 2012). All suffer the endangerment of structures of meaning and identity, and all must mobilize social and cultural resources in their efforts to re-establish viable social groups and communities and to restore adequate levels of material and cultural life (Bennett and McDowell 2012).

However different the driving forces and policies, moving the community to a new location does not solve the problem. They may have stopped moving, but that is just the beginning of another process, resettlement, which in all too many cases ends up becoming a second crisis. Some displacements will involve sudden rapid-onset events that evoke at initial stages strategies akin to emergency management such as evacuation and temporary shelters. Other cases integrate the displaced into existing communities and resemble the resettlement of political refugees. Still other forms will be planned mitigation projects drawing on models from DFDR, community development and urban planning (de Sherbinin, Castro et al. 2011). Some relocations may involve several of these forms of displacement and resettlement over time. Finally, some relocations will constitute simply mass migrations, evoking very little formal institutional response. The challenge of displacement and resettlement thus requires inputs from many fields, ranging from emergency management to economic development and research from social, scientific and management disciplines.

In some circumstances, because events and processes associated with displacement and resettlement involve different time/space scales (lasting longer, encompassing wider areas, crossing ecological, jurisdictional and national boundaries, affecting heterogeneous populations), they will require multiple strategies and inter- and multi-national efforts and cooperation. At the same time, resettlement may involve masses of people, but responses will need to address culturally and socially defined constituent population groups. In most cases, solutions must be durable. There is often little hope of return.

The social scientific literature on displacement and resettlement over the last half century has become clustered around three themes: civil and military conflicts, disasters, and development projects. The majority of the literature on refugee resettlement focuses on individual and family resettlement rather than the resettlement of entire communities, although some large refugee populations, such as the Baha'i or Soviet Jews, have been resettled as communities. The literature on DFDR generally deals with resettlement of communities affected by large-scale infrastructure projects (Scudder and Colson 1982; Cernea and Guggenheim 1993; Cernea and McDowell 2000; de Wet 2006; Vandergeest, Idahosa and Bose 2007; Oliver-Smith 2009; Scudder 2009; Penz, Drydyk and Bose 2012). Post-disaster resettlement research has received somewhat less attention, in part due to the emphasis placed on reconstruction in situ. However, since the 1980s the resettlement of communities located in high-risk zones for disaster risk reduction (DRR) has gained greater attention (Perry and Mushkatel 1984; Oliver-Smith 1991; Correa 2011a, 2011b; Ferris 2011). These bodies of research are being complemented by a growing concern regarding internally displaced persons (IDPs) (Deng and Cohen 1999; Koser 2007).

DFDR

In DFDR, Scudder developed an approach to describe and analyze the process of involuntary dislocation and resettlement based on the concept of *stress* (Scudder 1981; Scudder and Colson 1982). The Four Stage Framework emphasizes how most "resettlers" can be expected to behave during each of the four stages, passage through which must be completed if the resettlement project is to be successful (Scudder 2009).

The resettlement process itself is represented as occurring in four stages, which Scudder (2009) labels as follows:

1. Planning for resettlement before physical removal;
2. Coping with the initial drop in living standards that tends to follow removal;
3. Initiating economic development and community-formation activities; and
4. Handing over a sustainable resettlement process to the second generation of resettlers and to non-project authority institutions.

Scudder and Colson posited that: physiological stress, psychological stress and socio-cultural stress (referred to as multi-dimensional stress), are experienced as affected people pass through the displacement and resettlement process. Physiological stress is seen in increased morbidity and mortality rates. Psychological stress, seen as directly proportional to the abruptness of the relocation, has four manifestations: trauma from the uprooting process, guilt about having survived, grief for a lost home and anxiety about an uncertain future. Socio-cultural stress is manifested as a result of the economic, political, and cultural effects of relocation such as inadequate material support, loss of power, loss of identity and culture. At roughly the same time, Cernea began to develop his now well-known Impoverishment Risks and Reconstruction (IRR) approach to understanding (and mitigating) the major adverse effects of displacement and the resettlement process; the IRR identifies eight basic risks associated with displacement (1996, 2000). Cernea models displacement risks by deconstructing the "syncretic, multifaceted process of displacement into its identifiable, principal and most widespread components," including landlessness, homelessness, joblessness, marginalization, food insecurity, increased morbidity, loss of access to common property resources, and social disarticulation (Cernea 2000, 19–29). All these risks follow the displacement process with the threat of a second calamity that entails such risks that can translate directly into losses. Cernea's IRR model is designed to predict, diagnose and resolve the problems associated with DFDR.

More recently, de Wet has asked why resettlement so often goes wrong (2006). He sees two broad approaches to responding to the question. The first which he calls the "Inadequate Inputs" approach, argues that resettlement projects fail because of a lack of appropriate inputs: national legal frameworks and policies, political will, funding, pre-displacement research, careful implementation and monitoring that can help to control and mitigate losses. De Wet favors an "Inherent Complexity" approach. He argues that there is a complexity in resettlement that is inherent in "the interrelatedness of a range of factors of different orders: cultural, social, environmental, economic, institutional and political—all of which are taking place in the context of imposed space change and of local level responses and initiatives" (de Wet 2006, 190). These interlinked, transformational changes are also influenced by and respond to impositions from external sources of power as well as the initiatives of local actors. Therefore, the resettlement process emerges out of the complex interaction of all these factors in ways that are not predictable and that do not seem amenable to a linear-based, rational planning approach. The fact that authorities are limited in the degree of control they can exercise over such a complex process creates a space for resettlers to take greater control over the process. The challenge thus becomes the development of policy that supports a genuine participatory and open-ended approach to resettlement planning and decision making (de Wet 2006).

Downing and Garcia-Downing (2009) argued that insufficient attention has been paid to the risks of psycho-socio-cultural (PSC) impoverishment inflicted by displacement. Few projects consider or attempt to mitigate this risk. The Downings argue that, in the PSC realm, it is highly improbable that a pre-displacement routine culture may be recovered, let alone be restored. However, some relative success of PSC recovery may be achieved in how well the transformed routine culture answers the primary questions of the displaced compared to the pre-displacement culture. Primary questions include: Who are we? Where are we? And how do we relate to one another? The applied question thus becomes, "What can be done to facilitate the new routine culture so that it adequately addresses the primary cultural questions faced by displaced peoples?"

Post-disaster resettlement

By and large the field of disaster management has, until quite recently, been focused predominantly on emergency management. However, over the past quarter-century, the issue of DRR is beginning to influence the ways disasters are conceptualized and to add to the social toolkit used to respond to both disaster risk and impact. Resettlement has actually been employed by responsible authorities in disaster recovery for centuries. However, such efforts rarely met with success. Post-disaster resettled populations often abandon the new settlements and return to previous home sites for a wide variety of environmental, economic, social and psychological motives (Oliver-Smith 1977, 1991). Part of the blame for these failures was due to the failure in design, construction, implementation and delivery of the resettlement project itself.

Resettlement in the framework of DRR involves a complex planning process similar, but not identical, to DFDR (Oliver-Smith 2009; Correa 2011a). In some cases, disasters and other environmental disruptions will force people to migrate as individuals and families, similar to political refugees, with few community-based resettlement efforts on their behalf. However, in other cases community based resettlement was undertaken for disaster-affected people in projects that involved planning processes similar to DFDR, but usually only when no risk mitigation was possible.

Issues of organization and design can be identified as significant in the success or failure of DIDR projects. Poor choice of site for resettlement is one of the most frequently mentioned causes of resettlement failure. Poor design or layout of the settlement has also resulted in failure. Cost cutting and other economic reasons often produce monotonous, uniform designs for resettlement sites. Housing design and inferior materials can result in rejection by intended beneficiaries. These problems derive from a lack of consultation with, and participation by, the affected people. This lack is generally due to a disparagement of local knowledge and culture on the part of policy makers and planners (Oliver-Smith 1991; Correa 2011a, 2011b).

The disarticulation of spatially and culturally based patterns of self-organization, social interaction and reciprocity constitutes a loss of essential social ties that affect access to resources, compounding the loss of natural and man-made capital. Thus, in displacement and resettlement, people's adaptations to the social disarticulation produce new dynamics that influence their access and control over resources, often leading to a process of further impoverishment. Therefore understanding the role of social institutional processes, such as governance or social networks, in resettlers' adaptive strategies is crucial for identifying the socio-culturally specific nature of the impoverishment risks (Cernea 2000), thus helping to explain why displacement and resettlement so often result in greater impoverishment of affected households (McDowell 2002).

Cross-cutting issues

Although there are clear distinctions (particularly in the initial stages of displacement) between people who flee over international borders to escape persecution or death, and those displaced internally by disasters or development projects, over the past twenty-five years, the fields of development-induced displacement, refugee studies and disaster research have revealed that displaced peoples share many similar challenges (Hansen and Oliver-Smith 1982; Hansen 1993; Cernea 1996, Oliver-Smith 2005; Turton 2006). The social effects of all three forms of displacement present similar challenges and generate similar responses over the long term in affected peoples. Ultimately, regardless of the cause of dislocation, displaced people, disaster victims and refugees are confronted with a complex, cascading sequence of events which are reflected in a wide variety of cross-cutting issues such as gender, class and ethnicity, public health and resistance, to name only a few.

To note a few examples, gender is a powerful construct differentially affecting social, cultural, economic and political relations between men and women in all resettlement processes (Koenig 1995; Fordham and Enarson 1998; Colson 1999; Martin 2004; Mehta 2009). Class and ethnicity have been clearly identified as impacting all resettlement processes differentially, proving to be key markers of vulnerability (Oliver-Smith 2010; Cernea and Guggenheim 1993). Resistance to resettlement also demonstrates the importance of place in the construction and maintenance of identity, and questions the right of governments to uproot and resettle (Oliver-Smith 1991, 2010; Correa 2011a, 2011b). Public health concerns, including mental health, also prove to be key issues in successful outcomes for all resettlement processes (Kedia 2009).

Since the poor and the marginalized are almost always the most affected, displacement and resettlement are invariably intertwined with the question of development. Systemic forms of vulnerability and exposure and their tragic outcomes are frequently linked to unresolved problems of development (Maskrey 1989; Wisner *et al.* 2004; Lavell and Oppenheimer 2011). Since resettlement should focus on durable solutions, humanitarian responders must be prepared to coordinate effectively with development actors to ensure successful resettlement outcomes. If displacement results only in warehousing affected populations in "temporary" or otherwise permanent camps (rural slums), or the dispersal of affected populations to poverty stricken slums, the process of resettlement will compound the trauma and human rights violations of uprooting and consign them to long-term misery.

Therefore, resettlement projects must be configured as development projects. The projects must include the appropriate investments to enable people to become active and self-sufficient members of resilient communities. However, to date, relatively few nations have either the necessary legislation or the administrative structure and capacity to competently undertake the task of resettling displaced populations. Generally speaking, a mix of public agencies, with a wide array of environmental, social, and economic responsibilities, is assembled and charged with planning and implementing resettlement, frequently creating projects with serious internal contradictions and conflicting agendas, and little if any consultation with affected people (Cernea 2000).

Existing guidelines

A global regime of principles and organizations pertinent to displacement and resettlement has taken shape around such specific issues as the environment and the rights of indigenous peoples. Discussion of guidelines for refugee protection is found elsewhere in this volume, so we turn directly to a discussion of guidelines for DFDR and DIDR.

DFDR

Increasing media attention, public and non-governmental organization (NGO) recognition of human rights violations, and resistance by local peoples, in relation to large development projects financed by the World Bank stimulated efforts to formulate a set of resettlement policy guidelines within the Bank (cf. Fox and Brown 1998; Clark, Fox and Treakle 2003). The first resettlement guidelines in 1980 drafted by Cernea, named Operational Directive (OD) 4.30: Involuntary

Resettlement (World Bank 1990), is the strongest policy the bank has yet produced. OD 4.30 called for minimizing resettlement, an improvement or restoration of living standards, earning capacity and production levels, resettler participation in project activities, a resettlement action plan, and valuation and compensation for assets lost (World Bank 1990, 1–2). Shortly thereafter, the Organization for Economic Cooperation and Development (OECD) and the regional development banks developed similar guidelines. In addition, the Food and Agriculture Organization of the United Nations (FAO) also published guidelines that focused on the environmental impacts of resettlement projects in the humid tropics (Burbridge, Norgaard and Hartshorn 1988).

Pressure from borrower nations, which claimed the OD 4.30 guidelines violated national sovereignty, resulted in the 2001 iterations, now referred to as Operational Policy and Best Practice (OP/BP 4.12), which neglected to cover a wide range of cultural and psychological impacts (Scudder 2005, 281). The recent February 2011 revision also still places too little emphasis on the development dimensions of DFDR. There is also little question that, deficient though they may be in many respects, the guidelines have helped to improve resettlement planning and implementation, and to reduce the numbers of people affected by projects. However, as Scudder points out, even the World Bank would not claim that, generally speaking, income-earning capacity and living standards of displaced peoples have been restored (2005, 278). There are a number of other guidelines, including Indigenous Peoples (OD 4.20), Environmental Assessment (OD 4.01), Project Supervision (OD 13.05), Disclosure of Information (Best Practice (BP) 13.05) and Management of Cultural Property (OPN 11.03), that can be used to protect the interests of displaced and resettled communities.

Other initiatives over the past fifteen years that address the rights and protections for people threatened with DFDR, include the World Commission on Dams (WCD) (2000), the Equator Principles (2003), the Extractive Industries Review (2004) and the International Hydropower Association (2010). Each addressed the displacement and resettlement impacts of specific forms of development (dams, mines, etc.), some of them reflecting sector interests, and developed adequate and appropriate responses to remedy the problems. The WCD, as the most far-reaching, established a set of guidelines for good practice based on the five values of equity, efficiency, participatory decision making, sustainability and accountability (WCD 2000), most of which have not been adopted or followed.

DIDR

Disaster management performance standards have been established in a variety of institutional settings, such as the Sphere Project (2000) to address needs during the emergency period. However, over the past quarter-century DRR has grown in influence, adding to the social toolkit used to reduce disaster impacts. Attention is now turning to the option of resettlement, both for durable and sustainable reconstruction, and for reduction of exposure of communities in risk zones that are not possible to mitigate. Over the past several years, two notable efforts—the Inter-Agency Standing Committee (IASC) Operational Guidelines on the Protection of Persons in Situations of Natural Disasters (IASC Operational Guidelines) (2006, 2010), and two volumes from the World Bank, *Populations at Risk of Disaster: A Resettlement Guide* (Correa 2011a) and *Preventive Resettlement of Populations at Risk of Disaster* (Correa 2011b)—have attempted to address the issue of disaster related resettlement in terms of human rights and good practice guidelines.

The IASC Operational Guidelines adopt a human rights-based approach to help protect populations threatened or afflicted by disasters. They are intended to complement existing guidelines on humanitarian standards in disasters. The guidelines are organized by thematic grouping: protection of life, protection of rights related to food, health, etc., protection of rights related to housing and livelihoods, protection of rights related to freedom of movement and religion, throughout the time phases of the disaster. The IASC Operational Guidelines highlight areas where these rights are also threatened by the resettlement process. However, they do not provide a set of measures, guidelines or good practices in resettlement to ensure that these rights are safeguarded in and by the resettlement process, where, in fact, they are frequently violated.

Recent World Bank volumes help make up for this deficit. While not formally recognized as guidelines per se, the two volumes, together with the World Bank *Involuntary Resettlement Sourcebook* (World Bank 2004), constitute a major source of knowledge on the implementation of resettlement. The first volume covers the task of DRR resettlement, recognizing the complexity of the process and the heterogeneity of affected populations, laying out a holistic understanding and providing a step-by-step approach to the challenges of resettlement. The guide identifies clearly the goals, purposes, activities, expected results, and monitoring and evaluation steps that need to be taken, recognizing that, while plans are organized in a linear fashion, project implementation rarely follows that linearity since many unanticipated issues emerge.

The second volume, composed of case studies, demonstrates an awareness that resettlement is a holistic process that entails much more than housing provision, but is fundamentally about the reconstruction of community. The holistic perspective is manifested in the goal of maintaining as much as possible the social networks of the resettled population that are vital in accessing resources essential to the rebuilding of community. Sensitivity to the importance of socio-cultural issues, particularly in housing options, house form and urban design, is also a vital component in post-resettlement adaptation to the new environment. The linking of resettlement to skill building and economic opportunities will also be a key element in the long-term viability of any resettled population. Analysis of the legal and institutional frameworks and responsibilities illustrates the administrative complexity and challenges that DRR resettlement presents, as well as the importance of the credibility of government, enhanced through the care taken and the participation encouraged in the resettlement process itself.

Working toward durable solutions

Emerging guidelines

According to the International Federation of Red Cross and Red Crescent Societies (IFRC):

> there are no well recognized and comprehensive legal instruments which identify internationally agreed rules, principles and standards for the protection and assistance of people affected by natural and technological disasters. As a result, many international disaster response operations are subject to ad hoc rules and systems, which vary dramatically from country to country and impede the provision of fast and effective assistance—putting lives and dignity at risk.
>
> *(IFRC 2004, 1)*

However, increasing concern regarding IDPs displaced by wars, disasters and development projects and their rights, now includes recognition that climate change effects also impact people's human rights and welfare (Koivurova 2007; McInerny-Lankford, Darrow and Rajamani 2011). The Conference of the Parties to the UN Framework Convention on Climate Change (UNFCCC) encouraged "measures to enhance understanding, coordination and cooperation with regard to climate change induced displacement, migration and planned relocation" (UNFCCC 2010, part II(14)f.). Despite this there are still no nationally or internationally binding agreements or treaties that guarantee the rights of people who have been uprooted by causes such as climate change, environmental disruption, disasters or development projects. Although widely recognized as an international standard, and certainly helpful in guiding NGOs and other aid organizations in assisting IDPs, the UN Guiding Principles on Internal Displacement (Guiding Principles) have not been agreed upon in a binding covenant or treaty, although the African Union Convention for the Protection and Assistance of Internally Displaced Persons in Africa (Kampala Convention) provides a continental instrument that binds governments to provide legal protection for the rights and well-being of those forced to flee inside their home countries due to conflict, violence, natural disasters or development projects.

There is a vigorous debate about how to categorize people displaced by environmental and climate change—as refugees, IDPs, forced migrants or migrants—and what protections to afford them, if any (Bakewell 2011; Leighton 2011; McAdam 2011). Drawing on the Guiding Principles, recent efforts have been made to craft legal instruments for the protection of people displaced by climate change. Resettlement is among the forms of protection that are proposed, but concrete guidelines have yet to be formulated. Biermann and Boas (2010) advocate the construction of a new global governance architecture for the protection and voluntary resettlement of people displaced by sudden or gradual alterations in their natural environment by sea level rise, extreme weather events and drought and water scarcity. They advocate for a new Protocol on Recognition, Protection and Resettlement of Climate Refugees to the UNFCCC. The Nansen Principles are equally clear in declaring the need for addressing displacement, but do not address resettlement in any specificity (2011).

Currently proposals that specifically address the challenges of the resettlement process for climate change-affected peoples are appearing. For example, de Sherbinin, Castro *et al.* (2011) address displacement from direct impacts of climate change such as sea level rise, coastal erosion, drought, desertification and intensified coastal storms, among others, and from mitigation and adaptation projects, suggesting that the lessons learned from DFDR provide a useful template for improved policy and practice options for affected people. Berringer (2011) divides the larger rubric of climate phenomena into sudden impact and slow onset, each paired with a set of general tools (short-term aid, development strategies, planned resettlement) and strategies for durable solutions drawn from DFDR and DIDR.

In 1935, the US created an independent agency, the Resettlement Administration, for people displaced by the Great Dust Bowl, the devastating Mississippi flood of 1927 and the Great Depression that existed until 1947 (Maldonado *et al.* 2013). However, Bronen (2008, 2011), working with the community of Newtok in the Alaskan arctic, documents the present confusing and contradictory institutional environment impeding the voluntary resettlement of this community, which is threatened by coastal and riverbank erosion caused by melting permafrost due to climate change (Bronen, Chapter 11 in this volume). Since the US has no single federal program that proactively provides operational guidance and funding for the relocation of communities, Bronen calls for the creation of an adaptive governance response based on human rights doctrine. She asserts that a relocation strategy framework must contain two primary organizational instruments: a relocation policy framework and an adaptive governance structure. The relocation policy framework defines the human rights principles and objectives that frame the steps that governmental and non-governmental agencies must use to implement a resettlement process, including the determination of when and where resettlement must take place, the organizational arrangements between agencies, the protection of the human rights of affected peoples and their full participation, and the funding mechanisms. The Adaptive Governance Framework refers to the institutional arrangements that govern natural resources and can respond to rapid ecosystem changes. It includes amendments to the existing hazard mitigation and post-disaster recovery legislation to cover climate change induced processes, creating a relocation institutional framework which clearly delineates stakeholders and roles, the role of local governance, land acquisition processes, and decision making procedures and other tasks and responsibilities involved in relocation and resettlement. Bronen's recommendations are similar to those of Correa (2011a, 2011b) in stressing the need for an institutional framework to map and guide the necessary legal tasks and responsibilities in displacement and resettlement.

Noting the necessity of joining humanitarian, development and human rights actors with climate change experts to develop general principles to protect the rights of people resettled due to climate change, Ferris (2012) also recommends consulting the Guiding Principles, the IASC Operational Guidelines, the IASC Framework for Durable Solutions, and the World Bank Guidelines (4.30; 4.12) as well as mining the fields of DFDR for policy and practice options. She emphasizes the general guidelines from DFDR studies: that resettlement should always be considered a last resort, should always be adequately funded, well planned ahead of time, with a focus on land provision, should consider the rights of affected communities, and should be based on lessons from prior experiences (2012, 17–22). Based on these principles, Ferris outlines 23 protection principles and guidelines for climate change driven displacement and resettlement, organized by the definition of rights and responsibilities, protection and human rights, the prevention of risks of impoverishment and monitoring mechanisms. Among the related issues that Ferris notes as important are the legal policy bases for planned resettlement and the land and property issues, both of which echo the recommendations of Correa (2011a, 2011b) and Bronen (2008, 2011). She further recommends that a consultative process be established to develop specific protection principles and concrete guidelines that will be useful to all stakeholders, including affected peoples, development and humanitarian actors, and governments who may be obligated to consider resettlement as an adaptation to climate change. Most recently, the Peninsula Principles on Climate Displacement within States have been drawn up to build the foundations for a new normative framework to address the rights of people displaced by climate change effects (2013).

There is no agreement as yet on guidelines for anticipatory or preventive resettlement (that is, resettlement in *advance* of significant impacts), or indeed by what criteria such resettlement might be deemed necessary. Ferris (2012) points out that the lack of a clear internationally accepted definition of uninhabitability of a region and the likelihood that such conditions are due to multiple factors, make it difficult to determine both causality and responsibility. Furthermore, it is unclear whether residents of a risk-prone area should be moved in advance of potential impacts, even given uncertainties concerning timing and magnitude, or whether it is best to wait until after a major disaster occurs (Kelman 2008). It is not clear that an authority's responsibility to avoid loss of life supersedes its responsibility to protect property. The Chinese government policy of ecological migration could be viewed as a form of anticipatory resettlement, where residents in marginal environments are induced to move to locations with better economic and ecological prospects. Yet its impacts have been controversial (*The Economist* 2012). In general, when confidence in government is low there is likely to be more community resistance (Oliver-Smith 2010). In such cases, there is a need to reconcile the ethics of policies that remove people from high-risk areas with the potential that they will undermine historical freedoms and long-standing cultural patterns of settlement, mobility and livelihood (Johnson 2012). This risk that vacated lands might be appropriated for financial gain or that resettlement might be used as a tool against politically marginalized peoples suggests that criteria and guidelines are needed, lest anticipatory resettlement open a Pandora's box of unwanted outcomes. Yet even in the best of circumstances, it may be difficult to muster the necessary political will or resources in the absence of a major disaster, and indeed residents may be very reluctant to leave an area even if the probability of a disaster occurring is high.

Emerging resettlement strategies

Although there are no completely new strategies for resettling those who have been displaced by development, disasters or discord, given likely increases in the numbers of people affected by climate and conflict risks, it is worth reviewing some potential future strategies for relocating people in ways that increase economic opportunities while protecting human rights.

One strategy is resettlement through urban migration. Traditionally, in DFDR rural communities that lose land to a large project are provided new lands in other rural areas to enable them to continue their rural livelihood strategies. Some resettlers may in fact migrate to urban areas, but that is on an ad hoc basis, and independent of any government assistance. This strategy may no longer make sense for a number of reasons. Competition over land resources in rural areas is increasing, such that large swaths of unoccupied or thinly settled lands are largely a thing of the past. Natural resource-dependent livelihoods are therefore likely to be increasingly precarious under climate change, and economic opportunities largely lie in urbanized areas. So, from a livelihood improvement perspective, resettling populations in urban areas could improve economic prospects for resettlers, providing that adequate infrastructure, training and employment opportunities are made available. However, as Fagen notes (Chapter 16 in this volume), migrants in crisis situations tend to move to cities that are already seriously stretched, enduring high unemployment and lack of infrastructure.

Under its "ecological migration" strategy, the Chinese government is using a combination of incentives and coercion to relocate rural populations from environmentally marginal areas, mostly in the western provinces, into urban centers. For example, resettlement associated with a water transfer scheme from Shaanxi province to Beijing is linked to recurrent droughts that have affected Beijing's water supply (*The Economist* 2012). In many cases rural residents are provided with subsidized housing options in urban areas, but because most are not initially provided with urban household registrations under the *Hukou* system, they have difficulty obtaining urban services. These rural-to-urban resettlement schemes have yet to be studied to any extent, and so the impacts on resettlers are not fully known.

Another strategy is international resettlement in ethnic enclaves. With each wave of regional conflict and political upheaval, developed countries have absorbed refugees through organized resettlement programs. The US, under the Refugee Resettlement Act (1980), has resettled more than 1.6 million refugees (Singer and Wilson 2007), often in "ethnic enclaves" (Portes and Manning 1986) under the assumption that co-ethnics will help one another to acculturate and get established economically. In many cases the resettlement has occurred in communities with little recent experience with international migration or cultural understanding of ethnic minorities

(Gaber *et al.* 2004). For the most part international resettlement has been effective, affording refugees a better life with greater opportunities. Yet it is also costly, so for this reason the humanitarian community tends to favor repatriation and local integration into the country of first asylum (Dwyer 2010). Currently only 1 percent of all refugees are resettled internationally. The degree to which this strategy will succeed in the future will depend in large part on the availability of funds and the willingness of developed countries to receive displaced populations owing to factors other than conflict or human rights concerns. Although it cannot be expected that traditional refugee recipient countries will open their arms to potentially large numbers of environmentally displaced peoples, or so-called "climate refugees," the new strand of discussion on Loss and Damage within the UNFCCC opens the door for discussions of compensation for damages from climate change owing to developed country emissions, which may include international resettlement. This possibility has been raised for the small island states of the South Pacific that have close historical or cultural ties to larger countries, such as Tuvalu and Kiribati in relation to Australia and New Zealand, and the Marshall Islands in relation to the US (Kelman 2008).

Conclusions

The subject of future resettlement is a complex one, and by necessity we have not been able to address all aspects. Given anticipated levels of climate related displacement, the question of available land for resettlement will become crucial in both urban and rural contexts. Adequate and available land must be identified for resettlement sites. Procedures for establishing ownership and clear legal title, both traditional and formal, must be established. Legal instruments and procedures must be developed to acquire such properties and to establish the compensation of owners of land acquired for resettlement sites. Financing, therefore, also becomes a central issue. While the normative frameworks we have described for protecting human lives while also guaranteeing human rights is the gold standard, we must recognize that the governments of most developing countries, where significant climate impacts are projected to occur (and indeed are already occurring), may have the fewest resources to prepare and implement them. Developed countries will be reluctant to assume the costs of resettlement, even given responsibility for past emissions.

Resettlement is a complex social process. At its best, resettlement should support and nourish the coping and adaptation processes that enable a population to regain the functionality and coherence of a viable community, resilient enough to deal with social and environmental stressors within a range of variation. Central to these tasks are the issues of rights, poverty, vulnerability and other forms of social marginality that are intrinsically linked to displacement. While the field of displacement and resettlement studies has achieved significant advances over the last half century, deficiencies in planning, preparation, and implementation of involuntary resettlement projects have produced far more failures than successes. Indeed, as currently practiced, it is questionable whether resettlement could be categorized as a form of protection.

Nevertheless a key element to progress in the field will continue to be the recognition that the displaced must be seen as active social agents with their own views on rights and entitlements, which have to be considered in any displacement and in the planning and implementation of resettlement projects. If, in fact, the uprooted are resettled in some systematic way, the quality of the resettlement project itself may play a major role in the capacity of the community to recover. In many cases such projects are really about reconstructing communities after material destruction and often profound social disarticulation and trauma. Reconstructing and reconstituting community is an idea that needs to be approached with a certain humility and realism about the limits of our capacities. Such humility and realism have not always characterized the planners and administrators of projects dealing with uprooted peoples. Planners often perceive the culture of uprooted people as an obstacle to success, rather than as a resource (Oliver-Smith 2005).

While the focus of this paper has been on the past and future of organized resettlement in the context of global change, it must be recognized that most of those displaced by disasters and conflict, climate induced or not, will likely be left to fend for themselves. Humanitarian relief, when provided, has rarely been sufficient to make up for lost assets or to establish displaced persons in new locations. Potentially large-scale self-generated population movements will provide little solace to long-range planners and those who care about social justice, but ignoring this likelihood is unrealistic. We must also recognize that past precedent suggests (Foresight 2011; de Sherbinin *et al.* 2012) that migrants who choose their own destinations—wherever they fall on the continuum from forced to voluntary—are as likely to move *into* areas of climate risk as out of them.

Note

1 Resettlement has been considered by some a subset of the larger category of *planned relocation*, which is a term more commonly used in the humanitarian community. According to Ferris (2012, 10), "relocation is much less ambitious than resettlement in that it does not necessarily imply restoration of living standards and livelihoods."

References

Bakewell, O. (2011) "Conceptualising Displacement and Migration: Processes, Conditions, and Categories," in Koser, K. and Martin, S. (eds.), *The Migration–Displacement Nexus: Patterns, Processes, and Policies*, Oxford: Berghahn Books.

Bennett, O. and McDowell, C. (2012) *Displaced: the Human Cost of Development and Resettlement*, New York: Palgrave Macmillan.

Berringer, A. (2011) "Possible Frameworks for Climate Change IDPs: Disaster and Development Induced Displacement and Resettlement Models and their Integration," *International Journal of Climate Change: Impacts and Responses*, 2(4): 89–99.

Biermann, F. and Boas, I. (2010) "Preparing for a Warmer World: Towards a Global Governance System to Protect Climate Refugees," *Global Environmental Politics*, 10(1): 60–68.

Birkmann, J., Garschagen, M., Fernando, N., Tuan, V., Oliver-Smith, A. and Hettige, S. (2012) "Dynamics of Vulnerability: Relocation in the Context of Natural Hazards and

Disasters," in Birkmann, J. (ed.), *Measuring Vulnerability to Natural Hazards* (2nd edn.), Tokyo: United Nations University Press.

Black, R., Bennett, S., Thomas, S. and Beddington, J. (2011) "Migration as Adaptation," *Nature*, 478: 447–449.

Borras, S.M. and Franco, J.C. (2012) "Global Land Grabbing and Trajectories of Agrarian Change: A Preliminary Analysis," *Journal of Agrarian Change*, 12(1), DOI:10.1111/j.1471-0366.2011.00339.x.

Bronen, R. (2008) "Alaskan Communities' Rights and Resilience," *Forced Migration Review*, 31: 30.

Bronen, R. (2011) "Climate Induced Community Relocations: Creating an Adaptive Governance Framework Based on Human Rights Doctrine," *NYU Review of Law and Social Change*, 35: 356–406.

Bryssea, K., Oreskes, N., O'Reilly, J. and Oppenheimer, M. (2013) "Climate Change Prediction: Erring on the Side of Least Drama?" *Global Environmental Change*, available online at: http://dx.doi.org/10.1016/j.gloenvcha.2012.10.008.

Burbridge, P., Norgaard, R.B. and Hartshorn, G.S. (1988) "Environmental Guidelines for Resettlement Projects in the Humid Tropics," FAO, Environment and Energy Paper 9.

Cernea, M. (1990) "Poverty Risks from Population Displacement in Water Resource Projects," Development Discussion Paper No. 355, Harvard Institute for International Development.

Cernea, M. (1996) *Eight Main Risks: Impoverishment and Social Justice in Resettlement*, Washington, DC: World Bank Environment Department.

Cernea, M. (1997) "The Risks and Reconstruction Model for Resettling Displaced Populations," *World Development*, 25(10): 1569–1588.

Cernea, M. (2000) "Impoverishment Risks, Safeguards, and Reconstruction: A Model for Population Displacement and Resettlement," in Cernea, M. and McDowell, C. (eds.), *Risks and Reconstruction: Experiences of Resettlers and Refugees*, Washington, DC: World Bank.

Cernea, M. and Guggenheim, S. (1993) *Anthropological Approaches to Resettlement*, Boulder, CO: Westview Press.

Cernea, M.M. and McDowell, C. (2000) *Risk and Reconstruction: Experiences of Settlers and Refugees*, Washington, DC: World Bank.

Clark, D., Fox, J. and Treakle, K. (eds) (2003) *Demanding Accountability: Civil Society Claims and the World Bank Inspection Panel*, Lanham, MD: Rowman & Littlefield.

CNA (2011) *National Security and the Threat of Climate Change*, Alexandria, VA: CNA Corp.

Colson, E. (1971) *The Social Consequences of Resettlement: The Impact of the Kariba Resettlement Upon the Gwembe Tonga*, Manchester: Manchester University Press.

Colson, E. (1999) "Gendering Those Uprooted by 'Development'," in Indra, D. (ed.), *Engendering Forced Migration: Theory and Practice*, New York: Berghahn Books.

Correa, E. (2011a) *Populations at Risk of Disaster: A Resettlement Guide*, Washington, DC: World Bank.

Correa, E. (2011b) *Preventive Resettlement of Populations at Risk of Disaster*, Washington, DC: World Bank.

de Sherbinin, A., Castro, M. and Gemenne, F. (2010) "Preparing for Population Displacement and Resettlement Associated with Large Climate Change Adaptation and Mitigation Projects," Background Paper for the Bellagio Conference on Preparing for Population Displacement and Resettlement Associated with Large Climate Change Adaptation and Mitigation Projects, Bellagio, Italy, November 2–6.

de Sherbinin, A., Warner, K. and Ehrhart, C. (2011) "Casualties of Climate Change," *Scientific American*, 64.

de Sherbinin, A., Castro, M., Gemenne, F., Cernea, M.M., Adamo, S., Fearnside, P.M., Krieger, G., Lahmani, S., Oliver-Smith, A., Pankhurst, A., Scudder, T., Singer, B., Tan, Y., Wannier, G., Boncour, P., Ehrhart, C., Hugo, G., Pandey, B. and Shi, G. (2011) "Preparing for Resettlement Associated with Climate Change," *Science*, 334: 456–457.

de Sherbinin, A., Levy, M., Adamo, S., MacManus, K., Yetman, G., Mara, V., Razafindrazay, L., Goodrich, B., Srebotnjak, T., Aichele, C. and Pistolesi, L. (2012) "Migration and Risk: Net Migration in Marginal Ecosystems and Hazardous Areas," *Environmental Research Letters*, 7, 045602.

de Wet, C. (2006) "Risk, Complexity and Local Initiative in Involuntary Resettlement Outcomes," in de Wet, C. (ed.), *Towards Improving Outcomes in Development Induced Involuntary Resettlement Projects*, Oxford and New York: Berghahn Books.

Deng, F. and Cohen, R. (1999) "Masses in Flight: The Global Crisis of Internal Displacement, and: The Forsaken People: Case Studies of the Internally Displaced," *Human Rights Quarterly*, 21(2): 541–544.

Displacement Solutions (2013) "The Peninsula Principles on Climate Displacement with States," Geneva: Displacement Solutions. http://displacementsolutions.org/peninsula-principles/.

Downing, T. and Garcia-Downing, C. (2009) "Routine and Dissonant Cultures: A Theory about the Psycho-socio-cultural Disruptions of Involuntary Resettlement and Ways to Mitigate Them without Inflicting Even More Damage," in Oliver-Smith, A. (ed.), *Development and Dispossession: The Crisis of Development, Forced Displacement and Resettlement*, Santa Fe: SAR Press.

Dwyer, T. (2010) "Refugee Integration in the United States: Challenges and Opportunities," Church World Service Immigration and Refugee Program.

Economist, The (2012) "Shifting the Problem: A Massive Resettlement Project in Northern China is not All it Seems," March 24.

Ferris, E. (2011) "Planned Relocations, Disasters and Climate Change," prepared for Conference on Climate Change and Migration in the Asia-Pacific: Legal and Policy Responses, Sydney, November 10–11.

Ferris, E. (2012) *Protection and Planned Relocations in the Context of Climate Change*, UNHCR Legal and Protection Policy Research Series, Geneva: UNHCR.

Field, C.B., Barros, V., Stocker, T.F., Qin, D., Dokken, D.J., Ebi, K.L., Mastrandrea, M.D., Mach, K.J., Plattner, G.K., Allen, S.A., Tignor, M. and Midgley, P.M. (eds.) (2012) *Managing the Risks of Extreme Events and Disasters to Advance Climate Change Adaptation. A Special Report of Working Groups I and II of the Intergovernmental Panel on Climate Change*, Cambridge: Cambridge University Press.

Fjelde, H. and von Uexkull, N. (2012) "Climate Triggers: Rainfall Anomalies, Vulnerability and Communal Conflict in Sub-Saharan Africa," *Political Geography*, 31: 444–453.

Fordham, M. and Enarson, E. (1998) *The Gendered Terrain of Disaster: Through Women's Eyes*, Miami: Laboratory for Social and Behavioral Research, Florida International University.

Foresight Project on Migration and Environmental Change (2011) *Final Project Report*, London: Government Office for Science.

Fox, J.A. and David Brown, L. (eds.) (1998) *The Struggle for Accountability: The World Bank, NGOs, and Grassroots Movements*, Cambridge, MA: MIT Press.

Gaber, J., Gaber, S., Vincent, J. and Boellstorff, D. (2004) "An Analysis of Refugee Resettlement Patterns in the Great Plains," *Great Plains Research*, 14: 165–183.

Hansen, A. (1993) "African Refugees: Defining and Defending Their Human Rights," in

Cohen, R., Hyden, G. and Nagan, W. (eds.), *Human Rights and Governance in Africa*, Gainesville: University Presses of Florida.

Hansen, A. and Oliver-Smith, A. (eds.) (1982) *Involuntary Migration and Resettlement: The Problems of Dislocated Peoples*, Boulder, CO: Westview Press.

Hansen, J., Sato, M. and Ruedy, R. (2012) "Increasing Climate Extremes and the New Climate Dice," unpublished paper released August 10, available online at: http://twileshare. com/uploads/James_Hansen_2012_DiceDataDiscussion1.pdf.

Hsiang, S.M., Meng, K.C. and Cane, M.A. (2011) "Civil Conflicts are Associated with the Global Climate," *Nature*, 476: 438–441.

IASC (InterAgency Standing Committee) (2006) *IASC Operational Guidelines on the Protection of Persons in Situations of Natural Disasters*, Washington, DC: Brookings-Bern Project on Internal Displacement.

IASC (2010) *IASC Framework on Durable Solutions for Internally Displaced Persons*, Washington, DC: Brookings-Bern Project on Internal Displacement.

International Federation of Red Cross and Red Crescent Societies (IFRC) (2004) World *Disasters Report*, Geneva: IFRC.

Johnson, C.A. (2012) "Governing Climate Displacement: The Ethics and Politics of Human Resettlement," *Environmental Politics*, 21(2): 308–328.

Kedia, S. (2009) "Health Consequences of Dam Construction and Involuntary Resettlement," in Oliver-Smith, A. (ed.), *Development and Dispossession: The Crisis of Development Forced Displacement and Resettlement*, Santa Fe: SAR Press.

Kelman, I. (2008) "Island Evacuation," *Forced Migration Review*, 31, October.

Koenig, D. (1995) "Women and Resettlement," in Gallin, R. and Ferguson, A. (eds.), *The Women and International Development Annual, Volume 4*, Boulder, CO: Westview: 21–49.

Koivurova, T. (2007) "International Legal Avenues to Address the Plight of Victims of Climate Change: Problems and Prospects," *Journal of Environmental Law and Litigation*, 22: 267–299.

Koser, K. (2007) "The Global IDP Situation in a Changing Humanitarian Context," UNICEF Global Workshop on IDPs, Brookings Institute.

Lavell, A. and Oppenheimer, M. (2011) "New Dimensions in Disaster Risk, Exposure, Vulnerability, and Resilience," *Managing the Risks of Extreme Events and Disasters to Advance Climate Change Adaptation*, Intergovernmental Panel on Climate Change: Ch. 1.

Leighton, M.T. (2011) "Desertification and Migration," in Johnson, P., Mayrand, K. and Paquin, M. (eds.), *Governing Global Desertification: Linking Environmental Degradation, Poverty and Participation*, Burlington, VT: Ashgate.

Maldonado, J.K., Shearer, C., Bronen, R., Peterson, K. and Lazrus, H. (2013) "The Impact of Climate Change on Tribal Communities in the US: Displacement, Relocation, and Human Rights," in Maldonado, J.K., Pandya, R.E. and Colombi, B.J. (eds.), *Climate Change and Indigenous Peoples in the United States: Impacts, Experiences, and Actions, Climatic Change Special Issue*, 120(3): 601–614.

Martin, S. (2004) *Refugee Women*, Lanham, MD: Lexington Books.

Maskrey, A. (1989) *Disaster Mitigation: A Community Based Approach*, London: Oxfam.

McAdam, J. (2011) "Swimming Against the Tide: Why a Climate Change Displacement Treaty is Not the Answer," *International Journal of Refugee Law*, 23(1): 2–27.

McDowell, C. (2002) "Involuntary Resettlement, Impoverishment Risks, and Sustainable Livelihoods," *Australasian Journal of Disaster and Trauma Studies* 2, available online at: http://www.massey.ac.nz/~trauma/issues/2002-2/mcdowell.htm.

McInerny-Lankford, S., Darrow, M. and Rajamani, L. (2011) *Human Rights and Climate Change: A Review of International Legal Dimensions*, Washington, DC: World Bank.

Mehta, L. (2009) *Displaced by Development: Confronting Marginalisation and Gender Injustice*, New Delhi: Sage Publications.

Nansen Principles (2011) Available online at: http://www.regjeringen.no/upload/UD/Vedlegg/Hum/nansen_prinsipper.pdf.

New, M., Liverman, D., Schroder, H. and Anderson, K. (2011) "Introduction. Four Degrees and Beyond: The Potential for a Global Temperature Increase of Four Degrees and its Implications," *Philosophical Transactions of the Royal Society A*, 369: 6–19.

Oliver-Smith, A. (1977) "Traditional Agriculture, Central Places and Post-disaster Urban Relocation in Peru," *American Ethnologist*, 3(1): 102–116.

Oliver-Smith, A. (1991) "Success and Failures in Post-disaster Resettlement," *Disasters*, 15(1): 12–24.

Oliver-Smith, A. (2005) "Communities after Catastrophe: Reconstructing the Material, Reconstituting the Social," in Hyland, S. (ed.), *Community Building in the 21st Century*, Santa Fe: School of American Research Press.

Oliver-Smith, A. (2009) "Disasters and Diasporas: Global Climate Change and Population Displacement in the 21st Century," in Crate, S.A. and Nuttall, M. (eds.), *Anthropology and Climate Change: From Encounters to Actions*, Walnut Creek, CA: Left Coast Press.

Oliver-Smith, A. (2010) *Defying Displacement: Grass Roots Resistance and the Critique of Development*, Austin, TX: University of Texas Press.

Penz, P., Drydyk, J. and Bose, P. (2012) *Displacement by Development: Ethics, Rights and Responsibilities*, Cambridge: Cambridge University Press.

Perry, R. and Mushkatel, A. (1984) *Disaster Management: Warning, Response and Community Relocation*, Westport, CT: Quorum Books.

Portes, A. and Manning, R. (1986) "The Immigrant Enclave Theory and Empirical Examples," in Olzak, S. and Nagel, J. (eds.), *Competitive Ethnic Relations*, Orlando, FL: Academic Press.

Schellnhuber, H.J., Cramer, W., Nakicenovic, N., Wigley, T. and Yohe, G. (2006) *Avoiding Dangerous Climate Change*, Cambridge: Cambridge University Press.

Scudder, T. (1981) "What it Means to be Damned: The Anthropology of Large-scale Development Projects in the Tropics and Subtropics, *Engineering & Science*, XLIV(4): 9–15.

Scudder, T. (2005) *The Future of Large Dams*, London: Earthscan.

Scudder, T. (2009) "Resettlement Theory and the Kariba Case: An Anthropology of Resettlement," in Oliver-Smith, A. (ed.), *Development and Dispossession: The Crisis of Development Forced Displacement and Resettlement*, Santa Fe: SAR Press.

Scudder, T. and Colson, E. (1982) "From Welfare to Development: A Conceptual Framework for the Analysis of Dislocated People," in Hansen, A. and Oliver-Smith, A. (eds.), *Involuntary Migration and Resettlement*, Boulder, CO: Westview Press.

Singer, A. and Wilson, J.H. (2007) "Refugee Resettlement in Metropolitan America," *Migration Information Source*.

Sphere Project (2000) *The Sphere Project: Humanitarian Charter and Minimum Standards of Disaster Response*, Oxford: Oxfam.

Trenberth, K.E. (2012) "Framing the Way to Relate Climate Extremes to Climate Change," *Climatic Change*, 115: 283–290.

Turton, D. (2006) "Who is a Forced Migrant?" in de Wet, C. (ed.), *Development-induced Displacement: Problems, Policies, and People*, Oxford: Berghahn Books.

UNFCCC (UN Framework Convention on Climate Change) (2010) *Enhanced Action on Adaptation*, available online at: http://unfccc.int/resource/docs/2010/cop16/eng/07a01.pdf#page=4.

UNOCHA (UN Office for the Coordination of Humanitarian Affairs), IDMC (International Displacement Monitoring Centre) and NRC (Norwegian Refugee Council) (2009) *Monitoring Disaster Displacement in the Context of Climate Change*, Geneva: IDMC.

Vandergeest, P., Idahosa, P. and Bose, P.S. (2007) *Development's Displacements: Ecologies, Economies and Cultures at Risk*, Vancouver: University of British Columbia Press.

Walker, P., Glasser, J. and Kambli, S. (2012) "Climate Change as a Driver of Humanitarian Crises and Response," report issued by the Feinstein International Center, Tufts University.

WCD (World Commission on Dams) (2000) *Dams and Development: A New Framework for Decision Making*, London: Earthscan.

Wisner, B., Cannon, T., Blaikie, P. and Davis, I. (2004) *At Risk: Natural Hazards, People's Vulnerability and Disasters*, London: Routledge.

World Bank (1990) "Operational Directive 4.30: Involuntary Resettlement," *The World Bank Operational Manual*, Washington, DC: World Bank.

World Bank (2004) *Involuntary Resettlement Sourcebook*, Washington, DC: World Bank.

Yenotani, M. (2011) *Displacement Due to Natural Hazard-Induced Disasters. Global Estimates for 2009 and 2010*, Oslo: International Displacement Monitoring Centre (IDMC) and Norwegian Refugee Council (NRC).

PART III
At-Risk Populations

13

PROTECTING NON-CITIZENS IN SITUATIONS OF CONFLICT, VIOLENCE AND DISASTER

Khalid Koser

Introduction

In recent years, significant numbers of non-citizens have been displaced by conflict, violence and disasters, in countries where they reside and work, and in some cases where they may become "trapped" in transit. They include migrant workers (both regular and irregular), but also asylum-seekers, refugees and stateless persons. Non-citizens have been displaced for example by invasion in Lebanon in 2006, xenophobic violence in South Africa in 2008, revolution in Libya in 2011, civil war in Côte d'Ivoire in 2010–11, flooding in Thailand also in 2011, and more recently, by ongoing conflict in Syria.

Displacement usually results in vulnerability, for example as a result of a lack of shelter, loss of property and access to livelihoods, a lack of access to services, and discrimination. As a result, the displaced often are more likely than those who are not displaced to become victims of gender-based violence, be separated from family members, be excluded from education, and be unemployed. While there has been very little research on the experiences of non-citizens during humanitarian crises, it is reasonable to suppose that many of them may be even more vulnerable to displacement, and suffer its consequences more acutely, than citizens. Reasons include that they may not speak the local language or understand the culture, they may lack job security, and they may lack a social safety net. In some of the recent examples of non-citizens caught up in crises, they have been affected as bystanders, whereas in other cases they have been deliberately targeted. Equally, it may be harder for displaced non-citizens to resolve their displacement, especially if they are unable or unwilling to return to their country of origin, and they may face specific challenges in regaining property, employment, and identification cards in the country where they have been displaced. Even where they can return to their countries of origin, they may face significant reintegration challenges there, too.

Such vulnerabilities are likely to be heightened for irregular migrants, whose lack of legal status limits their rights in many countries, who may not be willing to access assistance from the state experiencing crisis even where it is available, and for whom return home is often problematic. Special attention is also required to the situation of asylum-seekers, refugees and stateless persons, whose rights as enshrined under international law may be hard to guarantee during crises.

As in other examples of "crisis migration," in many cases non-citizens who become displaced may fall into "protection gaps." The rights of non-citizens caught in crises are not explicitly stated in existing laws, conventions, or standards. Instruments that cover displacement do not deal with non-citizens; while those that cover non-citizens do not deal with displacement. Neither is responsibility for protecting and assisting non-citizens during crises clearly ascribed. In Lebanon and Libya, developed nations evacuated their own citizens; whereas in South Africa, Thailand and Côte d'Ivoire, and currently in Syria, where most of those affected were citizens of states that either lacked the political will or the capacity to assist them, a variety of agencies tried to assist. These included local and international non-governmental organizations (NGOs), and international organizations, especially the International Organization for Migration (IOM) and the UN High Commissioner for Refugees (UNHCR).

The displacement of non-citizens is likely to become more common in the future. For example, the expansion of Chinese interests in Sub-Saharan Africa is already resulting in large numbers of migrants working in unstable states (Duchâtel and Bates 2012). Climate change may make many of the developing states where more migrants are moving to work susceptible to an increasing frequency of natural disasters—Thailand is a recent case in point. This combines with targeted violence against immigrants as a result of rising xenophobia in many countries around the world at the moment. And as poor and developing countries continue to export migrant workers, assistance and protection during times of crisis is likely to continue to fall significantly on the international community, as such sending countries may lack the capacity effectively to protect their own citizens overseas. At the same time, it is worth noting recent initiatives by major labor-exporting countries such as Bangladesh and Philippines to protect the rights of their migrant workers abroad—for example, within the framework of the Colombo Process.

Against this background, this chapter has three main objectives. The first is to review the existing legal, normative, and institutional and operational frameworks that apply to non-citizens during times of displacement. The second is to describe and analyze recent examples of the displacement of non-citizens and responses to their displacement—and here the paucity of data and evidence in most cases is worth noting. The third is to consider lessons learned to inform the development of good practice and guiding principles.

The legal, normative and institutional framework

Legal and normative framework

Human rights and humanitarianism proscribe discrimination on the basis of nationality (and legal status), and so the nationality of people affected by crises should be irrelevant. Although they do not explicitly make reference to migrants, the core treaties of human rights law clearly extend to all migrants in all situations—irrespective of legal status (Grant 2005). The International Covenant on Civil and Political Rights (ICCPR) states in Article 2.1 that:

> Each State Party to the present Covenant undertakes to respect and to ensure to all individuals within its territory and subject to its jurisdiction the rights recognized in the present Covenant, without distinction of any kind, such as race, color, sex, language, religion, political or other opinion, national or social origin, property, birth or other status.

International Humanitarian Law (IHL) does explicitly refer to non-citizens in Section II of Part II of the Fourth Geneva Convention entitled "Aliens in the Territory of a Party to Conflict," but applies only to situations of armed conflict, and is not applicable for example in natural disaster or other crisis situations. Article 4 stipulates that:

> Persons protected by the Convention are those who, at a given moment and in any manner whatsoever, find themselves, in case of a conflict or occupation, in the hands of a Party to the conflict or Occupying Power of which they are not nationals.

Yet the rights of non-citizens during humanitarian crises or displacement are not explicitly enumerated either in international treaties or standards that protect the rights of people who are displaced, or in those that protect the rights of migrants.

For non-citizens displaced inside the country where they live or work, the instrument with closest relevance is the Guiding Principles on Internal Displacement (Guiding Principles). These define internally displaced persons (IDPs) as follows:

> . . . persons or groups of persons who have been forced or obliged to flee or to leave their homes or places of habitual residence, in particular as a result of or in order to avoid the effects of armed conflict, situations of generalized violence, violations of human rights or natural or human-made disasters, and who have not crossed an internationally recognized State border.
>
> *(UNOCHA 1998, 2)*

This definition incorporates a non-exclusive list of examples of causes of displacement that includes natural disasters, and thus covers a wide range of crisis situations.

It is not, however, clear whether the Guiding Principles extend to non-citizens, and there are no recent examples of states or relevant organizations interpreting the Guiding Principles in this way. Certainly they are not explicitly mentioned either in the definition or elsewhere in the Guiding Principles, but an argument may be made that upon being displaced, non-citizens are also leaving ". . . their homes or places of habitual residence" and thus can be defined as IDPs (Principle 29(1)). The extent to which this argument applies to short-term or temporary migrant workers is debatable; and neither is it clear whether the Guiding Principles apply to irregular migrants who are unlikely to be able to demonstrate formally that they have a home or place of habitual residence. Furthermore, the Guiding Principles enumerate the rights to which all citizens—whether or not displaced—are entitled. But in most cases non-citizens are legally entitled to fewer rights than citizens.

Even if it can be argued that the Guiding Principles apply to non-citizens, it is important to acknowledge that the protection regime for all IDPs remains relatively weak. The Guiding Principles comprise an expert document that is not legally binding. About thirty countries have developed their own national laws and policies on internal displacement, although the scope of these laws and policies varies significantly, with some for example limited to armed conflict only. The African Union Convention for the Protection and Assistance of Internally Displaced Persons in Africa (Kampala Convention) is the only regional binding instrument on IDPs. Yet even where national and regional laws and policies are in place, implementation is often challenging and incomplete (Koser 2008).

For non-citizens displaced across an international border, the most relevant instrument is the 1951 Convention relating to the Status of Refugees (and its subsequent Protocol) (1951 Refugee Convention) and regional extensions. Article 1A(2) provides the following definition of a refugee:

> A person who owing to a well-founded fear of being persecuted for reasons of race, religion, nationality, membership of a particular social group or political opinion, is outside the country of his nationality and is unable or, owing to such fear, is unwilling to avail himself of the protection of that country; or who, not having a nationality and being outside the country of his former habitual residence as a result of such events, is unable or, owing to such fear, is unwilling to return to it . . .

There are at least three scenarios where this definition might apply to non-citizens displaced across borders. The first is where they have already been recognized as a refugee in the country from which they are subsequently displaced. Crossing an international border involuntarily to a third country would not affect the legal status of a refugee, although such movement needs to be distinguished from "secondary movements" where refugees move to third countries for reasons unrelated to the 1951 Refugee Convention—for example, to find work. The second is where non-citizens have applied for asylum in the country from which they are subsequently displaced, and the outcome of their application is pending. A number of registered asylum-seekers and refugees were displaced into Tunisia during the recent crisis in Libya. The third is where non-citizens have not yet made an application for refugee status, but may satisfy the criteria.

Displacement may be the catalyst to apply for refugee status, for example by asylum-seekers who have recently arrived and not yet submitted a claim for asylum; for irregular migrants who have thus far avoided engaging with authorities; or for migrant workers who cannot return to their country of origin. A number of Zimbabweans displaced by xenophobic violence in South Africa successfully applied for asylum rather than return to Zimbabwe, for example, while some Somalis and Eritreans who had been working without authorization in Libya have also applied for asylum once they were displaced into Tunisia. But for the most part, non-citizens displaced across an international border would not be recognized as refugees.

The 1951 Refugee Convention applies mainly in situations of conflict or political unrest and violence where there is individualized persecution, and certainly not to humanitarian crises triggered by other events and processes such as those influenced by environmental change, pandemics or nuclear accidents. Non-citizens who may be displaced across a border, for example by natural disaster or as a result of the effects of climate change, fall into a "protection gap" that they share even with citizens of the country from which displacement occurs. Efforts at the multilateral level to fill this gap are focusing on the development and consolidation of normative principles that can inform regional or national laws and policies on environmental migration. The Nansen Principles, for example, build on existing norms in international law, and identify the responsibility of local, national, and international actors (Nansen Conference 2011). A recommendation of this chapter is that where new standards are developed on displacement, such as the Nansen Principles, they make explicit reference to the situation of displaced non-citizens.

Turning to instruments that provide protection for migrants: neither of the two International Labour Organization (ILO) instruments relating to migrant workers, nor the UN International Convention on the Protection of the Rights of All Migrant Workers and their Families (ICMW), mention migrants in crisis; and they do not consider provisions for their displacement. They do nevertheless enumerate a comprehensive list of protections for migrants and their families, including provisions barring the arbitrary expulsion of migrant workers. ILO Convention No.97 (1949) articulates the principle of equal treatment with national workers, for example regarding working conditions, trades union membership, accommodation and legal proceedings, but is not explicit concerning involuntary departure from (and subsequent return to) work or accommodation, and neither does it apply to irregular migrant workers. ILO Convention No.143 (1975) again is not explicitly applicable, but does impose an obligation on states ". . . to respect the basic human rights of all migrant workers," confirming its applicability to irregular migrant workers (Article 1).

ICMW explicitly covers irregular migrants, but does not specifically mention migrants in crisis situations. This omission has been described as a significant gap in a Convention that purports to be overwhelmingly comprehensive on migrant workers' rights (Jureidini 2011). It is also worth noting that the Convention has still not been ratified by any major destination countries for migration.

One of the consequences of the fact that the rights of non-citizens during crises are not explicitly stated in international laws, conventions or relevant standards is that national laws and policies that often draw on international guidelines are equally mute on the rights of non-citizens during crises. In an extensive evaluation of the humanitarian response to the displacement of foreign nationals in South Africa in 2008, for example, it was noted that the Disaster Management Act of 2002 did not provide sufficient guidance on the rights of non-citizens during disasters (FMSP 2009).

Institutional responsibilities

Responsibilities for protecting and assisting non-citizens are not clearly assigned. Reflecting human rights law and IHL, the Guiding Principles clearly assign primary responsibility to the state where the displacement takes place. Principle 3(1) states that, "National authorities have the primary duty and responsibility to provide protection and humanitarian assistance to internally displaced persons within their jurisdiction." The same assumption is made in the Inter-Agency Standing Committee (IASC) Operational Guidelines on the Protection of Persons in Situations of Natural Disasters (IASC 2011).

Yet it has been suggested that governments of the country of origin also have a legal, as well as civil and moral, responsibility to protect their own citizens abroad (Jureidini 2011). The Vienna Convention on Consular Relations (1963) conceives consular functions to include helping and assisting their nationals (Article 5e) as well as "protecting in the receiving State the interests of the sending State and of its nationals, both individuals and bodies corporate, within the limits permitted by international law" (Article 5a). At the same time the Convention places responsibility on the host state to facilitate contact. Article 36(1)(a) stipulates that:

> consular officers shall be free to communicate with nationals of the sending State and to have access to them. Nationals of the sending State shall have the same freedom with respect to communication with and access to consular officers of the sending State.

Whether it is the sending or receiving state that has primary responsibility for protecting non-citizens in times of crisis, often neither has the political will and/or capacity to provide adequate protection and assistance, in which case attention turns to the international community. But here, responsibility is equally unclear.

No UN agency has a mandate to protect or assist IDPs. UNHCR has a mandate focused on asylum-seekers, refugees and stateless people—important categories of non-citizens—but not migrant workers. The IOM is the world's leading migration agency, but is outside the UN system and does not have a protection mandate; although it is worth mentioning IOM's recently developed Migration Crisis

Operational Framework (MCOF) intended to provide a "migration lens" on crises. As a result, in most of the case studies below, where migrant workers have not been evacuated by their own countries, they have been assisted in an unpredictable manner by a combination of IOM, UNHCR, the International Committee of the Red Cross (ICRC) (in conflict situations), and local and international NGOs. In the case of responses by the international community more broadly, political and strategic interests may be relevant—for example, in explaining the contrasting attention paid to the crisis in Libya and its consequences as compared with Côte d'Ivoire, as may be funding and donor priorities.

Finally, in a number of the cases considered in the next section, private companies have also played a significant role in organizing the evacuation of their employees from crisis situations. While the financial clout and organization of corporations is an advantage, most lack standard operating procedures (SOPs) to handle crises, and many do not have risk assessment units or senior security officer positions. Although not directly on point, the 2011 Guiding Principles on Business and Human Rights has some relevance. Principle 7(b) states that:

> Because the risk of gross human rights abuses is heightened in conflict-affected areas, States should help ensure that business enterprises operating in those contexts are not involved with such abuses, including by providing adequate assistance to business enterprises to assess and address the heightened risks of abuses, paying special attention to both gender-based and sexual violence.

Case studies

Lebanon, 2006

The crisis in Lebanon resulted from an invasion by Israel against Hizbollah forces, and lasted thirty-three days from July 12 to August 14, 2006. At the time there were hundreds of thousands of foreign nationals living and working in Lebanon. A non-exhaustive list drawn mainly from press reports includes: Sri Lanka (80,000), Canada (50,000), Philippines (30,000), Australia (25,000), US (25,000), UK (22,000), France (20,000), Bangladesh (20,000), Egypt (15,000), India (12,000), Sweden (7,000), Denmark (4,100), Nepal (4,000), Venezuela (4,000), Germany (2,600), Greece (2,500–5,000), Russia (1,500), Romania (1,200), Armenia (1,200), Ukraine (1,200), Poland (329), Moldova (240), Mexico (216 wishing to be evacuated), Bulgaria (207 wishing to be evacuated), Iran (200), Ireland (161), Cyprus (102), Croatia (58), Slovakia (56), Peru (50), Kazakhstan (31), and Malaysia (1). To this should be added Italy, Ethiopia, Sudan, Ghana, Vietnam, Cameroon, Seychelles and Madagascar for all of which countries numbers were unknown (Jureidini 2011).

In total, some 70,000 foreign nationals were evacuated from Lebanon by various, mostly richer, countries with significant populations present. While these evacuations prioritized the nationals of the relevant countries, they also extended to other nationals. Thus the Australian Defence Force (ADF) evacuated 5,000 Australians but also 1,350 other "approved foreign nationals"; the French sea and air mission, "Opération Baliste," included Lebanese residents with dual Lebanese and French passports as well as nationals of other countries including European and US citizens; the Indian Navy also evacuated Nepalese and Sri Lankan nationals; less than half of those evacuated by the UK were British citizens, and the rest were made up of some fifty other nationalities; while Greece also evacuated European Union (EU) and US citizens.

Poorer countries with significant numbers of citizens working in Lebanon at the time of the invasion, like Sri Lanka, the Philippines and Ethiopia, had neither the financial means nor organizational capacity to arrange for large-scale evacuations of their citizens. In some cases bus convoys were organized across local borders, mainly to Syria. But on the whole they relied heavily on IOM to pay for and arrange relief, accommodation, transportation and repatriation for their nationals. IOM received around €11 million from the EU, US$1 million from the US, US$2 million from Belgium and AUD600,000 from Australia for the evacuation of migrants from developing countries. Under IOM auspices, more than 13,000 "Third Country Nationals" were evacuated, often in collaboration with the Catholic NGO Caritas.

Responsibility for protecting and assisting non-citizens who were not evacuated was not specifically assigned, but they were assisted alongside nationals and IDPs in Lebanon by the considerable Lebanese NGO and civil society network, comprising more than 6,000 organizations. These delivered health, food and nutrition, water and sanitation, logistics, legal assistance, shelter, and common services to affected people, including IDPs, and including non-citizens whether displaced or not. The role of the Lebanese government was mainly one of coordination, and it established a Higher Relief Council as the main body to coordinate relief activities.

In terms of international efforts, an appeal for humanitarian aid was issued in late July by the UN Office for the Coordination of Humanitarian Affairs (UNOCHA) following a consultative process within the IASC and with the assistance of the Red Cross and Red Crescent Movement. A regional Task Force was established to liaise with the Israeli authorities. The United Nations Relief and Works Agency for Palestine Refugees in the Near East (UNRWA), the agency responsible for Palestinian refugees (but that does not have a protection mandate) continued its work with some 400,000 Palestinian refugees in Lebanon, particularly in Ein El Helwa, the largest camp in the south, which was most exposed to the conflict. The World Health Organization (WHO), the Swiss Development Cooperation (SDC), World Vision and Médecins Sans Frontières (MSF) provided various healthcare facilities and supplies, assisting hospitals throughout the main affected areas of the country, and continued after the war assisting those returning or attempting to return to their homes.

For the duration of the conflict, as many as 10,000 people arrived at the Syrian border each day, including tourists from the Gulf States and many poor Palestinians. UNHCR sent emergency mobile teams to the main transit routes between

Lebanon and Syria. The government of Syria and the Syrian Arab Red Crescent Society (SARC) took a lead role in registering, accommodating and assisting the most vulnerable of the people displaced there from Lebanon.

South Africa, 2008

Unlike the other cases considered in this chapter, the displacement of foreign nationals in South Africa in May 2008 resulted from violence targeting them specifically, and they exclusively were displaced, rather than as part of a wider cross-section of the population.

The violence started in Gauteng Province on May 11, and by May 22 had spread to Durban, KwaZulu-Natal, Free State, North West and Limpopo Provinces. It was targeted at foreign nationals living mainly in townships and informal settlements in urban areas within the affected provinces. Widespread robbery and looting of foreign-owned businesses, and theft of personal property took place, either from premises still occupied by foreigners or from their residences after they had fled. Victims included nationals of Bangladesh, Burundi, the Democratic Republic of Congo (DRC), Kenya, Malawi, Mozambique, Nigeria, Pakistan, Somalia and Zimbabwe, as well as South Africans from minority language groups. Sixty-two deaths were reported, of which about one-third were of South Africans. Estimates of the total number of people displaced ranged from 80,000 to 200,000 (FMSP 2009).

The humanitarian response to the displacement has been characterized as falling into three main phases. During the first phase, large-scale repatriation to countries of origin took place, often self-organized by foreign nationals themselves, and especially to neighboring countries. The Mozambican and Zimbabwean Embassies also provided a repatriation service to their nationals, and IOM helped provide safe passage. It was reported that, by May 27, some 27,500 Mozambicans had returned home.

At the same time, however, significant deportation also took place, despite assurances by the Department of Home Affairs that the victims of violence would not be deported regardless of their immigration status. According to one estimate, 17,000 Zimbabweans were deported during June and July 2008. Shelter and basic welfare was provided on an ad hoc basis to those who remained in situ and those who were displaced internally. During these first days of the displacement the key actor was civil society, as the South African government geared up structures provided for in the Disaster Management Act.

During the second phase, the displaced were centralized in government shelters, processes were established to register them, and a combination of solutions was pursued, including repatriation home or return and reintegration to the place in South Africa from which displacement occurred. At this stage the government had assumed primary responsibility, assisted by UNHCR and other UN agencies as well as civil society. A particular challenge during this phase concerned documentation and registration. Efforts were made to replace lost documentation, fast track new asylum and refugee documentation, and provide shelter access, aid distribution registration and temporary permits.

By the third phase, lasting to about the end of September 2008, most of the government shelters were closed, and the emphasis was on return and reintegration to places of residence and work in South Africa. Reintegration assistance was provided by the government, various UN agencies and a range of civil society actors. Particular challenges arose around ensuring the cessation of violence and justice mechanisms for the perpetrators of the violence, documentation, regaining a livelihood basis, and trauma (FMSP 2009).

The key findings of an evaluation of the humanitarian response by the Forced Migration Studies Programme (FMSP) were: a lack of experience and established systems, a lack of government leadership especially in the early phases of the crisis, the fragmentation of civil society, and confusion regarding the rights of displaced foreigners. The evaluation recommendations concerned: greater communication, better consultation and participation, more coordination, better information collection and management, more effective emergency preparedness and contingency planning, and an internal evaluation of the response.

Côte d'Ivoire, 2010–2011

While it is clear from press and agency reports that the 2010–11 civil war in Côte d'Ivoire impacted—and in some cases deliberately targeted—migrants working there, data, research and evidence on the experiences of these migrant workers and the responses of the government, countries of origin and the international community remains scarce. In some ways this gap in information, certainly as compared to that available for Lebanon, Libya and South Africa, for example, reflects the relative lack of attention generally paid to the Côte d'Ivoire crisis by the international community.

After Independence, President Félix Houphouët-Boigny instituted a regional policy that encouraged the free movement of people within West Africa to Côte d'Ivoire. As a result the share of foreigners living there increased from 5 percent of the population in 1950 to 26 percent by 1998; indeed the World Bank estimated in 2009 that Côte d'Ivoire was one of the top 12 destinations for international migrants worldwide (UNDP 2011). In 2010, before the onset of violence following the November elections, some three million migrants were estimated to be working in Côte d'Ivoire, mainly originating in Mali and Burkina Faso, but also including significant numbers from the other neighboring countries, Ghana, Guinea and Liberia, as well as further afield, including Benin, Mauritania, Niger and Togo.

The 2010–11 Ivorian crisis began after Laurent Gbagbo, President since 2000, was proclaimed winner of the Presidential election in November 2010. The opposition, and a number of countries, organizations and leaders around the world, claimed that the election had in fact been won by the opposition leader Alassane Ouattara. Sporadic violence followed, and Ouattara's forces launched a military offensive that besieged key targets in the capital city Abidjan. On April 11, 2011, Gbagbo was arrested and sent for trial by the International Criminal Court (ICC) in The Hague in November of that year.

Migrant workers were impacted in at least three ways. First, some were specifically targeted by Gbagbo supporters, apparently on the suspicion that they were supporting the opposition forces. This targeted violence continued for at least a few months after Gbagbo was arrested. Second, they were caught up generally in the violence and political instability, and were particularly susceptible as foreigners, as poor people, and as people often working in the informal sector and sometimes without legal authorization. Third, the livelihoods of many migrant workers were disrupted by the violence and unrest.

It is estimated that the civil war in Côte d'Ivoire displaced up to one million people from their homes, and the effects on migrant workers specifically need to be understood within this wider displacement context. It appears that some migrant workers from Mali and Mauritania initially sought shelter at their Embassies in Côte d'Ivoire, before apparently being evacuated by their countries of origin in cooperation with UNOCHA. Conditions in these Embassies were dire, with reports of inadequate living conditions, and without proper access to food, water, and medical care. Presumably those with access to the Embassy mainly comprised the elite, and it is reported that many other migrant workers from these and other countries were displaced inside the country without immediate assistance. In January 2011 the governments of Mauritania, Mali, Burkina Faso and Liberia requested the assistance of IOM to evacuate their citizens. IOM's appeal was severely underfunded, and disrupted by violence. UNHCR provided sporadic assistance to asylum-seekers, refugees and IDPs in Côte d'Ivoire, Ghana, Liberia, Burkina Faso, Guinea, Benin, Nigeria, Mali, Sierra Leone and Togo. Neighboring Mali, Liberia and Burkina Faso also drew up contingency plans to receive refugees from the crisis.

But reports suggest that by far the most common response by migrant workers and other non-citizens displaced by or caught up in the crisis was to return home by themselves. By the peak of the crisis in 2011, it was estimated that 50,000 migrant workers had returned home, mostly without direct assistance from governments or agencies. In part this can be viewed as a traditional coping strategy—it is reported that during Côte d'Ivoire's 2002–03 civil war, about 200,000 migrant workers returned to Burkina Faso and Mali, most of whom subsequently returned again after the civil war. While there is no clear evidence on the fate of those migrant workers who returned home in 2010–11, it may be assumed that the majority also returned to Côte d'Ivoire once the violence had subsided.

Libya, 2011

When the Libyan Civil War began in early 2011 there were an estimated 2.5 million migrant workers in Libya, including a workforce drawn from all over the world associated with the oil industry, as well as significant numbers of sub-Saharan Africans often working in the informal sector and Asian migrants working in construction. It has been estimated by IOM that up to 1.5 million of these migrant workers did not have legal status (UNHCR 2011a).

According to IOM statistics, by June 22, 2011, around half a million migrant workers had left Libya (UNHCR 2011b); 253,957 migrant workers had crossed the border into Tunisia, comprising 60,942 Tunisians and 193,015 third country nationals. A further 183,334 had entered Egypt, comprising 105,821 Egyptians and 77,513 third country nationals. Of the 73,618 who had entered Niger, the majority (69,859) were nationals of Niger, but 3,759 were third country nationals. There are no disaggregated data for entry into Libya's other neighboring countries, but it is assumed that the majority of those entering Algeria (24,050), Chad (43,795) and Sudan (2,800) were nationals of those countries.

Some 60,000 of these evacuated third country migrant workers were subsequently flown home on around 300 flights, coordinated either by their home governments, their employers, or by the international community—as indicated below, there were subsequent criticisms that these international efforts were inefficiently coordinated. Significant numbers returned to Bosnia-Herzegovina, Canada, China, Croatia, Greece, India, Jordan, Lebanon, Macedonia, Morocco, the Netherlands, Nigeria, South Korea, Syria, Turkey, the UK, the US and Vietnam, with the largest number—an estimated 35,000—to China alone.

At least some migrant workers also became internally displaced. It has been estimated that, during the peak of the crisis, there were around 150,000 IDPs in Libya, around 58,000 of them in IDP settlements and camps. Significant internal displacement was reported in Ajdabiya, Derna and Tubruq. A proportion of IDPs in Libya were migrant workers. For example, of those migrant workers evacuated by boat from Misrata to Benghazi (Press TV 2011), not all were subsequently evacuated from the country, and some remained stranded in Benghazi (IOM 2011).

Another category of concern during the crisis was around 3,500 asylum-seekers and 8,000 refugees registered by UNHCR in Libya before the uprising, the majority originating in Iraq, Eritrea, Ethiopia, Palestine, Somalia and Sudan (UNHCR 2011a), as well as some stateless persons (Tuaregs and Sahrawis) (Fiddian-Qasmiyeh 2012). It remains unclear what happened to these people. At least some appear to have escaped Libya: UNHCR reported on 22 June that there were 920 people of concern to the agency at the Saloum border crossing with Egypt; and more than 3,000 refugees and asylum-seekers in Choucha camp on the Tunisian border (UNHCR 2011b). But the origin of these people is unclear. Some may indeed have been asylum-seekers and refugees who were registered in Libya before the uprising, but others may have been migrant workers—and in particular irregular migrants—who crossed the border and claimed asylum.

During the uprising there was very limited access for international organizations in areas held by government forces, and in particular in Tripoli. There was no international presence on the borders with Chad or Niger, where some migrant workers are thought to have been stranded on the Libyan side of the border, and security clearance was not easy for agencies working in eastern Libya. As a result, humanitarian assistance inside Libya was sporadic and geographically limited. IOM evacuated migrant workers from Misrata to Benghazi by boat, and then onwards to the Tunisian border by bus. UNHCR worked with the World Food Program (WFP) and local NGOs, including the Libyan Committee for Humanitarian Aid and Relief (LCHR), to provide shelter, food and non-food items to IDPs in and around Benghazi.

The response of neighboring states to Libyans as well as third country nationals displaced across the border was generally positive. The UN High Commissioner for Refugees, António Guterres, publicly praised the governments of Egypt and Tunisia for keeping their borders open, despite the massive influx of migrants, which placed considerable pressure on public services (*Lancet* 2011). There were serious fuel, milk and water shortages in Tunisia as a result of increased demand from the migrant population, and also compounded by an increase in cross-border smuggling into Libya.

To an extent the pressures on Tunisia and Egypt were mitigated through significant international assistance in supporting migrants fleeing Libya. The evacuation of third country nationals from the Egyptian and Tunisian borders heralded probably unprecedented cooperation between UNHCR and IOM. The two agencies also cooperated in processing "boat arrivals" in Lampedusa and Malta, and in supporting migrants and refugees at border camps. The ICRC assisted people within Libya affected by the conflict, including IDPs and migrant workers. It provided tracing services, and emergency shelter to border camps and IDP camps, as well as to "trapped" populations.

Following the downfall of Gaddafi, there has been international support for resolving some of the consequences of the displacement during the Libyan crisis. UNHCR has announced a series of training workshops for the Department of Justice of the Transitional National Council in Libya, and in collaboration with Mercy Corps for the Libyan Red Crescent Society (LRCS), on durable solutions for IDPs (UNHCR 2011b). It is not clear whether these initiatives will also consider the particular circumstance of non-citizens who have been internally displaced. Housing, land and property issues remain an unresolved challenge for perhaps 70,000 people who remain displaced inside Libya. This number is thought to include some Sub-Saharan Africans (UNHCR 2012).

UNHCR also continues to provide support to the asylum-seekers, refugees, and others of concern to the agency in camps on the borders. Most of the refugees on the border are from countries such as Eritrea, Somalia and Sudan where they cannot easily be returned, and so UNHCR is focusing its efforts on finding resettlement places for them in third countries.

Many Egyptian and Tunisian workers, as well as most of those from third countries who left Libya and were not evacuated home, are reported to have now crossed the borders back into Libya. It is not clear to what extent they have been able to start work again, or what challenges they may have faced—for example, in regaining property. There are particular concerns about many of the Sub-Saharan Africans caught up in the crisis. During the conflict some suffered abuse at the hands of both the rebels and Gaddafi loyalists, under the accusation of being "foreign mercenaries." Since the end of the conflict there have been reports that the National Transitional Council (which governed Libya for a ten-month period after the end of conflict) continued to discriminate against Sub-Saharan Africans, for example in terms of access to justice.

Thailand, 2011

Data and analysis on the displacement that occurred in Thailand as a result of the 2011 floods are not readily available, and this brief section therefore largely relies on web-based reporting.

It is estimated that there are about three million migrant workers in Thailand, mainly from Cambodia, the Lao People's Democratic Republic and Burma; many of whom have irregular status. Migrant workers from Burma are estimated to make up between 5 and 10 percent of Thailand's entire labor force. Restrictions—for example, relating to movement between regions—apply to many of those even with legal status in Thailand.

The floods of 2011 affected as much as one-fifth of Thailand; and the Thai Labour Ministry estimated at the time that about one million migrant workers lived and worked in flood-affected areas. Aid workers estimated in November 2011 that as many as 600,000 migrant workers were stranded by floodwaters. While the approximately 150,000 refugees from Burma in camps on the Thai–Burma border were largely unaffected, some of the 2,000 asylum-seekers and refugees in Thailand were reported to have been affected.

Restrictions on movement between provinces reportedly resulted in many migrant workers staying put even in flood-affected areas, because of concerns about losing their work permits, about arrest or deportation, or at least obstacles to future employment in Thailand. Many therefore rode out the floods, despite harsh conditions and health risks. Furthermore many of the factories and production plants where they had worked were closed down as a result of the floods, meaning they had no income.

The UN system in Thailand coordinated relief efforts with the government, and assistance was provided to Thais and migrant workers alike in flood-affected areas, including by the Thai Red Cross, MSF and Save the Children. These relief efforts were not specifically targeted at migrants; indeed some migrants claimed that they had been deliberately excluded from relief efforts. The ILO conducted an analysis on livelihoods and employment and mobilized some resources for protecting migrant workers and ensuring their access to relief supplies; while UNHCR provided cash grants to displaced people affected by the floods.

At the same time an estimated 120,000 migrant workers were reported to have crossed the border back into Burma (98,237 between September 1 and November 10, according to IOM). In some cases they were reported to have experienced extortion and abuse by immigration officials and police along the way. It was reported that fees ranging from 12,000 to 15,000 baht (400 to 500 US dollars) were being charged at border crossings. Others were deported, having illegally left their provinces of employment. Irregular migrants were also deported, despite pleas by migrant rights groups for a temporary amnesty.

After the floods, many migrant workers who had returned to Burma tried to return to Thailand to resume working there. In the absence of any formal program, many apparently paid smugglers to get them back to Thailand. A joint plan was being discussed between the governments of Burma and Thailand to regularize the re-entry of those who returned to Burma during the floods, but it is not clear to what extent this has been effectively implemented. Equally there were concerns businesses may not be ready to re-employ migrant workers, at least on the previous scale, as they recovered from the floods.

Lessons learned

There are a number of lessons to learn from the preceding analysis and case studies, which in turn can inform the development of effective practice and guiding principles.

Non-citizens may have particular vulnerabilities during and after displacement

The various case studies covered here lend some weight to the assertion in the Introduction that non-citizens may be more vulnerable to displacement than nationals, may suffer its consequences more acutely, and may find achieving durable solutions more challenging. In the case of Lebanon, invasion affected an extant refugee population of some 400,000 Palestinian refugees, who were already in a particularly vulnerable situation prior to the conflict. The same is currently taking place in Syria. In South Africa non-citizens were displaced specifically because they were targeted on the basis of their nationality; some were deported on the basis of their immigration status, and for many, justice mechanisms were absent upon reintegration in South Africa. In Libya and Thailand, many of those displaced both inside and outside the country were irregular migrants, and their lack of legal status compounded their vulnerability, for example to discrimination. In Côte d'Ivoire, many of the affected migrants were from very poor Sub-Saharan African countries to which they could not turn for assistance in evacuation. There is very little research on the extent to which those who return to their countries of origin permanently managed to reintegrate successfully.

The rights of non-citizens during displacement are not explicitly enumerated

The analysis offered in the first part of this chapter is that the rights of non-citizens affected and displaced by conflict and other events and processes are inherent in human rights law and IHL, and probably also in laws, conventions and standards that apply to the displaced and to migrant workers. But nowhere are they explicitly stated. It may be argued that this is not a problem, as in all the cases cited above at least some displaced non-citizens were provided with basic protection and assistance, alongside nationals. But normative principles are important because they provide the foundation to respond to the specific protection and assistance needs of affected people—for example, by informing the development of national laws and policies. As indicated above, there may be grounds to assume that non-citizens have specific needs. Equally as non-citizens—and particularly if in an irregular situation—their rights are legally fewer than those of displaced nationals.

Origin countries have different capabilities to assist in times of crisis

The Vienna Convention on Consular Relations notwithstanding, it is clear from the case studies above that not all countries have the capability to assist their citizens when they are affected by humanitarian crises while abroad. In the case of Burma and Zimbabwe, the political will to assist citizens displaced in Thailand and South Africa respectively was largely absent; indeed, some Zimbabwean migrant workers applied for asylum from Zimbabwe upon displacement. In other cases poorer countries have simply lacked the capacity to assist.

Even the involvement of richer states has not been without controversy. In the aftermath of the evacuation of Canadian citizens from Lebanon, for example, a Canadian Senate Report expressed concerns that:

> The issue of available resources also highlights the need for greater coordination between Canada and its international allies, including the United Kingdom, the United States, Australia, and France. The attempt by many countries to unilaterally secure the necessary transport resources for evacuating their nationals created a situation of competition amongst countries for available resources, whether this competition was intended or not. Countries with greater military resources than Canada were able to begin the evacuation of those in greatest need at an earlier time than Canada. There was also a subsequent event where a vessel chartered by Canada for evacuation from the port of Tyre left at half capacity. To improve effectiveness and efficiency, Canada should work with officials bilaterally and through the G8 and NATO to devise strategies to ensure that in future cases of mass evacuation, Canada and its allies cooperate to the maximum extent possible to secure and evacuate their respective nationals.
>
> *(Standing Senate Committee on Foreign Affairs and International Trade 2007)*

Coordination among international agencies is unpredictable

In different ways, each of the cases above illustrates the ability of international agencies to cooperate in times of crisis—cooperation between IOM and UNHCR during the Libyan crisis has been cited as a particular success. But such cooperation has certainly not been systematic, and neither is it predictable nor can it be guaranteed in future crises, although IOM's new MCOF may provide a platform for better cooperation. Part of the reason is that the rights of non-citizens during displacement are not enumerated, and neither are responsibilities. One implication is that on the whole, international organizations have assisted non-citizens alongside other displaced persons, without discerning where non-citizens may have particular needs. An additional problem is that in the case of internal displacement in particular, the institutional framework is weak even for displaced citizens, let alone non-citizens. The inter-agency "cluster approach" has certainly improved the capacity of the international community to respond to situations of internal displacement and other humanitarian crises, but by no means is it comprehensive. As indicated in this chapter, one variable includes international commitment and attention to the crisis situation in question—compare, for example, Libya and Côte d'Ivoire.

National responses are often inadequate

The case studies above have illustrated that even where states have developed specific laws and policies relating to disasters, crises and displacement, a lack of clarity about the rights of non-citizens at the level of international laws, conventions and standards translates to the national level, too. Furthermore, as highlighted by the evaluation of the humanitarian response to the displacement in South Africa in 2008, many affected states lack the basic capacity to implement laws and policies during periods of crisis.

Recommendations

Research and evaluation

1. More research, including better data collection, is required to establish the extent to which non-citizens face particular vulnerabilities at all stages of displacement. Comparing the above case studies, it is noticeable that there has been very little research on the crises in Côte d'Ivoire and Thailand in particular.
2. Internal evaluations are needed on the humanitarian response to the displacement of non-citizens, such as that conducted by the FMSP in South Africa. There is a need to share experiences among both origin and host states that have responded to the displacement of non-citizens in recent years.

International standards

3. Where new standards on displacement are developed or updated, they should make explicit reference to the rights of displaced non-citizens and responsible

parties. The Nansen Principles are a case in point; as is the current work of the International Law Commission (2012) on standards on the "expulsion of aliens" and the "protection of persons in the event of disasters."

4. There may be a case to refer the issue of displaced non-citizens to the UN Special Rapporteur on the Human Rights of Migrants and the UN Special Rapporteur on the Human Rights of Internally Displaced persons.

National laws and policies

5. More states should be encouraged to develop national laws and policies on the rights of IDPs, including non-citizens.
6. Existing national laws and policies should be updated to make explicit reference to the rights of displaced non-citizens. In this regard it is likely that states will be most willing to extend rights to those in a regular situation.

Capacity development in host states

7. Greater national capacity is required to protect and assist IDPs—including non-citizens—during crisis, ranging from establishing a response framework, to a clear allocation of responsibilities, and consultation with affected populations. The Brookings-Bern Project Framework for National Responsibility provides a good overview of capacity requirements (Brookings-Bern 2005).

Countries of origin

8. Countries of origin with large overseas worker populations should develop SOPs for the protection of migrant workers during crises, including detailed information on in situ protection measures, relocation, evacuation and repatriation procedures.
9. An international emergency fund should be considered, for access by countries of origin to evacuate their citizens during crises.
10. Pre-departure training that is offered to migrant workers should include contingency planning for crisis situations.
11. Micro-insurance schemes should be considered to assist migrants cope with emergency situations.
12. Consular capacity should be developed to protect migrant workers.

Corporations

13. Corporations that employ significant numbers of overseas nationals should develop SOPs on protecting and evacuating workers, establish risk assessment units, and establish senior chief security officer positions.

International cooperation

14. Contingency planning should take place at a bilateral and regional level to ensure effective cooperation between states during evacuations of non-citizens from crisis situations.
15. A coordinating mechanism should be established to ensure effective cooperation between relevant international agencies to assist and protect non-citizens displaced internally and across borders during crisis situations.
16. IOM's MCOF should provide a platform for greater international cooperation.

Conclusion

At the time of writing, concerns are being expressed about the security of migrant workers and Iraqi and Palestinian refugees in Syria. IOM estimates that at least 15,000 migrant workers in Syria require evacuation assistance, while UNHCR has reported that 360,000 Palestinians, as well as 94,000 Iraqi and other refugees, need immediate humanitarian support (IOM 2013). As observed at a recent Brookings-LSE Project on Internal Displacement roundtable, the ongoing unrest in Syria makes it difficult to discuss durable solutions for Iraqis who remain there (Brookings 2012).

It has been reported that some 400,000 Palestinians have been affected by the violence. There were more than 60,000 Palestinian refugees from Syria in Lebanon in June 2013, and more than 6,000 in Jordan, which closed the border to Palestinian refugees from Syria in January 2013. There are also thought to be tens of thousands of Palestinians displaced inside Syria, with violence breaking out in some of the camps—including the largest, Yarmouk—last summer, causing thousands to flee.

The very real threat of the displacement of non-citizens in Syria reinforces the observation made in the Introduction to this chapter that this phenomenon is likely to become more common. Any guiding principles developed in response to crisis migration should take explicit account of the situation of non-citizens during these migrations.

References

Brookings (2012) "Roundtable on Improving Prospects for Durable Solutions for Iraqi Internally Displaced Persons and Refugees," available online at: http://www.brookings.edu/events/2012/02/24-iraq-displacement.

Brookings-Bern (2005) "Addressing Internal Displacement: A Framework for National Responsibility," available online at: http://www.brookings.edu/fp/projects/idp/20050401_nrframework.pdf.

Duchâtel, M. and Bates, G. (2012) "Overseas Citizen Protection: A Growing Challenge for China," Stockholm International Peace Research Institute, available online at: www.sipri.org/media/newsletter/essay/february12.

Fiddian-Qasmiyeh, E. (2012) "Invisible Refugees and/or Overlapping Refugeedom? Protecting Sahrawis and Palestinians Displaced by the 2011 Libyan Uprising," *International Journal of Refugee Law*, 24(2): 263–293.

Forced Migration Studies Programme (FMSP) (2009) *Humanitarian Assistance to Internally Displaced Persons in South Africa: Lessons Learned Following Attacks on Foreign Nationals in May 2008*, Witswatersrand: University of Witswatersrand.

Grant, S. (2005) "Migrants Human Rights: From the Margins to the Mainstream," *Migration Information Source*, March, available online at: http://www.migrationinformation. org/feature/display.cfm?id=291.

Inter-Agency Standing Committee (IASC) (2011) *IASC Operational Guidelines on the Protection of Persons in Situations of Natural Disasters*, Washington, DC: Brookings-Bern Project on Internal Displacement.

International Law Commission (2012) *Report of the International Law Commission*, sixty-fourth session.

IOM (2011) "IOM Evacuation of Migrants from Libyan Port of Benghazi Resumes," available online at: http://reliefweb.int/node/393449.

IOM (2013) "Emergency Appeal: Syria Crisis, January–June 2012," available online at: http://www.iom.int/files/live/sites/iom/files/Country/docs/IOM_Syria_Crisis_ Response_Appeal_January-June_2013.pdf.

Jureidini, R. (2011) "State and Non-state Actors in Evacuations during the Conflict in Lebanon, July–August 2006," in Koser, K. and Martin, S. (eds.), *The Migration–displacement Nexus: Patterns, Processes, and Responses*, New York: Berghahn: 197–215.

Koser, K. (2008) "Gaps in IDP Protection," *Forced Migration Review*, 31: 21.

Lancet (2011) "The Arab Uprisings and Health," 378(9796): 1050, September 17.

Nansen Conference (2011) "The Nansen Conference: Climate Change and Displacement in the 21st Century," Oslo, Norway, June 5–7, available online at: http://www.unhcr. org/4ea969729.pdf.

Press TV (2011) "Libya IDP Crisis Escalates," May 25, available online at: http://www. presstv.ir/detail/181600.html.

Standing Senate Committee on Foreign Affairs and International Trade (2007) "The Evacuation of Canadians from Lebanon in July 2006: Implications for the Government of Canada," May, available online at: http://www.parl.gc.ca/content/sen/committee/ 391/fore/rep/rep12may07-e.htm.

UNDP (2011) "The Conflict in Côte d'Ivoire and its Effects on West African Countries: A Perspective from the Ground," Issue Brief, July.

UNHCR (2011a) "Update No. 1 on the Humanitarian Situation in Libya and Neighbouring Countries," March 2, available online at: http://www.unhcr.org/4d7788729.html.

UNHCR (2011b) "Update No. 30 on the Humanitarian Situation in Libya and Neighbouring Countries," June 22, available online at: http://www.unhcr.org/4e0201a09.html.

UNHCR (2012) *Housing, Land and Property Issues and the Response to Displacement in Libya*, Geneva: UNHCR.

UNOCHA (1998) *Guiding Principles on Internal Displacement*, New York: UNOCHA.

14

"TRAPPED" POPULATIONS

Limits on mobility at times of crisis

Richard Black and Michael Collyer

Introduction

Amid widespread policy interest in the numbers of individuals likely to be displaced by humanitarian crises, there is a growing weight of evidence that particular drivers, such as environmental change, may actually prevent rather than encourage movement[1] (Black *et al.* 2011; Gray and Mueller 2012b). There are obvious humanitarian reasons to be concerned about situations in which individuals are unable to move to escape danger, as such immobility both magnifies their vulnerability and may inhibit the access of humanitarian actors. Yet a research agenda investigating why populations might become "trapped" is still very much in its infancy, and faces a number of challenges. This chapter considers these challenges using the limited information on trapped populations in a variety of contexts globally. It sets out a justification for a research agenda on how and why people become trapped in times of crisis, and considers how research could progress in this currently very limited but potentially highly significant field. The chapter concludes with a consideration of the policy implications of this agenda.

Conceptually, there have been some attempts to shift the focus of migration studies away from the 3 percent of the world's population that are international migrants to the 97 percent that are not (cf. Hammar *et al.* 1997). The suggestion that immobility rather than mobility is what needs explaining has generated very little response in the academic literature, probably due to the challenge in explaining a behavior common to such a vast number of people in anything other than very general terms. Yet a consideration of trapped populations is somewhat different, as it must distinguish between ability, desire and need to move. There are two groups of people who are barely reflected in the theoretical literature and frequently absent from policy considerations in acute crisis situations: those who wish to move but can't and those who don't wish to move but probably should if they are not to endanger others.

This presents obvious methodological challenges. Distinguishing those who wish to move (or need to do so in times of crisis) but remain in situ from those who do not wish to move is likely to be extremely difficult, not least because people's judgment about whether it is necessary to move will likely change over quite short periods of time. A need to move is most obvious in acute crises, which therefore receive a disproportionate share of our attention in this chapter, although slower onset disasters such as drought are equally significant in preventing mobility. Yet, even when the immobile are in a small minority amongst a generally mobile population there can be no assumption that they are unable to move. The methodological difficulty is to identify a desire (or need) to move that is distinct from movement and explain why potential movement does not occur. Only then can assumptions about the size and nature of the trapped population be properly evaluated.

Much of the limited literature devoted to immobility is concerned with various forms of policy as a barrier to international movement (Harris 2002). Involuntary immobility is the inevitable result of migration policy in most instances, given that only 10 percent of countries in the world report that they wish to increase outward migration, and just nine percent wish to increase inward migration (UN 2013a). Indeed, many among the 16.5 percent of governments that have a stated overall concern to reduce immigration would view reduced mobility as an indication of successful policy (UN 2013a). Immobility is occasionally an explicit policy objective, as in detention, but it is almost always a consequence of tougher immigration controls. Warnings of the humanitarian consequences of such restrictions have been raised for decades in relation to their impact on potential refugee movement (ECRE 2002) and much more recently in terms of the detrimental effect on development strategies (UNDP 2009). It is clear that policy has a differential impact on immobility, and that those whose movement is prevented or facilitated share certain characteristics. The same could be assumed of other controls on movement.

The justification for a concern with the immobile is therefore that particularly vulnerable populations will be trapped. Detailed analysis of events such as Hurricane Katrina suggest that this was the case, as residents of care homes or those without their own transport were amongst those left behind (Cutter and Emrich 2006). Yet the challenge is to avoid imposing any kind of hierarchy of vulnerability onto our emerging understandings of these events. Policy approaches to internally displaced persons (IDPs) have been criticized for privileging a population that is undoubtedly highly vulnerable but at least has some kind of access to resources in comparison to those who have not been able to move at all (Hathaway 2007). Trapped populations are likely to be more vulnerable than IDPs, just as IDPs typically have less access to financial or social capital than those who migrate internationally. Still, movement creates additional concerns for IDPs, just as international travel changes the nature of vulnerability for international migrants, and this justifies their special categorization.

This chapter aims to review what we know of trapped populations. Given the potential size of such populations, at least over the next few decades, what we know is surprisingly little, even incorporating the larger literature on those trapped by policy initiatives. Yet the potentially extreme vulnerability of the involuntarily immobile justifies greater attention to this group. It also justifies some attempt to extrapolate existing information to gain some understanding of how those who are trapped might respond to progressively more severe crises or shocks (involving both "natural" and "human induced" disasters), and how these responses could be supported.

"Earthbound compulsion": theorizing immobility

In 1939, the anthropologist Fei Xiaotong set out to explain why the Chinese peasantry were so immobile, coining the wonderful phrase "earthbound compulsion" to describe the attachment to the land that contributed to the very low rates of internal migration seen in China at the time (Fei 1939). The Chinese have become far more mobile since then but at a global level migration is still an unusual behavior. It is estimated that 214 million people—only about 3 percent of the world's population—live outside of the country of their birth (UN 2009). Recent analysis of 2005 census data suggests that a further 763 million, 9 per cent of world's population, remained in their country but outside their region of birth, a rough definition of internal migration (UN 2013b). Available data therefore suggest that as many as 88 percent of the world's population are immobile—at least to the extent that mobility is statistically legible at this very general level of approximation.

This empirical observation was the starting point for a well-cited critique of existing migration theory by Tomas Hammar and colleagues (1997). They argued that existing theories of migration focus on the question of why people leave, whereas greater attention should be devoted to why they stay (Hammar and Tamas 1997). Their collection concludes with a reframing of migration theory around three concepts: migratory space, local assets and cumulative causation (Faist 1997). This nuanced account was undoubtedly a step forward in explaining the full range of mobility decisions. Yet, despite the authors' declared intentions, the apparent focus is still an explanation of mobility in ways that allow for the existence of immobility rather than an explanation of immobility.

One of the earliest attempts to explicitly theorize immobility was Guy Standing's (1981) Marxist analysis in the *Journal of Peasant Studies*. Standing sought to explain the historically highly varied experiences of mobility of those living in poverty, writing against the assumptions of dominant neo-classical accounts, which saw mobility of the very poor and unemployed as an inevitable result of spatial discrepancies in wages. Standing, in contrast, relates mobility to the dominant mode of production. His account begins with the communal mobility of pastoralist societies. This mobility is curtailed during the transition to various forms of feudal system, where the opportunity to exploit the labor of peasant farmers or plantation workers requires that they are prevented from generating the surplus necessary to be able to move elsewhere. A second mobility transition occurs with the growing influence of industrial capitalism, which depends on a mobile population of surplus labor, initially through rural to urban migration. Even thirty-five years ago, Standing noted a further shift, as state policy introduces new restrictions on mobility:

> Though essential for capital, migration on the scale it is reaching in many parts of the world is posing an increasing threat to the stability of the global capitalist system, notably by transforming a rapidly growing proportion of the world's poor into a visible and desperate surplus population. In many social formations the State has recognised the threat posed by this process and has attempted to take remedial action.
>
> *(Standing 1981, 202)*

The new mobility restrictions Standing notes were primarily a response to contemporary concerns around massive rural to urban migration in poorer countries. This was also the beginning of the progressive tightening of controls on international migration, which have become one of the most obvious limits to global mobility. The decline of international refugee protection, caused by the gradual closure of the territories of wealthier states, was a significant concern of the 1990s, highlighted by critical approaches to the "non-entrée regime" (e.g. Chimni 2000). By the end of the decade Zygmunt Bauman outlined his now celebrated argument that mobility, not capital, had become the key differentiating factor of the globalized world (Bauman 1998).

The function of migration policy in restricting mobility is now widely commented on, particularly in relation to detention and deportation, which have become a more common focus for the analysis of state violence (De Genova and Peutz 2010). The 2009 Human Development Report cited migration controls as a significant barrier to the international development potential offered by people on the move (UNDP 2009). Similar observations are now relatively common and a critical theorization of these processes is emerging (Harris 2002). For example, the papers in Deirdre Conlon's (2011) recent special issue of *Gender, Place and Culture* focus on experiences of waiting in the migration process, highlighting a particularly clear example of mostly state-imposed immobility.

Poverty is a further significant factor in immobility that emerges from Standing's early account. Despite a structuralist framework that appears to allow little more room for individual action than the neoclassical theories it set out to critique, Standing is attentive to intra-household dynamics, highlighting that the poorest households may be too poor to finance migration of household members to nearby urban areas, or unable to spare the labor. This offers a clear contrast to neoclassical assumptions that the poorest would have to migrate. Similarly, we may infer that the urban "visible and desperate surplus population" that provokes restrictive state responses have at least partially become "trapped" in urban areas.

Although the vast migration and development literature has mostly neglected immobility (with certain exceptions, e.g. Skeldon 1997), poverty has provided an important focus for a more limited interest in the selectivity of migration (Kothari 2003). Uma Kothari develops an analysis of six forms of capital (social, cultural, human, economic, geographical and political) to argue that "a lack of capitals can both require and limit movement" (Kothari 2003, 648). In her terms, "staying put" may be a positive choice but may also result from various forms of exclusion. An analysis of exclusion, using these six capitals, highlights how they may combine with a variety of mobility impacts. In more recent research into the "unfreedoms" of plantation and contract labor, Kothari has continued to investigate "mobility, movement, displacement and confinement . . . A study of those who move to become immobile further challenges dichotomous and generalized understandings whereby unfreedom connotes restrictions on movement, while freedom implies ability to move" (2013, 5).

Carling (2002) draws many of these strands together. He identifies six widely accepted reasons for immobility: lack of development, risk averseness of potential migrants, location specific advantages that would be lost with migration, cumulative immobility (the corollary of cumulative causation), anti-migrant discrimination at destination and migration policy (2002, 9). Yet he also raises the theoretical problem of distinguishing between not wanting and not being able to migrate, and critiques existing approaches to migration for largely failing to allow the possibility of *involuntary* immobility. Using empirical material from Cape Verde, he draws up a new model based on aspiration and ability. Ability is modeled by what Carling calls the "immigration interface," an approach that has much in common with Kothari's (2002) focus on various forms of capital as a way of investigating exclusion. Analysis of aspiration focuses on the "emigration environment" which is "the macro-level approach to asking why people want to emigrate, . . . the historical, social, economic, cultural or political setting that encourages migration or not" (2002, 16–17). Aspiration and ability may obviously be intertwined, influencing each other in important ways, but the clear analytical separation Carling attempts to draw between the two is a substantial contribution to theorizing immobility.

Given the context of labor migration in his Cape Verdean case study, Carling's model excludes conflict. Yet conflict is a further factor, which may disrupt existing patterns of mobility and prevent further migration taking place. For example, it could be argued in relation to conflicts in the 1990s in Bosnia, Sri Lanka, Somalia and elsewhere, that those in most humanitarian need were precisely those unable to flee from conflict and violence, rather than those who moved to become refugees or IDPs. Recognizing this, international actors sought to establish "safe havens" within these countries, where both in situ and internally displaced populations could benefit from UN protection and assistance, although in practice these zones did not always remain "safe," as was illustrated most obviously in the town of Srebrenica (Hyndman 2003).

More broadly, Lubkemann argues that refugee studies focuses so consistently on movement that it "renders the involuntarily immobilized invisible" (2008, 456). He draws on detailed investigation of a drought-prone rural area of Mozambique during the civil war to identify the gender-differentiated impacts of conflict. A predominantly male group with established patterns of labor migration to neighboring South Africa were able to benefit economically from forced migration, whereas the disproportionately female group left behind were prevented by the intensification of violence from engaging in their usual small scale mobility in response to the prolonged drought of the early 1980s, and their impoverishment increased.

> Those who moved the least ultimately suffered most dramatically from the war's effects on migration precisely because their baseline mobility strategies were profoundly disrupted through forced immobilization. The most detrimental effects of the war were thus the novel impediments to mobility that rendered inoperable well-established (mobility-dependent) strategies for household reproduction and for coping with ecological insecurity.
>
> *(Lubkemann 2008, 468)*

Lubkemann labels this experience of forced immobility "displaced in place," arguing that it may be more analytically relevant than the more common focus on forced mobility.

This combination of conflict-related violence and chronic or acute environmental disturbance, which Hyndman labels "dual disasters" (2011), appears to be increasingly common (see e.g. Lindley, Chapter 8 in this volume). Investigation of the relationship between migration and the environment has been dominated by discussions of how many people may be forced to migrate (Gemenne 2011). Yet the immobilizing effects of environmental change have also recently started to be observed, most clearly in the 2011 Foresight report of the UK's Government Office for Science (Foresight 2011). Drawing on a very broad range of evidence, this report focused on a more complex set of migration drivers and investigated how they would be affected by environmental impacts. It concluded that it was possible that migration might become less rather than more prevalent in the context of climate change, with one of the report's key insights being to highlight the greater vulnerabilities of populations "trapped" by environmental change.

This conclusion of the Foresight report drew on at least some empirical evidence (e.g. Zolberg, Suhrke and Aguayo 1989), which has been bolstered since its publication. For example, drawing on multivariate analysis of historical data from Ethiopia and Bangladesh, Gray and Mueller (2012a, 2012b) also argue that environmental changes such as droughts or floods have relatively limited impacts on long-term migration. In both cases, migration occurs predominantly over short distances and some movements may be restricted entirely. They found that, in Ethiopia, drought increases men's labor migration but restricts the migration of women for marriage (2012a). In Bangladesh, they argue that "although mobility can serve as a post-disaster coping strategy, it does not do so universally, and disasters in fact can *reduce* mobility by increasing labor needs at the origin or by removing the resources necessary to migrate" (2012b, 6003). This echoes Standing's (1981) argument, and suggests that it is the most marginalized who are most likely to be trapped. This picture is reinforced by Black and colleagues, writing in *Nature*, who find that "The greatest risks will be borne by those who are unable or unwilling to relocate, and may be exacerbated by maladaptive policies designed to prevent migration" (Black *et al.* 2011, 447).

The combination of multiple constraints on opportunities for mobility is likely to compound the impact of enforced immobility. This is just as true for the combined effects of environmental disasters and restrictive migration policy in Bangladesh as it is for conflict-related violence and drought in Mozambique. The range of terms used in this section, from "involuntary immobility" to "staying put" and "earthbound compulsion" to "displaced in place," all highlight the same range of issues: the greatest burden falls on those who are least able to cope. We continue with the term from the Foresight report—"trapped"—since it is concise and contains a degree of emotive force that demands a response. The use of this term highlights the issue of *need* to migrate, which is not always present in previous discussions of enforced or involuntary immobility.

The limited research into trapped populations, briefly reviewed here, highlights a range of causes of immobility, though it is typical for multiple constraining factors to exert a cumulative impact. Yet as most of this research emphasizes, the trapped are usually overlooked, given the more common focus on mobility. A shift in focus to those who are trapped challenges both theoretical and practical approaches to mobility and crisis, which prioritize movement. The difficulties faced by trapped populations are typically compounded by policy interventions, sometimes explicitly, as in the case of migration policies, but sometimes inadvertently, in the case of certain humanitarian initiatives, such as safe havens. Finally, there is a humanitarian concern: those who have lost control of the decision to move away from potential danger have inevitably lost a lot more too and are therefore amongst the most vulnerable members of society.

This is reminiscent of Hathaway's critique of the category of IDPs (Hathaway 2007) though we do not wish to construct some kind of hierarchy of vulnerability, or suggest that IDPs are less worthy of attention. Just as the circumstances of refugees are linked to the nature of their international movement, IDPs have a particular set of vulnerabilities associated with their more limited mobility. Those who are denied access to mobility entirely, whether through lack of various capitals (cf. Kothari) and/or through other constraints such as conflicts, hazards or policies, are likely to have a similarly distinct set of vulnerabilities that are rarely acknowledged and hardly ever addressed. Having drawn this research together our aim is to explain why trapped populations have not received greater attention, and to highlight some of the challenges of addressing this.

Challenges to conceptualizing trapped populations

The key challenges to any conceptualization of trapped populations are the questions of volition, highlighted by Carling, and the additional consideration of need. Distinguishing between those who choose to stay and those who are forced to stay is essential if the notion of trapped populations is to have anything other than a very broad conceptual application. Carling (2002) suggests interpreting volition as a simple expression of a desire to move elsewhere. He argues that the likelihood that such a move could occur or the practical preparation made to turn this aspiration into a reality, such as visa applications, should not be considered since this would start to blur the lines between aspiration and ability, which his model sets out to separate. Yet these lines are perhaps inevitably blurred, as indeed are the lines between forced and voluntary mobility (Richmond 1993). In this context, an alternative basis for distinguishing involuntary immobility could be the *need* to move, based on some form of well-founded fear of the consequences if movement does not take place.

Carling's approach is broadly the one taken by the Gallup World Poll, initiated in 2005, which asks three related migration questions: "Would you like to move abroad?", "Do you plan to move in the next 12 months?" and "Are you currently making preparations to move?" (Esipova, Ray and Pugliese 2011). Existing polling data covers representative samples totaling 750,000 people in 150 countries, allowing estimation of a global total number of people who wish to move (internationally) but are unable to do so. Modeling these patterns at the global scale suggests that 630 million people, or 14 percent of the world's population, would like to migrate internationally, but only forty-eight million plan to move in the next twelve months and only nineteen million are making preparations to do so (Esipova *et al.* 2011, 20). Variations between countries are obviously substantial but, assuming the sample is representative, the key point is that the vast majority of the world's population (86 percent) have no wish to leave the country in which they are now living. This provides an indication of the level of migration aspirations, at least at the international level.

A crisis situation, such as a political emergency or environmental disaster, will obviously change the nature of these expressed intentions, such that "aspiration" is no longer sufficient to convey the urgency of an intention to migrate under these circumstances. Yet aspiration cannot simply be conflated with need, since even where there is a clear humanitarian need to migrate, some people will prefer to remain. Harry Randall Truman refused to leave his cabin on Mount St. Helens before the May 1980 eruption and enjoyed a brief period of fame through regular TV interviews in which he described his love of the mountain on which he had lived for most of his life. His inevitable death inspired songs, poems and a full-length memoir (Rosen 1981). Similar choices are presumably made in any crisis situation, though it is unusual for them to be so highly mediatized. For these reasons, Lubkemann (2008) is critical of the negation of agency inherent in the language of forced migration, since even in extreme circumstances mobility results from a positive choice.

In a humanitarian context, Carling's aspiration/ability approach should therefore be supplemented with the additional consideration of need. This fits the urgency suggested by the use of the term "trapped" and helps to further narrow the conceptualization of the term. Trapped populations are those people who not only aspire but also *need* to move for their own protection, but who nevertheless lack the ability. The remainder of this section considers ability, aspiration and need in turn, drawing on elements of the theorization of immobility considered in the previous section.

Influences on ability and aspiration to migrate

Ability to migrate is a function of access to resources, understood broadly in terms of various capitals (Kothari 2003). The conceptualization of trapped populations in the Foresight report (Figure 1.2, p. 29) considers vulnerability to environmental change to be inversely related to levels of capital and ability to move as directly related, so low levels of capital indicate both high vulnerability to crises and low ability to move away from crises. Different forms of capital may have a more direct influence on ability to move, such as financial capital, or access to transportation, or a less direct influence, such as involvement in social networks beyond the area immediately affected by the crisis. Migration policy seems to be a distinct issue. Although it could be argued that policy interventions such as migration controls are less of a barrier for those with the right sort of capital, this is stretching capitals analysis considerably. Linking policy with capital in this way also undermines Bauman's (1998) insight that access to mobility is replacing capital as the key differentiating factor in a globalized world. Policy interventions are therefore considered as a separate control on movement to resources.

In recent years, resource limitations on mobility have been most obvious in urban contexts. This results from a combination of the rapid global growth of urban populations, the concentration and isolation of large numbers of very poor, marginalized people in urban areas, and the uncertainty and inexperience of much humanitarian response given the challenges of operating in crisis-stricken urban areas (e.g. see Fagen, Chapter 16 in this volume).

The most striking recent example is the impact of Hurricane Katrina on New Orleans. Although preparation in rural and suburban coastal areas was generally good, the very poor disaster preparedness in the city of New Orleans is considered as one of the main reasons for the catastrophic impact of the crisis in the city itself, after the hurricane hit in August 2005: "Those with resources left in advance of the approaching hurricane; those without (largely the poor, African American, elderly or residents without private cars) remained, trapped in the rising floodwaters" (Cutter and Emrich 2006, 105). The resources required were very simple, according to Waller: "The difference between those who escaped with their lives and loved ones, and those who did not, often came down to access to a car and enough money for gas" (Waller 2005, 1). Those with friends and family elsewhere, with whom they could go and stay, were also more likely to leave. The overall impact of an evacuation scheme that relied on private transport was, as Cutter and Emrich (2006) argue, that the dangers of the crisis were disproportionately faced by the most vulnerable.

This is perhaps most surprising in the wealthiest country in the world, but is common to other urban disasters, such as the impact of the 2010 Haitian earthquake on the population of Port-au-Prince, where access to resources also dictated the nature of mobility in the immediate aftermath of the earthquake (Clermont *et al*. 2011; see also Ferris, Chapter 4 in this volume). Immobility is particularly apparent in the case of fast-onset disasters, such as floods or earthquakes, but it can result in similar levels of inequality in more chronic contexts.

The fate of the "visible and desperate surplus population" in urban areas, identified by Standing (1981, 202), has only become more significant. This includes places that would have surprised Standing, writing thirty years ago, such as Detroit, where the population has fallen drastically as those who can have moved elsewhere, leaving behind those whose homes or savings were not sufficient to get out of the city (Sugrue 2005). The sprawling informal settlements in cities in poorer parts of the world face similar problems of large numbers of very poor people, essentially trapped there. In some cases the political economy is strikingly different from the negative equity trapping people in Detroit. The homes in the infamous Mumbai slum of Dharavi, once Asia's largest, are partially protected from demolition under the 1995 Slum Rehabilitation Act, but their impoverished inhabitants cannot move as they cannot sell their homes, and the land on which they live is too valuable for them to give up (Arputham and Patel 2010). Nevertheless, they face serious public health risks due to poor sanitation and are more vulnerable to other disasters, such as the rain-induced landslide that killed at least 63 people in the slum of Lal Bahadur Shastri Nagar in northern Mumbai in July 2005 (*The Hindu* 2005).

Mobility itself is also a valuable resource, which is obviously unevenly distributed. Remittance income sent by migrants may provide additional resources for poor households. Even where migrant incomes are low, migration typically offers a more stable income source and so serves an important insurance role, particularly in situations of environmental, economic or political uncertainty. There is a vicious circle here, as lack of access to the resources and stability offered by migration becomes a further factor limiting access to mobility. Even migration over relatively short distances requires resources that the very poorest households may find insurmountable. For smaller households, particularly female-headed households, which are typically over-represented amongst the very poorest, migration of household members may be prevented by labor constraints. In either case, where mobility brings benefits, trapped populations are further marginalized

In research on the relationship between mobility, vulnerability and environmental fragility in Bolivia, Senegal and Tanzania, Tacoli found that "in all locations in all three countries, the most vulnerable households were unanimously identified as those not receiving remittances from migrant relatives" (2011, 17). Tacoli's research reinforces an understanding of migration as a form of adaptation to environmental change, which emerges strongly from the Foresight report, among other recent work (Black *et al*. 2011). Migration provides access to financial resources and in some cases reduces risk as a form of insurance or protection. In a limited number of cases particular patterns of migration have become so socially embedded that migration also serves an important cultural role. Jónsson's (2008) ethnography highlights the impact of restricted mobility on prospective Soninke migrants from the Senegal River Valley. Over the last century a culture of migration to France has developed, to the extent that it is seen as a virtual rite of passage for young Soninke men. Enhanced controls on international migration have resulted in a growing number of "unwilling non-migrants" (2008, 11) who find it difficult to marry, are stigmatized as lazy and sometimes openly mocked. As a result, according to Jóns-son, an entire system of social organization is beginning to disintegrate.

If migration is a resource, policy that limits or controls that migration contributes to trapping populations, whether deliberately or incidentally. It is possible that people may be trapped by controls on international migration as their preferred destination introduces controls that inhibit their travel (Carling 2002; Jónsson 2008). This is an important factor, particularly as access to international migration has become significantly dependent on accident of birth as much as resources. Yet, in most cases, those likely to be trapped would never have access to international mobility anyway. It is now well established that the dominant pattern of crisis migration involves temporary moves over short distances. Policy will therefore be most significant in trapping populations where it affects this type of movement. With the exception of populations immediately adjacent to borders, the enhanced controls on international migration are likely to have less impact than other, often non-migration-related policies.

Yet policy restrictions on internal mobility have often been substantial, and remain so in many countries even though they have generally become indirect or implicit rather than direct and explicit. Thus direct restrictions on mobility, such as those associated with the *hukou* system in China or the *ho khau* system in Vietnam, are now generally limited to more extreme totalitarian states such as North Korea. But in a number of countries, internal mobility is restricted in practice through limitations on access to public services for those who move internally, as for example in Sri Lanka, where new migrants from Tamil areas in the north of the country find it difficult or impossible to register with the authorities if they move to the capital, Colombo (Collyer 2011).

Limitations on migration and other policies may also combine to leave individuals genuinely trapped in the face of crises. The decision by the government of Burma not to allow international humanitarian aid into the country following the devastation caused by Cyclone Nargis in May 2010 provides one such example. The cyclone resulted in at least 130,000 deaths, a figure likely increased by the combination of restrictions on movement, and restrictions on access for humanitarian actors. The result was a cutting-off of the survivors from vital resources, which made them far more vulnerable in the fragile post-disaster situation (McLachlan-Bent and Langmore 2011). A further, much more well-intentioned example may be traced to UN High Commissioner for Refugees Sadako Ogata's declaration of a "right to remain" for crisis-affected populations in 1993, which was initially criticized as an implicit attempt to limit mobility and, crucially, restrict the right to seek asylum (Morel, Stavropoulou and Durieux 2012). This inspired a policy of "preventative protection" by the UN in the 1990s, initially in Somalia, then in Bosnia and Herzegovina where Srebrenica became the first "safe area" in 1993. In this context, although well intentioned, and seemingly an initiative to protect those who were trapped, this policy could be seen as punitive for those trapped not simply by "events" but as a direct or indirect consequence of policy itself.

To be trapped, individuals must not only lack the ability to move but also either want or need to move. Ability to migrate is clearly a complex and multi-faceted indicator that includes access to significant resources or capitals and a range of potentially relevant policies that may impede movement. In contrast, aspiration is arguably rather more straightforward. Carling imagines this as a binary variable, a simple "yes" or "no." This is a worthwhile simplification, though it may also be possible to conceptualize aspiration along a continuum from complete disinterest to tremendous enthusiasm or even desperation. This may be applied in new research or used as a basis for interpreting existing data sets on immobile populations in order to assess levels of aspiration to move. Alternatively, aspiration may be inferred from other behaviors—most obviously a limited or curtailed experience of mobility. Kothari (2013) argues that the ability to move should not be automatically associated with freedom, since individuals could move into a situation of "unfreedom," which we have called "trapped."

Being "trapped" on the move

It is not necessary for trapped individuals to have always remained in one place. Conditions that trap particular populations may arise at any stage in their migration process. Protracted refugee situations offer an obvious example of a partially mobile yet trapped population. This is particularly the case in refugee or IDP camps where mobility out of the camp is officially restricted. Individuals exercised a degree of mobility or short distance migration to reach the camp and although this usually provides an immediate solution to short-term protection needs, it also establishes conditions of complete dependence, depriving individuals of possible access to resources that would allow them to move on and effectively trapping them in the camp. Mobility once again becomes highly differentiated. Travel between the Dadaab camps in Kenya and Nairobi offers a clear example where travel is restricted to a mostly male population who are more dynamic, younger and healthy, and already have access to some resources to start with, often through remittances from migrant family members (Horst 2006). In most cases, camps are not particularly safe places to be, yet moving out remains very difficult. Examples of increased vulnerability in camps are sadly not hard to find. It may be through further political repression, as in the Manik Farm IDP camps in Sri Lanka in 2009 (Weiss 2011), disease, such as the cholera outbreak in Haiti in 2010 (Clermont et al. 2011), or increased risk of domestic abuse and violence seen in the vast Syrian refugee camps in Jordan in 2013 (CPGVE 2013).

Another example of those trapped on the move is that of people trapped at migrant destinations or in bordering countries due to conflict or natural disaster in country. The 2006 Lebanese war highlighted the significant presence of foreign nationals in the country; an estimated 15,000 US nationals were evacuated, while at least 90,000 Sri Lankan migrant workers were trapped, offered very limited assistance from their own government (Jureidini 2011). More recently, a major part of the humanitarian intervention following the 2011 conflict in Libya concerned the evacuation of third country nationals who had become trapped in neighboring countries, with no right to remain and no independent means to return to their home countries (see Koser, Chapter 13 in this volume).

Being trapped on the move may also result from a more individual migration project. The growing inaccessibility of some of the most attractive destinations combined with the greater accessibility of information about the attractive life-styles that may be found there is responsible for generating new patterns of mostly undocumented migration over the last decade or so. In many cases these involve long distances over land or sometimes sea, and very high risks (see e.g. Kumin, Chapter 15 in this volume). Attempts to reach Australia by sea from Southeast Asia are more frequently blocked (Mountz and Hiemstra 2012); it is now increasingly common for migrants from West Africa to stop in North Africa rather than reach Europe (Collyer and De Haas 2012); migrants aiming for the US from Central America or trying to reach Israel from Sudan risk horrendous human rights abuses from criminal gangs in Mexico or the Sinai (see e.g. Albuja, Chapter 6 in this volume). Alternative routes from East to South Africa, or crossing from Somalia to Yemen, are equally hazardous. The inevitable disjuncture of this type of movement increases vulnerabilities of migrants forced to wait for extended periods of time at particular nodes along the route, producing new risks of exploitation. Collyer (2010) has labeled these new patterns of movement "fragmented migration." They are characterized by the growing risk that migrants become trapped at particular points along the long journey, deprived of resources or blocked by new migration controls, unable to return home, and gradually becoming more desperate.

Influences on the need to migrate

Finally, the question of need is a fundamental one in relation to forced migration. The need to move is central to the 1951 Convention definition of a refugee and both a complex and contentious process that we do not wish to try to simplify here. Yet the need to move must be an important factor distinguishing trapped populations. Acute crisis situations produce an especial urgency to move but beyond the crisis immediate mobility needs diminish. Those "trapped" in the long term may nonetheless not be a subject of humanitarian need. For example, owner occupiers of originally expensive homes on the UK coastline may well be at risk of subsidence or coastal erosion. They may even be prevented from moving, since insurance is unlikely to cover everything and their homes are now worth much less than the price of an equivalent home elsewhere. Yet, despite their vulnerable residential location, a relocation program would be highly unlikely.

Most organized population movements have faced resistance from at least some of those required to move. Given the legacy of failure of forced relocations, from terrible human rights abuses of colonial times, to apparently more enlightened agrarian reforms (Scott 1998) and mass relocations for large development projects, such as dams (World Commission on Dams 2000), such resistance is entirely understandable. Suspicion of the intentions of government or international institutions in their efforts to move people is not restricted to those living under authoritarian regimes. A disproportionate share of the impact of such relocations typically falls on the poor, vulnerable and marginalized who may therefore have an especially good reason to question the motives behind such projects. Thus individuals may *need* to move, *and* be offered an opportunity to do so, under particular conditions but still refuse to leave. Under such circumstances, analysis should turn to the nature of information provided to individuals before moving. Such individuals must still be considered as trapped.

Ways forward: research and policy responses to trapped populations

There is only very limited research investigating the situation of those we have called "trapped" populations. Much of this has focused on immobility more broadly, rather than the specific difficulties of those trapped as a result of crisis situations. The additional consideration of "needing" to move introduced a specifically humanitarian focus to our understanding of "trapped." Research with those prevented from moving by environmental or political crises faces obvious difficulties or dangers, at least in the immediate aftermath of the crisis. This explains why much recent research on immobility and environmental crisis has relied on existing data sources (Gray and Mueller 2012a, 2012b) or has focused on trapped migrants some distance from the immediate source of instability or danger.

This research base can be expanded substantially. Having established that people are trapped in crisis situations, which now seems widely accepted, we need to be able to explain how individuals and families become trapped, and what that means. This requires further efforts to analyze existing large-scale data sources, but also depends on the kind of information that can best be gained through more qualitative focused work. We also need to know something about potential public policy responses to the humanitarian challenges posed by trapped populations. These questions need to be considered in relation to those who have been able to move short distances and are now trapped in a different context, and those for whom even leaving home presented an insurmountable challenge.

Research into the situation of those who are trapped in complete immobility presents the greatest difficulties. The Foresight report considers their difficulties as largely economic. As the review of resource constraints in the previous section demonstrated, this can take many forms; constraints may not be directly financial, and may include things such as access to transport or labor power, or they may be entirely non-economic, such as access to geographically distant social networks. Conceptually, this relationship between financial constraints, mobility and vulnerability needs to be further explored in order to expand this economic focus. The necessary qualitative work may have to be conducted indirectly, such as through interviews with those who *have* been able to move out of a crisis situation, focusing on the selectivity of migration and the reasons why people remained behind. Engagement with the Nansen Initiative's work on cross border migration would offer a clearly limited research focus. Betts' recent research into "survival migration" presents an established conceptual framework to ground an understanding of institutional responses to movement of this nature (Betts 2013) that could be expanded to highlight difficulties in identifying and reaching those trapped by the crisis.

A research focus on those trapped at some point along their journey presents fewer practical challenges and could link to a range of established work, such as Conlon's (2011) analysis of waiting or Mountz's recent research in island detention centers (Mountz 2011). There is also developing research into long, dangerous journeys, particularly in places like Morocco or Turkey. A broader, more global analysis of why individuals are trapped at particular points along long, dangerous journeys could explore commonalities in these experiences and help to identify politically acceptable, humanitarian solutions to the tremendous vulnerability faced by trapped migrants. In certain contexts, such as the Sinai or northern Mexico, research would again present significant challenges of safety but local NGOs are already engaged in responding to such situations and entry points for research may be available. In other areas, such as Morocco or South Africa, migrants themselves are organized and proactively campaigning for action. These activist movements themselves provide important examples for further research. Conceptually this research also requires a shift from standard assumptions in migration studies that view migration as an unproblematic transition from origin to destination. The situation of trapped migrants requires an expansion in imaginations of the journey into an experience that may last many years. Collyer's (2010) work on "fragmented migration" offers a suitable conceptual framework for this work.

Practical policy responses are not obvious, though there are existing points of engagement. Existing regional agreements, most recently the African Union Convention for the Protection and Assistance of Internally Displaced Persons in Africa (Kampala Convention), which entered into forced in December 2012, provide clear indications of political will and strategies to respond to displacement. The Kampala Convention clearly recognizes rights of the internally displaced. It considers war and political instability but also makes specific mention of "climate change" as one of the possible reasons for displacement (in Article 5.4), recognizing the possibility for "dual disasters" where conflict is exacerbated by natural disaster. Yet the convention makes no mention of those unable to move.

The Guiding Principles on Internal Displacement have inspired several similar attempts to reach international agreement around non-binding principles that could form the basis for more concrete steps forward. The 2012 Nansen Initiative on disaster-induced cross border displacement draws inspiration from the Nansen Principles. The final Nansen Principle focuses on "National and international policies and responses," which include "planned relocation," which should be implemented "on the basis of non-discrimination, consent, empowerment, participation and partnerships with those directly affected" and, finally, "without neglecting those who may choose to remain." Choosing to remain is obviously substantially different from being unable to move, though given the long history of failure of relocation schemes, sketched in the previous section, refusal to participate may in some cases be the most sensible choice and such people may share many characteristics with the trapped. The Peninsular Principles on Climate Displacement within States, published in August 2013, are the most recent attempt to set out a normative framework for responding to climate-related displacement (Displacement Solutions 2013). The principles are fundamentally rights based, and again emphasize the need for participation and consent of the displaced, though they acknowledge the need for forcible relocation "in exceptional circumstances when necessary to protect public health and safety."

New relocation programs are one of the most common solutions advocated to assist trapped populations, yet given our currently limited state of knowledge, and the history of failure of relocation projects, this seems premature. This history is acknowledged by de Sherbinin and colleagues in a recent article in *Science* (de Sherbinin *et al.* 2011). They argue that relocation is already a necessary response to climate change and will continue to be so. The policy response they advocate is to devise high-quality relocation programs. Even though relocation is taking place, mostly in informal ways, the Foresight report highlighted how individuals frequently relocated to equally vulnerable areas, in poor-quality, flood-prone areas of cities, for example. Organized relocations offer one possible solution to this, but it is a short step to forced relocation. While we have such limited information on trapped populations, the policy goal should be to avoid situations in which people are unable to move when they want to, not to promote policy that encourages them to move when they may not want to. The best way to avoid such problematic issues as forcible displacement is to ensure that everyone has the ability to move if they wish to, and up-to-date information allowing them to make an informed choice.

While it is difficult to imagine exact details of such policies, it does seem clear that they must not be restricted to national-level initiatives. The variability of crises and responses is so great that regional initiatives, such as the Kampala Convention, must be combined with sub-national developments, particularly at city level. Given the significance of cities to the examples reviewed in this chapter—from acute crises in New Orleans or Port-au-Prince to more chronic political economic problems trapping large numbers of poor people in Detroit or Mumbai—city-level initiatives must be part of the solution. Policies focused on enabling mobility and providing timely access to relevant information can be more easily targeted at local level.

In conclusion, policy responses should be informed by the nature of the problem. The problem is *not* people being in the wrong place in relation to climate change or other crises. The problem is people being in the wrong place and being unable to do anything about it. There are plenty of examples in rich countries where relatively well-off people live in vulnerable places, but they have the resources to insure themselves against risk and would be able to move away if they deemed it absolutely necessary. The example considered in the previous section of the expensive properties perched on top of rapidly eroding cliffs along the south coast of the UK illustrate this. The most urgent issue is to identify how existing responses can reduce the likelihood of individuals being trapped in crisis situations. At present our understanding of the mechanics of trapped populations is too limited to suggest any clear policy measures to reduce their vulnerability or enable them to move when they feel they need to. Advancing understandings of the reasons behind their immobility may help existing policy responses begin to take account of their situation.

Note

1 We use the terms "movement," "mobility" and "migration" very widely in this chapter, though with distinct meanings. "Migration" is any move over a significant administrative boundary of more than three months. "Mobility" is any more local and/or short-term movement. "Movement" is a more general, generic term that encompasses both.

References

Arputham, J. and Patel, S. (2010) "Recent Developments in Plans for Dharavi and for the Airport Slums in Mumbai," *Environment and Urbanisation*, 22: 501–504.
Bauman, Z. (1998) *Globalization: The Human Consequences*, New York: Columbia University Press.
Betts, A. (2013) *Survival Migration: Failed Governance and the Crisis of Displacement*, Cornell University Press.
Black, R., Bennett, S.R.G., Thomas, S.M. and Beddington, J.R. (2011) "Migration as Adaptation," *Nature*, 478: 447–449.
Carling, J. (2002) "Migration in the Age of Involuntary Immobility: Theoretical Reflections and Cape Verdean Experiences," *Journal of Ethnic and Migration Studies*, 28(1): 5–42.
Child Protection and Gender-based Violence in Emergencies Sub-Working Group (CPGVE) (2013) Findings from the Inter-agency Child Protection and Gender Based Violence Assessment in the Za'atari Refugee Camp, available online at: http://www.data.unhcr.org/syrianrefugees/download.php?id=2306?.
Chimni, B.S. (2000) "Globalisation, Humanitarianism and the Erosion of Refugee Protection," Refugee Studies Centre Working Paper No. 3, University of Oxford.
Clermont, C., Sanderson, D., Sharma, A. and Spraos, H. (2011) *Urban Disasters: Lessons from Haiti. Study of Member Agencies Responses to the Earthquake in Port au Prince, Haiti, January 2010*, Disaster Emergency Committee.
Collyer, M. (2010) "Stranded Migrants and the Fragmented Journey," *Journal of Refugee Studies*, 23(3): 273–293.
Collyer, M. (2011) "When does Mobility Matter for Migrants to Colombo?" in Martin, S. and Koser, K. (eds.), *The Migration–displacement Nexus: Patterns, Processes and Policies*, Oxford: Berghahn.
Collyer, M. and De Haas, H. (2012) "Developing Dynamic Categorisations of Transit Migration," *Population, Space and Place*, 18: 468–481.
Conlon, D. (2011) "Waiting: Feminist Perspectives on the Spacings/Timings of Migrant (Im)mobility," *Gender, Place and Culture: A Journal of Feminist Geography*, 18(3): 353–360.
Cutter, S.L. and Emrich, C.T. (2006) "Moral Hazard, Social Catastrophe: The Changing

Face of Vulnerability along the Hurricane Coasts," *Annals of the American Academy of Political and Social Science*, 604: 102–112.

De Genova, N. and Peutz, N. (2010) *The Deportation Regime: Sovereignty, Space and the Freedom of Movement*, Durham: Duke University Press.

De Sherbinin, A., Castro, M., Gemenne, F., Cernea, M.M., Adamo, S., Fearnside, P.M., Krieger, G., Lahmani, S., Oliver-Smith, A., Pankhurst, A., Scudder, T., Singer, B., Tan, Y., Wannier, G., Boncour, B., Ehrhart, C., Hugo, G., Pandey, B. and Shi, G. (2011) "Preparing for Resettlement Associated with Climate Change," *Science*, 334: 456–457.

Displacement Solutions (2013) *Peninsula Principles on Climate Displacement Within States*, available online at: http://displacementsolutions.org/peninsula-principles/.

Esipova, N., Ray, J. and Pugliese, A. (2011) *Gallup World Poll: The Many Faces of Global Migration*, IOM Migration Research Series 43, Geneva: IOM.

European Council on Refugees and Exiles (ECRE) (2002) "Setting Limits: Research Paper on the Effects of Limits on the Freedom of Movement of Asylum Seekers Within the Borders of European Union Member States," London: ECRE.

Faist, T. (1997) "From Common Questions to Common Concepts," in Hammar, T., Brochmann, G., Tamas, K. and Faist, T. (eds.), *International Migration, Immobility and Development: Multidisciplinary Perspectives*, Oxford: Berg.

Fei, X. (1939) *Peasant Life in China*, London: Routledge.

Foresight: Migration and Global Environmental Change (2011) *Final Project Report*, London: Government Office for Science.

Gemenne, F. (2011) "Why the Numbers Don't Add Up: A Review of Estimates and Predictions of People Displaced by Environmental Changes," *Global Environmental Change*, 21S: S41–S49.

Gray, C.L. and Mueller, V. (2012a) "Drought and Population Mobility in Rural Ethiopia," *World Development*, 40(1): 134–145.

Gray, C.L. and Mueller, V. (2012b) "Natural Disasters and Population Mobility in Bangladesh," *Proceedings of the National Academy of Sciences*, 109(16): 6000–6005.

Hammar, T. and Tamas, K. (1997) "Why Do People Go or Stay?" in Hammar, T., Brochmann, G., Tamas, K. and Faist, T. (eds.), *International Migration, Immobility and Development: Multidisciplinary Perspectives*, Oxford: Berg.

Hammar, T., Brochmann, G., Tamas, K. and Faist, T. (eds.) (1997) *International Migration, Immobility and Development: Multidisciplinary Perspectives*, Oxford: Berg.

Harris, N. (2002) *Thinking the Unthinkable: The Immigration Myth Exposed*, New York: IB Tauris.

Hathaway, J.C. (2007) "Forced Migration Studies: Could We Agree Just to 'Date'?" *Journal of Refugee Studies*, 20(3): 349–369.

Hindu, The (2005) "200 Feared Trapped in Landslide," July 29, available online at: http://www.hindu.com/2005/07/29/stories/2005072907311300.htm.

Horst, C. (2006) "*Buufis* amongst Somalis in Dadaab: The Transnational and Historical Logics behind Resettlement Dreams," *Journal of Refugee Studies*, 19(2): 143–157.

Hyndman, J. (2003) "Preventative, Palliative or Punitive? Safe Spaces in Bosnia-Herzegovina, Somalia and Sri Lanka," *Journal of Refugee Studies*, 16(2): 167–185.

Hyndman, J. (2011) *Dual Disasters: Humanitarian Aid after the 2004 Tsunami*, Sterling, VA: Kumarian Press.

Jónsson, G. (2008) "Migration Aspirations and Immobility in a Malian Soninke Village," University of Oxford, International Migration Institute, Working Paper 10.

Jureidini, R. (2011) "State and Non-state Actors in Evacuations during the Conflict in Lebanon, July–August 2006," in Martin, S. and Koser, K. (eds.), *The Migration–displacement Nexus: Patterns, Processes and Policies*, Oxford: Berghahn.

Kothari, U. (2002) "Migration and Chronic Poverty," Institute for Development Policy and Management, Working Paper 16, University of Manchester.

Kothari, U. (2003) "Staying Put and Staying Poor?" *Journal of International Development*, 15: 645–657.

Kothari, U. (2013) "Geographies and Histories of Unfreedom: Indentured Labourers and Contract Workers in Mauritius," *Journal of Development Studies*, DOI:10.1080/00220388.2013.780039.

Lubkemann, S.C. (2008) "Involuntary Immobility: On a Theoretical Invisibility in Forced Migration Studies," *Journal of Refugee Studies*, 21(4): 454–475.

McLachlan-Bent, A. and Langmore, J. (2011) "A Crime against Humanity? Implications and Prospects of the Responsibility to Protect in the Wake of Cyclone Nargis," *Global Responsibility to Protect*, 3: 37–60.

Morel, M., Stavropoulou, M. and Durieux, J.F. (2012) "The History and Status of the Right not to be Displaced," *Forced Migration Review*, 41: 5–7.

Mountz, A. (2011) "The Enforcement Archipelago: Detention, Haunting and Asylum on Islands," *Political Geography*, 30: 118–128.

Mountz, A. and Hiemstra, N. (2012) "Spatial Strategies for Rebordering Human Migration at Sea," in Wilson, T.M. and Donnan, H. (eds.), *A Companion to Border Studies*, Oxford: Wiley-Blackwell.

Richmond, A. (1993) "Reactive Migration: Sociological Perspectives on Refugee Movements," *Journal of Refugee Studies*, 6(1): 7–24.

Rosen, S. (1981) *Truman of St. Helens: The Man and His Mountain*, Seattle: Madrona Publishers.

Scott, J. (1998) *Seeing Like a State: How Certain Schemes to Improve the Human Condition have Failed*, Yale University Press.

Skeldon, R. (1997) *Migration and Development: A Global Perspective*, Harlow: Longman.

Standing, G. (1981) "Migration and Modes of Exploitation: Social Origins of Mobility and Immobility," *Journal of Peasant Studies*, 8(2): 173–211.

Sugrue, T.J. (2005) *The Origins of the Urban Crisis*, Princeton: Princeton University Press.

Tacoli, C. (2011) "Not Only Climate Change: Mobility, Vulnerability and Socio-economic Transformations in Environmentally Fragile Areas of Bolivia, Senegal and Tanzania," Human Settlements Working Papers 28.

UNDP (2009) *Human Development Report 2009. Overcoming Barriers. Human Mobility and Development*, New York: UNDP.

United Nations (2009) "Trends in Total Migrant Stock: The 2008 Revision," available online at: http://esa.un.org/migration.

United Nations (2013a) "International Migration Policies 2013," Population Division, available online at: http://www.un.org/en/development/desa/population/publications/policy/international-migration-policies-2013.shtml.

United Nations (2013b) "Cross National Comparisons of Internal Migration: An Update on Global Patterns and Trends," Population Division Technical Paper 2013/1, available online at: http://www.un.org/en/development/desa/population/.

Waller, M. (2005) "Auto-mobility," *Washington Monthly*, October/November.

Weiss, G. (2011) *The Cage: The Fight for Sri Lanka and the Last Days of the Tamil Tigers*, London: Random House.

World Commission on Dams (2000) *Dams and Development: A New Framework for Decision Making*, London: Earthscan.

Zolberg, A., Suhrke, A. and Aguayo, S. (1989) *Escape from Violence*, Oxford: Oxford University Press.

15

POLICY ADRIFT

The challenge of mixed migration by sea

Judith Kumin

Introduction

On January 30, 2007, the Spanish maritime rescue service responded to a distress call from the *Marine 1*, a cargo ship stranded off the coast of West Africa. This was the start of an incident of what has come to be known as "mixed" migration (Wouters and Den Heijer 2010). It involved at least a dozen countries, two international organizations, several non-governmental organizations (NGOs), and 369 migrants[1] from Africa and Asia who had paid smugglers to take them to the Canary Islands.[2] They had embarked at Conakry (Guinea) and ended up detained, under Spanish police guard, in an abandoned fish processing plant in the Mauritanian city of Nouadhibou.

When the Spanish rescue service reached the *Marine 1*, it determined that although the ship was in Senegal's Search and Rescue area, the nearest port was Nouadhibou. It towed the vessel toward Nouadhibou, but Mauritania refused to allow the *Marine 1* to enter port. After long negotiations, a deal was reached (see Consejo General del Poder Judicial 2007, 6–7). Mauritania would allow the ship to dock, and the passengers to disembark, in exchange for a promise from Spain to take them out of the country within four hours. Spain would position aircraft in Nouadhibou for this purpose. The agreement stipulated that the Africans would be sent back to Conakry, while those Asians who volunteered to return home would be flown directly from Nouadhibou, and any who refused voluntary return would be repatriated via Spain or another third country. The agreement made no mention of screening for protection needs, nor of conditions of reception. No role was assigned to the UN High Commissioner for Refugees (UNHCR), while the International Organization for Migration (IOM) would assist the Spanish police and representatives of the countries of origin to interview the Asian passengers only.

The migrants disembarked on February 12. There were 334 Asians (299 claimed to be from India or Pakistan, twenty-two from Burma, ten from Sri Lanka and three from Afghanistan) and thirty-five Africans (twenty-three said they were from Guinea or Côte d'Ivoire, ten from Sierra Leone and two from Liberia). The four-hour deadline for removing all passengers from the country proved illusory, but the next day Spanish officials put the Africans on a plane bound for Guinea. When the aircraft was unable to land there owing to an uprising in Conakry, Spain rerouted it to Praia (Cape Verde), where the passengers were held in a local police lock-up.[3] On March 2, after relative calm had returned to Conakry, Spain deposited the Africans there and gave them some cash to help them to return home.

On February 14, the 35 passengers who claimed to be from Burma, Sri Lanka and Afghanistan were flown to the Canary Islands. Once on Spanish territory, fourteen asked for asylum while twenty-one reportedly opted to repatriate (one to Sri Lanka and twenty to Bangladesh, though they had earlier claimed to be from Burma). Spain found nine asylum applications inadmissible and deported the claimants to Sri Lanka, against UNHCR's advice. By April 24, when (then) President Musharraf of Pakistan made an official visit to Madrid, only twenty-three persons were still in Mauritania. The others had returned to India or Pakistan, with arrangements made by IOM, financed by Spain.[4] The remaining twenty-three migrants vehemently opposed return. Soon thereafter, four were taken in by Portugal, apparently at the personal request of the UN High Commissioner for Refugees, six were transferred to an immigration detention center in the Spanish enclave of Melilla, and the rest were deported to Pakistan.

Spain intervened to rescue the passengers on the *Marine 1*, but also to prevent them from reaching Spanish territory. It detained 369 persons on the territory of Mauritania, from where most were sent (by or with the help of Spain) to their countries of origin or to third countries. Screening to identify protection needs does not appear to have taken place, other than on the basis of (claimed) nationality. The migrants were held under conditions that would not have been permissible in Spain, and were denied the administrative and judicial recourse they would have enjoyed, had they landed in Spain. Efforts by Spanish NGOs to hold their government accountable for its actions failed.[5]

The *Marine 1* case illustrates how far states will to go to prevent irregular arrivals by sea, and the need for states to be equipped to deal with flows of people on the move for different reasons, with different protection needs. It highlights the tension between states' responsibilities for immigration control, on the one hand, and human rights protections, on the other, and shows that the line between rescue at sea and interception can be blurred. The case also raises the important question of states' accountability for extraterritorial actions. As the incident drew to a close, the UN High Commissioner for Refugees, António Guterres, wrote to Spanish Prime Minister Zapatero noting that, while Spain had averted a tragedy at sea, the process that determined the fate of the affected persons should be the "object of reflection" to "better prepare" for future situations.[6]

This chapter looks at the challenge of responding to mixed migration by sea in a rights-respecting manner. It explores the utility of the concept of mixed migration and specific problems related to rescue and interception at sea: where to disembark passengers, what arrangements should be in place for reception and determination of protection needs, and what solutions can be found for the persons concerned. The outcomes should not only respect the right to seek and enjoy asylum from persecution, but should also take account of the legitimate protection needs of "crisis migrants" who do not fall within the established frameworks for protecting refugees and victims of torture. The chapter points out that while "boat people" are often fleeing a situation of crisis, their mode of travel can generate a humanitarian crisis as well. It concludes that much more needs to be done to respond to irregular maritime migration in a way that protects fundamental rights and respects human dignity, but that the political will for this appears to be lacking.

The concept of "mixed" migration

Contemporary irregular migration is mostly "mixed" in nature, meaning that it consists of flows of people who are on the move for different reasons, but share the same routes, modes of travel and vessels. They cross land and sea borders without authorization, frequently with the help of people smugglers.

The IOM defines mixed flows as "complex population movements including refugees, asylum-seekers, economic migrants and other migrants" (IOM 2004, 42). UNHCR similarly notes that "mixed movements" can include migrants in an irregular situation as well as refugees, asylum-seekers and others with specific needs, such as trafficked persons, stateless persons, and unaccompanied or separated children (UNHCR 2011, Background paper). However, the dividing lines between these categories are not always clear, the groups are not mutually exclusive, and people often have more than one reason for leaving home. Also, while some organizations stress that "other migrants" includes people displaced due to climate change,[7] neither the term "other migrants" nor the phrase "migrants in an irregular situation" captures the extent to which mixed flows include people who have left home because they were directly affected or threatened by a humanitarian crisis and therefore need some type of protection, even if they do not qualify as refugees.

Mixed migration is not a new phenomenon. During the Cold War, for example, people left Eastern Europe for a host of reasons—political, economic, social and family related—as did the Vietnamese boat people and others commonly perceived as "refugees." What has changed is the scope and complexity of mixed migration, and the way countries of destination react to it. The proliferation of causes, the involvement of criminal enterprises, security concerns, and the sheer number of people on the move have led states to intensify their efforts to fight irregular migration, often applying blanket measures without any screening for protection needs. Where screening does take place, it generally serves only to identify refugees (and sometimes victims of torture) as defined by international or regional law, leaving other "crisis migrants" in limbo, along with persons who may be particularly vulnerable because of their age, gender or disability status.

What, then, is the utility of the notion of mixed migration? The concept featured prominently at UNHCR's Global Consultations on International Protection in 2001. This was a time when governments were increasingly preoccupied with irregular migration and abuse of asylum channels. UNHCR was keen to show understanding for states, notably those on which it depended for its funding. At the same time, it wanted to make clear that refugee protection and migration control were not mutually exclusive.

To be useful, however, the concept should be more than a device to "take the political steam out of the highly charged asylum debate" (Van Hear 2011, 4). It can help to advance refugee rights by emphasizing the need to identify refugees within broader flows, and this has essentially been UNHCR's strategy. But this approach carries the risk of delegitimizing those who do not qualify as refugees but nevertheless have protection needs, and can have a negative impact on how such persons are treated. One observer has warned of the danger of over-categorizing migrants (Linde 2011), noting that international law and institutions are highly developed for refugees but not for other "crisis migrants," such as those leaving situations of acute or slow-onset natural disasters, severe economic deprivation or generalized and indiscriminate violence. Although states remain reluctant to agree to new normative frameworks for the protection of vulnerable migrants who do not fall within the refugee definition, UNHCR and others, in particular the IOM, the International Federation of Red Cross and Red Crescent Societies (IFRC), the International Catholic Migration Commission (ICMC) and the Norwegian Refugee Council (NRC), have used the concept of mixed migration to highlight existing protection gaps.

In 2007, UNHCR issued a "Ten Point Plan of Action on Refugee Protection and Mixed Migration" (UNHCR 2007a) and the High Commissioner devoted his first "Dialogue on Protection Challenges" to the theme of Refugee Protection, Durable Solutions and International Migration. Participants at that Dialogue recognized the need for the international community to address existing protection gaps (UNHCR 2007b). The Ten Point Plan has been a useful platform, enabling UNHCR to invite governments and other actors to discuss regional approaches to mixed migration, including in parts of the world where such discussions had rarely taken place. In 2010, High Commissioner Guterres focused his "Protection Dialogue" more squarely on gaps in the international protection regime and soon thereafter emerged as one of the driving forces behind the "Nansen Initiative," to work toward a framework to address the protection needs of persons displaced across international borders by natural disasters, including by the effects of climate change (see Nansen Initiative n.d.).

Over the past decade, the terms "mixed migration" and "protection gaps" have entered the migration management vocabulary. Although governments are wary of accepting protection obligations beyond those pertaining to refugees and victims of torture, organizations working in the field of migration have started to look more closely at the profile of migrants and at their protection needs, including those which result from conditions in the migrants' countries of origin as well as those which arise in the course of their journey, and there is growing inter-agency cooperation with respect to vulnerable, non-refugee migrants. The Regional Mixed

Migration Secretariat (RMMS) in the Horn of Africa and Yemen is one example of a co-operative effort in a region where many mixed flows take place by sea. The RMMS aims to be an "independent agency acting as a catalyst, and where appropriate, as an 'agent provocateur' to stimulate forward thinking and policy development in the sector dealing with mixed migration. Its overarching focus and emphasis is on human rights, protection and assistance" (RMMS n.d., no page).

The particular challenge of boat migration

The unique feature of boat migration is that it almost always involves passage through waters that are not part of the territorial sea of any state. States increasingly see the "high seas" as an area to which they can extend their borders and border control measures, and are tempted by a variety of extraterritorial measures to prevent unauthorized arrivals. Some argue that their international legal responsibilities do not apply to actions taken outside their territory or territorial waters, essentially creating a zone where the rights of migrants are not protected—and where it is difficult to monitor the actions of states. It is on these "high seas" that state and private vessels frequently encounter boat people in distress.

Although comparative data is scarce, UNHCR identified the central Mediterranean as the "most deadly stretch of water" for refugees and migrants in relative terms in 2011 (UNHCR 2012a). But there is no overall tally of today's boat people, and some situations attract more attention than others. Attempts by Iranians, Afghans, Sri Lankans, Rohingya and others to reach Australia by boat, for instance, have frequently dominated Australian political debate. By contrast, the fact that more than 300,000 boat people, mostly Somalis and Ethiopians, landed in Yemen between 2010 and mid-2013 has received little publicity, to say nothing of the unknown number who perished en route (UNHCR 2010–2013). In the Indian Ocean, the waters around Mayotte have claimed countless lives (UNHCR 2012b), as people not only from Comoros and Madagascar but from the Democratic Republic of Congo, Rwanda, Burundi and elsewhere try to get to the French island.[8] Boats regularly founder off the west coast of Africa. In the Caribbean, Cubans, Haitians, Dominicans and others attempt to reach the US shores using small boats and makeshift rafts, while large, often barely seaworthy, vessels from Asia have travelled as far as the west coast of Canada and the United States.

Boat people, like other irregular migrants, are driven by a variety of push factors: from economic deprivation to political repression, from civil war to the chaotic aftermath of revolutionary change; from sudden onset natural disaster to the slower effects of climate change. Yet the arrival of boat people seems to trigger more emotional and extreme responses than is the case with overland migration. The UN High Commissioner for Human Rights has lamented that "[i]n many cases, authorities reject these [boat] migrants and leave them to face hardship and peril, if not death, as though they were turning away ships laden with dangerous waste" (Pillay 2009, no page). This is not new: unauthorized arrivals by sea confounded governments long before the term "boat people" was coined to describe those fleeing Vietnam after 1975. In 1914, when the Japanese steamship *Komagata Maru* arrived off the coast of British Columbia, carrying Sikh passengers from the Punjab region of India, Canada barred the ship from entering port and ordered it towed out to sea. In 1939 the *Saint Louis* tried to bring German Jews to safety in North America but was turned away by Cuba, the United States and Canada. In both cases, governments justified their actions, as they do today, as necessary to protect the integrity of state borders.

Rescue at sea and disembarkation

As governments have intensified their efforts to combat irregular migration, people-smugglers and migrants have resorted to ever more dangerous means of travel, using unseaworthy and overcrowded vessels, often without adequate food and water, experienced pilots or navigational aids. The result is situations of distress bearing little resemblance to those that the architects of the international law of the sea had in mind.

The duty to aid persons in distress at sea is a basic tenet of seafaring (UN Convention on the Law of the Sea 1982, Article 98), but international law is not explicit with respect to where persons rescued at sea should be disembarked. The greater the problems captains face to disembark persons picked up at sea, the more they hesitate to conduct rescues, and there have been many reports of ships ignoring migrant vessels in distress. At the height of the civil war in Libya in March 2011, for instance, the North Atlantic Treaty Organization (NATO) was accused of failing to aid a dinghy that had set out from Tripoli with seventy-eight people on board. After running out of fuel and drifting for fourteen days, the boat landed back on the Libyan coast with just nine survivors. They reported having been sighted by several vessels, which did not provide any help, and having been tossed a few packets of biscuits and bottled water by a NATO helicopter (Heller, Pezzani and Studio 2012).

It was traditionally assumed that persons rescued at sea would be fishermen or other seafarers who could be deposited at the next port of call, from where they would return to their home countries. Tension in the rescue at sea regime first arose in the 1970s when it was evident that Vietnamese boat people could not simply be sent back to their country. Coastal states in the region, determined not to be saddled with responsibility for the boat people, refused to let them disembark.

In the initial phases of the Vietnamese exodus, Western countries regarded the boat people ipso facto as refugees. This made it possible for UNHCR, in 1979, to set up the DISERO (Disembarkation Resettlement Offers) scheme, whereby Western countries guaranteed that they would resettle Vietnamese rescued and disembarked by ships flying flags of convenience. Problems remained with respect to those rescued by other ships, however, and in 1983 UNHCR floated a proposal for a broader scheme known as RASRO (Rescue at Sea Resettlement Offers), which was finally launched two years later. By that time, however, governments were starting to question the prima facie recognition and automatic resettlement of the boat people. Between 1980 and 1985, UNHCR's Executive Committee adopted no fewer than six Conclusions to encourage rescue and disembarkation in the Southeast Asian context (UNHCR ExCom Conclusions 20 (XXXI), 23 (XXXII), 26 (XXXIII), 31 (XXXIV), 34 (XXXV) and 38 (XXXVI)).

There have not been any collaborative measures to promote rescue at sea since that time, although rescue at sea would seem to lend itself to international cooperation, since both rescuing and coastal states find themselves with jurisdiction over the migrants essentially by chance. After the turn of the millennium, in the wake of high-profile incidents such as Australia's *Tampa* affair (2001) and the *Cap Anamur* case in the Mediterranean (2004),[9] UNHCR launched new efforts to secure arrangements for disembarkation of refugees and asylum-seekers rescued at sea, but expert meetings in Lisbon in 2002 and Athens in 2005, and a meeting of government representatives in Madrid in 2006 produced no concrete results.

The International Maritime Organization (IMO) was reluctant to become embroiled in disputes over disembarkation, although at UNHCR's urging, the two organizations produced a joint publication in 2006, explaining relevant recent developments in international law (International Maritime Organization and UNHCR 2006). These included the entry into force of the United Nations Convention on the Law of the Sea (UNCLOS) and amendments to the Safety of Life at Sea (SOLAS) and the Search and Rescue (SAR) Conventions, as well as implementing Guidelines issued by the IMO.[10] The Guidelines stipulate that the country in whose SAR region survivors are recovered is responsible for providing a "place of safety" for disembarkation, or for ensuring that such a place is provided. According to the Guidelines, a "place of safety" is a location where the survivors' safety or life is no longer threatened, where basic human needs can be met, and from where arrangements can be made for transportation to their next or final destination—leaving open whether it is also a place where refugee protection can be assured, and what the next or final destination might be (Strik 2012).

In 2007, UNHCR urged its Executive Committee to adopt a new Conclusion on rescue at sea (UNHCR 2007c). Member countries, anxious to avoid norm-setting in this sensitive area, resisted even further "soft" law, notwithstanding several notorious incidents, including one in the Mediterranean Sea in which twenty-seven survivors of a boat of more than fifty passengers clung for several days to a tuna pen towed by a Maltese trawler, while Malta and Italy debated who had responsibility for disembarkation (*Times of Malta* 2007).

In parallel, the European Commission tried to set rules for disembarkation of persons rescued at sea during operations coordinated by the European Union's (EU) border agency Frontex, but there was a lack of political will on the part of states to resolve this issue, even within the EU. First, the Commission convened a Working Group of Member States, Frontex, UNHCR, IOM and the IMO. The group met five times in 2007–08, but could not reach agreement. The main sticking points were human rights and refugee rights, and the identification of places for disembarkation. Then the Commission drafted Guidelines, which were approved in 2010 as a non-binding Annex to a Council Decision.[11] In 2012, that Decision was annulled on technical grounds by the Court of Justice of the European Union.[12] In 2013, the Commission made another attempt, putting forward a new draft Regulation that must be approved by Member States and the European Parliament.[13]

In 2011, with normative efforts unlikely to succeed, UNHCR proposed a non-binding "Model Framework" for cooperation on rescue at sea, intended to facilitate disembarkation and processing (UNHCR 2011). The Model Framework could be complemented by standard operating procedures for shipmasters. UNHCR further suggested that it could dispatch "Mobile Protection Response Teams" to help states with refugee status determination after disembarkation, coordinate resettlement and support the return of persons not in need of international protection. The proposal was put to states and other stakeholders at an expert meeting in Djibouti. The meeting put a spotlight on the migration crisis in the Gulf of Aden, but did not yield any concrete results, and no region has yet acted on the Model Framework proposal.

The US tried to negotiate a standing arrangement with countries in the East and Horn of Africa (Tanzania, Djibouti, Ethiopia, Kenya) for disembarkation of persons rescued at sea by US vessels in the region, but made no progress and the initiative has not been pursued. The IMO's 2011 decision to develop a draft regional agreement on rescue at sea in the Mediterranean had made little progress as of 2013 (IMO 2011). It remains to be seen whether parallel initiatives spearheaded by UNHCR in the Asia Pacific and Mediterranean regions will have more success.

From a human rights perspective, the lack of political will to resolve questions concerning rescue at sea and disembarkation, even within a regional context, is disturbing. Discussions led by UNHCR understandably have focused on measures to protect asylum-seekers and refugees rescued at sea, to secure respect for the principle of *non-refoulement* and to facilitate solutions for refugees. It is likely that the reluctance of states to make progress even on these narrower issues is linked to the fact that migrant vessels frequently carry not only refugees and bona fide asylum-seekers, but also individuals fleeing risks not covered by the refugee definition. There is little agreement on how to respond to people on the move who do not qualify as refugees yet cannot be returned to their countries of origin, whether for practical or protection-related reasons. In this context, cooperative arrangements for rescue at sea and disembarkation appear more elusive than ever.

Interception and state responsibility

Once boat people land, governments must care for them, determine any claims to refugee status, and face the uncertainty—and cost—of dealing with persons who are found ineligible for international protection. It has been reported, for example, that the EU agency Frontex spent €8.5 million in 2010 to repatriate just 2,038 irregular migrants (Fortress Europe 2011). Interception has thus, not surprisingly, become a preferred policy tool of governments that have the means to stop migrant vessels at sea and divert them to other countries (see Goodwin-Gill 2011 for a review of interception in the context of international refugee law).

UNHCR has defined interception as:

> All measures applied by a State, outside its national territory, in order to prevent, interrupt or stop the movement of persons without the required documentation crossing international borders by land, air or sea, and making their way to the country of prospective destination.
>
> *(UNHCR 2000, 10)*

The United States has long intercepted Cubans, Haitians, Dominicans and others in the Caribbean Sea (for an overview of US interception practice, see Legomsky 2008); Australia stops vessels heading for its territory from Indonesia and other departure points in Asia; Spain intercepts boats trying to reach the Canary Islands; Italy has intercepted boats in the Mediterranean (Hyndman and Mountz 2008).

Interception invariably results in lower levels of protection of fundamental rights than would have been available, had the migrants been allowed to continue to their destination. From the perspective of states, however, it is an attractive instrument not only because it prevents arrivals but also because it takes place largely beyond the public's view.

International law is not well developed with regard to interception. According to UNCLOS Article 110, states may board ("visit") a flagless vessel or one flying the flag of another country if they suspect that piracy, slavery or unauthorized broadcasting is occurring. The UN Protocol against the Smuggling of Migrants by Land, Sea and Air (UN 2000) gave states renewed authority to intercept, board and search vessels as part of the international effort to prevent smuggling and trafficking in human beings. States Parties are obliged to cooperate to the fullest possible extent to "prevent and suppress" the smuggling of migrants by sea (Article 7). The Protocol stipulates that they must also respect their obligations under international human rights law (Article 19), including the 1951 Convention relating to the Status of Refugees (1951 Refugee Convention), but does not provide guidance on how this balance is to be achieved.

There is broad consensus—recently reaffirmed by the European Court of Human Rights—that states are bound by their international human rights obligations wherever they assert their jurisdiction, including outside their territory or territorial waters (European Court of Human Rights 2012). The UN High Commissioner for Human Rights and the High Commissioner for Refugees emphatically agree with this position (UNHCR 2007d); the United States has remained quite isolated in maintaining that human rights and refugee law instruments do not apply when it acts beyond its territory.[14]

UNHCR's efforts to address the tension between human rights obligations and border control in the context of interception have focused on the distinction between refugees and asylum-seekers entitled to international protection and "other migrants who can resort to the protection of their country of origin" (UNHCR 2000, para. 34(a)). In 2003, after difficult negotiations, UNHCR's Executive Committee adopted a "Conclusion on Protection Safeguards in Interception Measures"

(UNHCR 2003) but declined to mandate UNHCR to develop Guidelines on the subject. The Conclusion simply reiterates the obvious: the human rights of intercepted persons should be respected, and interception should not result in asylum-seekers and refugees being denied access to international protection. It does not address interception on the high seas or the extraterritorial processing of intercepted persons, noting only that the state on whose territory interception occurs has "primary responsibility" for addressing any protection needs of intercepted persons—without specifying what this might cover.

As normative efforts stalled, interception without adequate protection safeguards continued. In 2009, in a highly publicized incident, Italy intercepted 750 boat people on the high seas in the Mediterranean and returned them to Libya. Some were transferred from Italian to Libyan vessels; others were taken to Libyan ports by Italian ships. NGOs were outraged and UNHCR condemned the practice.[15] The European Commission and other EU Member States remained largely silent, no doubt hoping that Italy's action would stem the flow of refugees, asylum-seekers and others who for years had been departing from the Libyan coast on small and medium-sized fishing vessels, bound for the Italian island of Lampedusa, just 160 nautical miles away. In 2008, some 30,000 people had landed there, representing sixty-one nationalities. Somalis and Eritreans, presumed to be refugees, accounted for 24 percent of arrivals, while Nigerians—one-third of whom were women—accounted for 19 percent, suggesting that this route was being used for trafficking for sexual exploitation (Tennant and Janz 2009).

UNHCR, while acknowledging the "mixed" nature of the flows, saw the Italian practice as a direct challenge to the international refugee protection system. That system would be devoid of meaning if a state could evade its responsibility by depositing asylum-seekers and refugees on the territory of a state not party to the 1951 Refugee Convention and lacking any refugee protection mechanisms. When Italy's action was challenged in the European Court of Human Rights, UNHCR intervened in the case, arguing that Italy had breached its 1951 Refugee Convention obligations by returning persons intercepted at sea, without any examination of their protection needs, to a country where there was no refugee protection system, and a clear risk of ill-treatment and *refoulement* (European Court of Human Rights 2011).

To justify the push-backs, Italy argued that it was putting into effect EU decisions on control of the external border, and implementing its 2008 bilateral agreement with Libya pertaining to the interception of migrant vessels in the Mediterranean and their return to Libya (European Court of Human Rights 2012, 93, 94; see also 2008 Italy–Libya Treaty on Friendship, Partnership and Cooperation, Article 19). It also referred to its obligations under the UN Protocol against the Smuggling of Migrants by Land, Air or Sea. However, neither the general principle of cooperation between states nor the above-mentioned Protocol exempts states from complying with international refugee protection norms.

In February 2012, the Grand Chamber of the European Court of Human Rights ruled that Italy had violated the European Convention on Human Rights and Fundamental Freedoms (ECHR) (European Court of Human Rights 2012). Even when states intercept vessels in international waters, the Court said, they must abide by their international obligations. It rejected Italy's arguments that it had not exercised "absolute and exclusive control" over the applicants and said that Italy could not circumvent its responsibilities under the ECHR by describing the events as rescue operations on the high seas (European Court of Human Rights 2012, 80). The Court's ruling confirms that the *non-refoulement* obligation applies on the high seas, and that states must take affirmative measures to ensure that intercepted migrants have access to protection.

This judgment may make European countries more cautious about how interception measures are conducted, but there is little indication of waning political interest in it as a tool of migration control. Even in the absence of empirical evidence that interception affects the "tipping point" at which people decide to leave their country, states believe it is a valuable deterrent. Moreover, as explained by the US Department of Homeland Security, "[i]nterdicting migrants at sea means they can be quickly returned to their countries of origin without the costly processes required if they successfully enter the [country]" (US Coast Guard n.d., no page).

For many years, the United States has intercepted Cubans, Haitians, Dominicans and others in the Caribbean, and refused to allow intercepted persons, including those demonstrated to be refugees, to enter the United States. It uses the US Naval Base at Guantánamo, Cuba, to house the refugees, while it tries to persuade other countries to admit them (Farber 2010).[16] In other words, even though the US maintains the formal position that its obligations under national and international refugee law do not extend to persons intercepted on the high seas, it refrains from forcibly repatriating those likely to face persecution or torture in their home countries, and seeks protection for them. To avoid the legal obligations that would flow from the label "refugee," it calls these persons "protected migrants."[17] Where potential countries of resettlement have asked for confirmation that candidates for resettlement qualify as refugees, UNHCR has conducted interviews at Guantánamo, and provided this certification.

Australia has implemented similarly aggressive policies to deter boat arrivals. Since 1992, it has used a variety of approaches ranging from non-reviewable, mandatory detention; to excising territory from the country's migration zone; to granting visas for temporary protection only; to interception coupled with offshore processing arrangements in Papua New Guinea and Nauru (the "Pacific Solution") and attempted processing arrangements with Indonesia and Malaysia (Foulkes 2012). Simply refusing to allow boats to land was excluded, not only because it would violate international law but also because boat people would scuttle their vessels in order to be rescued.

Like the United States, Australia has gone to great lengths to avoid bringing intercepted persons to its territory, where they would benefit from Australian legal protections. In one incident in 2009, an armed patrol vessel of the Australian Customs Marine Unit, the *Oceanic Viking*, stopped seventy-eight Sri Lankans in Indonesia's Search and Rescue zone. Australia wanted to disembark them on the Indonesian island of Bintan, where they would be held in a detention center financed by Australia. The Sri Lankans, who had previously been detained in Indonesia, demanded to be taken to Australia.

After a four-week stand-off, the passengers agreed to disembark in Indonesia. Under the terms of the never-published deal, UNHCR agreed to process the passengers' claims to refugee status, and to resettle those found to be refugees, though not necessarily in Australia. Even after UNHCR found all seventy-eight Sri Lankans to qualify as refugees, Australia pressed New Zealand, Canada, the US and European countries to take them in, apparently to signal that interception or rescue by Australia would not result in access to Australian territory, though the deterrent value of resettlement in Europe, North America or New Zealand is unclear and may explain later Australian moves to ensure that options open to refugees are less attractive.

Fresh from the *Oceanic Viking* experience, UNHCR set out its views on interception and the extraterritorial processing of international protection claims in a "Protection Policy Paper" (UNHCR 2010a). UNHCR reiterates that it is the existence of jurisdiction which triggers state responsibility under international human rights and refugee law, and affirms that claims for international protection presented by intercepted persons are "in principle" to be processed by the intercepting state within its territory. However, in a concession to Australia and the United States, and to European discussions about offshore processing,[18] the paper recognizes the utility of interception and extraterritorial processing as part of "a comprehensive or cooperative strategy to address mixed movements" and "to more fairly distribute responsibilities and enhance available protection space" (UNHCR 2010a, paras 3–4). Neither US processing at Guantánamo nor a new "offshore processing" law enacted by Australia in 2012 (Australia 2012) can be seen as meeting those conditions.

The new Australian law essentially revived the discredited "Pacific Solution," which had been implemented from 2001 to 2008. It authorized the transfer to countries designated by the Minister and Parliament as "regional processing" countries of asylum-seekers rescued at sea or intercepted by Australian vessels, and of individuals who have landed in Australia. Presented to the public as regional cooperation, it was in fact a way to legitimize bilateral arrangements with Nauru and Papua New Guinea, which were administered by Australia until 1968 and 1975, respectively. Their participation can be understood as yielding to Australian pressure or seeking to receive Australian benefits, or both. Nauru and Papua New Guinea are neither destination nor transit countries for asylum-seekers, and when the law was adopted, neither had procedures in place for the determination of refugee status. Yet the law allowed Australia to transfer responsibility for assessing asylum applications to these countries, where applicants could be held indefinitely.[19] This time, UNHCR declined to play an operational role, saying that, "Australia is not absolved of its legal responsibilities to protect people through all aspects of the processing and solutions" (Towle 2012, no page).

Under the 2012 arrangement, persons found to qualify as refugees were still to be allowed to settle in Australia, but in July 2013 Australia further hardened its stance toward boat people, and announced a new agreement with Papua New Guinea.[20] Under this arrangement, asylum-seekers arriving in Australia or intercepted en route are to be sent to Papua New Guinea not just for processing of their claims. Persons found to be refugees will no longer be permitted to settle in Australia; they are expected to settle in Papua New Guinea.

The policy shift came just weeks before scheduled elections in Australia, and was widely seen as a political move as well as an effort to deter boat people from setting out for Australia.

Reception, screening and solutions

Reception, screening and the search for solutions are often problematic in the context of persons intercepted or rescued at sea. In its Protection Policy Paper on interception and extraterritorial processing, UNHCR sets out some standards for reception, based on international human rights law, which are applicable to all new arrivals, not only those who seek protection as refugees. Reception arrangements must address basic needs and be consistent with the right to an adequate standard of living. Culturally appropriate meals, access to communication devices, space, privacy and security are required. Open centers are preferred; detention must be necessary, reasonable, proportionate and non-discriminatory. People with special needs (women, children, victims of torture and trauma) merit specific assistance (UNHCR 2010a, paras 23–29).

It is notoriously difficult to monitor the conditions under which intercepted persons are held, as one of the objectives of intercepting states is to limit access to legal protections. In the *Marine 1* case, it is clear than reception conditions were particularly poor. The Spanish government did not deny this, but claimed that the migrants were held in a facility owned by the Mauritanian state, over which Spain had no sovereignty, and that the means available to Mauritania were "very limited" (Consejo General del Poder Judicial 2007, 6, 8). Yet according to UNHCR's Protection Policy Paper, produced after the *Marine 1* incident, transfer of responsibility for intercepted persons by the intercepting state to another country requires "formal assurances" that the accepting country will respect essential protection standards, and the intercepting state must ensure that it does so (UNHCR 2010a, para. 35). In the case of Italy's push-backs to Libya, this was certainly not the case. Nor does it appear to be the case in the context of Australia's recent arrangements with Nauru and Papua New Guinea.

In December 2012, when UNHCR visited Nauru to assess Australia's latest offshore processing arrangement, nearly 400 asylum-seekers (from Afghanistan, Iran, Iraq, Pakistan and Sri Lanka) were held there. UNHCR's conclusions were devastating. It found the facilities "harsh and unsatisfactory," the detainees suffering from "a sense of isolation and abandonment" (UNHCR 2012c, 1, 12). UNHCR condemned the lack of information provided to the detainees and worried about the widespread depression, instances of self-harm and attempted suicide. No asylum procedures were in place four months after the agreement was concluded (UNHCR 2012c). Despite Australia's statements that its legal responsibilities for the asylum-seekers were extinguished at the time of transfer, UNHCR observed that Australia maintained de facto control over the detainees and considered that Australia remained responsible for ensuring a durable solution for those found to be refugees.

UNHCR's Protection Policy Paper stresses that "timely outcomes" are required for intercepted persons. Several refugees from the *Oceanic Viking* who had been transferred to Australia were still held in detention there as of mid-2012, a full three years after the incident.[21] On Nauru, UNHCR's monitoring team observed that the delays and uncertainty about processing arrangements were having a "significant and detrimental impact" on the mental health of asylum-seekers transferred from Australia (UNHCR 2012c, 2). Similarly, at Guantánamo, refugees may be detained indefinitely until a third country offers to resettle them, although the US refuses to call this detention, pointing out that the refugees may opt to return to their country of origin (of persecution) at any time.

Neither the Australian nor the US offshore processing arrangements make provision for persons who do not qualify as refugees, but have other vulnerabilities. The UNHCR Protection Policy Paper provides no guidance in this respect, stating simply that, "[f]or persons found not to be in need of international protection, resolution of their situation will generally consist of return to the country of origin" (UNHCR 2010a, para. 32). As the movement to define and secure the rights of persons who do not qualify as refugees but are fleeing other risks gathers steam, interception and offshore processing are likely to become even more attractive to states determined to limit their obligations. A US State Department spokesman, for instance, called Guantánamo Bay "an enormously valuable asset" for the US in case of a mass exodus following the 2010 earthquake in Haiti (Levine 2010, no page).

Conclusion

States need to come together to address irregular migration by sea in a cooperative, protection-oriented manner. Boat migration is a complex phenomenon, involving the intersection of several bodies of international law and thorny questions of jurisdiction. It affects countries of origin, of transit and of destination in all regions of the world. Despite its prevalence, states have failed to demonstrate the political will to work out an internationally accepted response to boat migration that would respect the sovereign right of states to control their borders, and protect the human rights and human dignity of the boat people. Instead, states experiment with ad hoc responses, with the balance between protection and control shifting as a function of domestic and external factors.

Irregular migration by sea almost always represents a response to a crisis. Why else would individuals take such terrible risks? It seems set to continue, as the drivers of migration multiply, other migration options are foreclosed and the steady intensification of migration control measures pushes migrants and people smugglers to take ever greater risks. Indeed, the very mode of travel frequently constitutes a humanitarian crisis, as evidenced by regular reports of tragedies at sea.

Inter-state agreements are needed to guarantee rescue at sea and safe disembarkation, as well as arrangements for reception and screening. States that practice interception at sea need to be held accountable for the protection of migrants' rights, and international organizations should be wary of participating in or otherwise lending their imprimatur to ad hoc arrangements that undermine state responsibility. There is no doubt that the mixed nature of the flows creates a real challenge, with states and international organizations only in the early stages of discussions about their responsibility for identifying and responding to protection needs going beyond those of asylum-seekers and refugees.

Notes

1 Throughout this chapter, the term "migrant" is used in a generic sense, to encompass people on the move for various reasons, including asylum-seekers, refugees and others.
2 In 2006, 32,000 boat people arrived in the Canaries, but the number dropped by 75 percent after Spain and the EU border agency Frontex launched intensive surveillance and interception efforts, and after Spain concluded return agreements with most of the sending countries in West Africa (see Carling and Hernandez-Carretero 2011).
3 In one of several inconsistencies, the UN Committee against Torture (see note 5 below) states that the Africans were taken to the Canary Islands before being flown to Guinea. Other sources, including UNHCR, UNDP and the Spanish NGO CEAR, confirm the Cape Verde episode.
4 IOM's Constitution stipulates that the organization engages only in *voluntary* returns. It is unclear whether the people held in Nouadhibou had any option other than return, and how voluntariness can be assessed in such a situation.
5 A first complaint by the NGO Collective for Justice and Human Rights was found inadmissible by Spain's Attorney General. The Spanish Committee for Aid to Refugees (CEAR) unsuccessfully petitioned the Administrative Court for transfer of the last twenty-three passengers to Spain. An appeal to the High Court also failed. Long after all passengers had left Mauritania, the Court held that Spain's actions in responding to a distress call in the search and rescue area of another country, and taking the passengers to the nearest safe port, did not give rise to any other rights or duties (Tribunal Supremo, Sala de lo Contencioso, Recurso 548/2008, judgment of 17.02.2010). The Collective for Justice and Human Rights turned to the UN Committee Against Torture, alleging violation by Spain of the Convention against Torture. The Committee found the complaint inadmissible, but affirmed that Spain had "maintained control over the persons on board the *Marine 1* from the time the vessel was rescued and throughout the identification and repatriation process that took place at Nouadhibou." The Committee reiterated that the jurisdiction of a State Party "refers to any territory in which it exercises, directly or indirectly, in whole or in part, de jure or de facto effective control . . ." (*J.H.A.v. Spain*, CAT/C/41/D/323/2007).
6 Information on the letter on file with the author.
7 For instance, in May 2009 the Danish Refugee Council revised its Policy on Mixed Migration to reflect concerns related to climate change.
8 Mayotte, a French *Département*, will become an "outermost region" of the European Union in 2014 and is thus seen as a "back door" to the EU.
9 The Norwegian freighter, *MV Tampa*, rescued some 400 Afghans at sea. Australia refused to let them disembark, triggering a political crisis and the so-called "Pacific Solution," whereby migrants rescued at sea would be diverted to Nauru for screening (see Bostock

2002). The *Cap Anamur*, chartered by a German NGO, picked up thirty-seven Africans from a sinking inflatable boat in the Mediterranean in June 2004, and spent three weeks trying to disembark them in Italy. The ship's captain and first officer were put on trial in Italy for encouraging illegal immigration, but were eventually acquitted (see BBC News 2009).

10 As of 2013, UNCLOS was ratified by 165 countries. The 1974 International Convention for the Safety of Life at Sea (SOLAR) and the International Convention on Maritime Search and Rescue (SAR) entered into force in 1980. Amendments to the SAR Convention, which entered into force in 2006, require Parties "to coordinate and cooperate to ensure that masters of ships providing assistance by embarking persons in distress at sea are released from their obligations with minimum further deviation from the ship's intended voyage; and arrange disembarkation as soon as reasonably practicable" (see IMO 2004, Annex 34).

11 Council Decision (2010/252/EU) of April 26, 2010, supplementing the Schengen Borders Code as regards the surveillance of the sea external borders (. . .).

12 European Court of Justice, Case C-355/10, Parliament v. Council, Judgment of the Court (Grand Chamber) of September 5, 2012. The Court ruled that the provisions of the Council Decision constituted a major development in the Schengen Borders Code system requiring the approval of the European Parliament.

13 Proposal for a Regulation establishing rules for the surveillance of the external sea borders (. . .), COM (2013) 197 final. With respect to disembarkation of persons intercepted in the territorial sea or contiguous zone of a Member State, disembarkation is to take place in that state. In the case of interception on the high seas, disembarkation is foreseen in the third country from which the boat departed or, if this is not possible, in the Member State "hosting" the Frontex operation. In the case of disembarkation following a rescue operation, this is to occur in a "place of safety." The Proposal is more explicit that the previous Decision with respect to guarantees that disembarkation respects human rights, though some observers believe these guarantees are still inadequate (see Meijers Committee 2013).

14 In *Sale v. Haitian Centers Council*, 509 U.S. 155 (1993), 113 S. Ct. 2549, the US Supreme Court held that the repatriation of Haitians intercepted on the high seas was not governed by international law.

15 The push-backs to Libya were widely reported and condemned, but summary returns were not new. In 2004–05, Italy had forcibly returned around 1,000 persons from Lampedusa to Libya without giving them the possibility to have their protection claims heard. Italy has also regularly returned Egyptians arriving by boat on the coast of the Apulia region, without examining any protection claims and without giving UNHCR access to the individuals (UNHCR 2010b).

16 Wikileaks has published US diplomatic cables reporting on US requests to other countries, including France, Latvia, Panama and Romania, to resettle "Protected Migrants" from Guantánamo.

17 There is no legal definition of "protected migrants," but the term appears in the State Department's online glossary: "Pursuant to an Executive Order of the President of the United States, an individual interdicted at sea who is determined to have a well-founded fear of persecution, or is more likely than not to face torture if he/she returns to his/her country of origin, and whom the US Government houses and cares for at its Migrant Operation Center on the Guantanamo Naval Base while it finds a third country in which to resettle him/her" (US Department of State 2001–2009).

18 In 2003, a UK government paper entitled "A New Vision for Refugees" proposed processing centers outside the EU to which asylum-seekers who had reached the EU could be sent. In 2004, German Interior Minister Otto Schily suggested the EU set up processing centers in North Africa where persons who might otherwise try to cross the Mediterranean to Europe could be held (see Garlick 2006).

19 The "Memorandum of Understanding between the Republic of Nauru and the Commonwealth of Australia, relating to the Transfer to and Assessment of Persons in Nauru,

and Related Issues" is available online at: http://www.minister.immi.gov.au/media/media-releases/_pdf/australia-nauru-mou-regional-processing.pdf; the agreement with Papua New Guinea is online at: http://www.minister.immi.gov.au/media/media-releases/_pdf/mou-between-png-australia-regional-processing.pdf.
20 BBC News, "Australia PM Kevin Rudd Defends PNG Asylum Deal," July 22, 2013.
21 Information provided verbally by UNHCR to the author in August 2012.

References

Australia, *Migration Legislation Amendment (Regional Processing and Other Measures) Act 2012*, No. 113 (2012).
BBC News (2009) "Italy Acquits Migrant Crew," 7 October.
Bostock, L.M.-J. (2002) "The International Legal Obligations owed to the Asylum Seekers on the MV Tampa," *International Journal of Refugee Law*, 14(2/3): 279–301.
Carling, J. and Hernandez-Carretero, M. (2011) "Protecting Europe and Protecting Migrants? Strategies for Managing Unauthorized Migration from Africa," *British Journal of Politics and International Relations*, 13: 42–58.
Consejo General del Poder Judicial (2007) *Buscador Jurisprudencia*, SAN 5394/2007, Audiencia Nacional, Sala de lo Contencioso, Madrid, Recurso 3/2007.
European Court of Human Rights (2011) *Submission by the Office of the United Nations High Commissioner for Refugees in the Case of Hirsi and Others* v. *Italy*, Application no. 27765/09, 29 March.
European Court of Human Rights (2012) *Hirsi Jamaa and Others* v. *Italy*, Application no. 27765/09, judgment of February 23.
Farber, S.R. (2010) "Forgotten at Guantánamo: The Boumediene Decision and its Implications for Refugees at the Base under the Obama Administration," *California Law Review*, 98(3): 989–1022.
Fortress Europe (2011) "Frontex, How Much Are You Costing Me?" available online at: http://fortresseurope.blogspot.co.uk/2011/07/frontex-how-much-are-you-costing-me.html.
Foulkes, C. (2012) "Australia's Boat People: Asylum Challenges and Two Decades of Policy Experimentation," *Migration Information Source*, July, available online at: http://www.migrationinformation.org/USFocus/display.cfm?ID=899.
Garlick, M. (2006) "The EU Discussions on Extraterritorial Processing: Solution or Conundrum?" *International Journal of Refugee Law*, 18(3/4): 601–629.
Goodwin-Gill, G.S. (2011) "The Right to Seek Asylum: Interception at Sea and the Principle of Non-Refoulement," *International Journal of Refugee Law*, 23(3): 443–457.
Heller, C., Pezzani, L. and Studio, S. (2012) "Report on the Left-to-die Boat," part of the European Research Council project "Forensic Architecture," Centre for Research Architecture, Goldsmiths, University of London, available online at: http://www.fidh.org/IMG/pdf/fo-report.pdf.
Hyndman, J. and Mountz, A. (2008) "Another Brick in the Wall? Neo- *refoulement* and the Externalization of Asylum by Australia and Europe," *Government and Opposition*, 43(2): 249–269.
International Maritime Organization (IMO) (2004) *Guidelines on the Treatment of Persons Rescued at Sea*, IMO Resolution MSC, 167(78), Annex 34.
IMO (2011) Facilitation Committee, Document FAL 37/6/1 of July 1.
IMO and UNHCR (2006) "A Guide to Principles and Practice as Applied to Migrants and Refugees," London and Geneva.
International Organization for Migration (IOM) (2004) "Glossary on Migration," Geneva, available online at: http://publications.iom.int/bookstore/free/IML_1_EN.pdf.

Legomsky, S.H. (2008) "The USA and the Caribbean Interdiction Program," *International Journal of Refugee Law*, 18(3/4): 277–295.

Levine, M. (2010) "US Preparing for Potential New Wave of Haitians Fleeing North," *Fox News*, January 15.

Linde, T. (2011) "Mixed Migration: A Humanitarian Counterpoint," *Refugee Survey Quarterly*, 30(1): 89–90.

Meijers Committee (2013) "Note on the Proposal for a Regulation Establishing Rules for the Surveillance of the External Sea Borders in the Context of Operations Coordinated by Frontex (COM(2013)197 final)," May 23.

Nansen Initiative (n.d.) "About Us," available online at: http://www.nanseninitiative.org.

Pillay, N. (2009) Statement of Ms. Navanethem Pillay, UN High Commissioner for Human Rights, at the 12th session of the Human Rights Council, September 15.

Regional Mixed Migration Secretariat (RMMS) (n.d.) "Homepage," available online at: http://www.regionalmms.org.

Strik, T. (2012) "Lives Lost in the Mediterranean Sea: Who is Responsible?" Committee on Migration, Refugees and Displaced Persons, Parliamentary Assembly of the Council of Europe Doc. 12893, April 5.

Tennant, V. and Janz, J. (2009) "Refugee Protection and International Migration: A Review of UNHCR's Operational Role in Southern Italy," UNHCR PDES/2009/5, September.

Times of Malta (2007) "International Shock over Tuna Pen Incident," May 29.

Towle, R. (2012) "Australia: 'Pacific Solution' Redux: New Refugee Law Discriminatory, Arbitrary, Unfair, Inhumane," Human Rights Watch press release, August 16.

UN (2000) Protocol against the Smuggling of Migrants by Land, Sea and Air, supplementing the United Nations Convention against Transnational Organized Crime, adopted November 15, 2000, entered into force 2004, UNTS, Vol. 2241, 507.

UNHCR (2000) "Interception of Asylum-seekers and Refugees: The International Framework and Recommendations for a Comprehensive Approach," EC/50/SC/CRP.17.

UNHCR (2007a) "Refugee Protection and Mixed Migration: A 10-point Plan of Action," Revision 1, January.

UNHCR (2007b) High Commissioner's Dialogue on Protection Challenges, available online at: http://www.unhcr.org/pages/4a12a6286.html.

UNHCR (2007c) "Proposals for an Executive Committee Conclusion on Rescue at Sea," January 16.

UNHCR (2007d) "Advisory Opinion on the Extraterritorial Application of Non-Refoulement Obligations under the 1951 Convention relating to the Status of Refugees and its 1967 Protocol," January 26.

UNHCR (2010a) "Protection Policy Paper: Maritime Interception Operations and the Processing of International Protection Claims: Legal Standards and Policy Considerations with Respect to Extraterritorial Processing," November.

UNHCR (2010b) Regional Representation for Southern Europe, "Access Denied: Interception at Sea and Returns to Libya Implemented by the Italian Government in 2009," June (unpublished paper).

UNHCR (2010–2013) "New Arrivals in Yemen Comparison," available online at: http://www.unhcr.org/4fd5a3de9.html.

UNHCR (2011) "Refugees and Asylum-seekers in Distress at Sea—How Best to Respond?" Background paper for Expert Meeting in Djibouti, November 8–10, available online at: http://www.unhcr.org/4ec1436c9.html.

UNHCR (2012a) "Mediterranean Takes Record as Most Deadly Stretch of Water for Refugees and Migrants in 2011," January 31, available online at: http://www.refworld.org/docid/4f2818452.html.

UNHCR (2012b) "Boat Tragedy off Mayotte," Press Briefing Note, October 9, available online at: http://www.unhcr.org/5073fe249.html.

UNHCR (2012c) "Mission to the Republic of Nauru, 3–5 December 2012, Report," available online at: http://reliefweb.int/report/nauru/unhcr-mission-republic-nauru-3-5-december-2012.

UNHCR Executive Committee (2003) "Conclusion on Protection Safeguards in Interception Measures," No. 97 (LIV), October 10.

US Coast Guard (n.d.) "Alien Migrant Interdiction," available online at: http://www.uscg.mil/hq/cg5/cg531/AMIO/amio.asp.

US Department of State (2001–2009) "Glossary," available online at: http://2001-2009.state.gov/g/prm/c26475.htm.

Van Hear, N. (2011) "Mixed Migration: Policy Challenges," the Migration Observatory at the University of Oxford, March 29, available online at: http://www.migrationobservatory.ox.ac.uk/node/850.

Wouters, K. and Den Heijer, M. (2010) "The Marine 1 Case: A Comment," *International Journal of Refugee Law*, 22(1): 1–9.

16

FLIGHT TO THE CITIES

Urban options and adaptations

Patricia Weiss Fagen

Introduction

"Crisis migrants," that is people for whom flight is the only option in the face of events and processes over which they have no control, have been altering demographic balances both within and beyond the borders of affected nations for centuries. The current crises likely to force people to leave their homes and regions are triggered by conflict, violence and repression, natural and man-made disasters, environmental degradation, climate change, loss of land and resources. A growing number of crisis migrants produced by all these factors, as this chapter will elaborate, are finding their way to cities in their own and in other countries. They are settling into the poorest segments of large and smaller cities, often outside the urban core, in informal settlements in peripheral areas where municipal authorities are only nominally in control, services are lacking, and conditions are precarious.

In fragile economies with weak institutions, the urban poor generally live amidst crime, deteriorating environmental conditions and high unemployment (for more on these themes, see *Disasters* 2012; IFRC 2012). The inability of national or local governments to address sexual abuse, drug traffic, and violent crime, combined with a lack of acceptable housing and overstretched public infrastructure, severely compromise the health, well-being and security of large numbers of urban dwellers. Humanitarian and long-term development strategies are very much overdue to improve the quality of urban life in the places where impoverished newcomers are increasingly concentrated. Adapting to urban life is challenging for all economically disadvantaged populations and new migrants are particularly vulnerable. Among these, forced migrants are especially at risk.

Newly arriving migrants and migrant families who have been able to plan their move to cities and to settle with other known people may reasonably expect to establish helpful safety nets and survival strategies even in the midst of poverty.

Moreover, planned migration almost always includes maintaining human and material ties in the places of origin. If one fails to adapt to the urban setting, it is therefore possible to return. In contrast, not only are those forced to flee due to conflict or criminal acts probably unable to count on support from others, they frequently face continuing threats to life, health, physical safety and subsistence in their new urban destinations. They seek economic opportunity, as all new migrants do, but remain traumatized by the events and processes that caused them to move.

More than half of the world's population now lives in urban settings. UN Human Settlements Programme (UN-HABITAT) estimates that in less than twenty years not only will 60 percent of the world population live in cities, but 93 percent of the additional growth will take place in the urban areas of developing countries.[1] African cities are growing especially rapidly. Africa's population is currently over one billion, with 40 percent living in urban areas that are disadvantaged by historically low levels of national economic growth. The numbers are rapidly increasing due to conflict, agrarian decline and climate change. Smaller, poorly resourced towns and cities are disproportionately impacted (UN-HABITAT 2008; 2010, 19; 2011, 21). Yet, however miserable the urban conditions, cities still offer—or are perceived to offer—better functioning and accessible institutions and greater access to social services than isolated, conflict-ridden or environmentally damaged rural or peri-urban areas. Urban spaces therefore will continue to draw more and more people.

Throughout modern history, rural residents have been leaving traditional agricultural lifestyles to seek non-agricultural employment. Yet, until about half a century ago, well past the initiation of the industrial revolution, rural livelihoods and village life still predominated nearly everywhere. Population shifts toward cities since the latter twentieth century are related to widespread agricultural modernization programs that promote commercial production amenable to export. The advantages of modernized agriculture notwithstanding, this global pattern, in combination with frequent political and environmental crises, has undermined small-scale agrarian economies in many countries and sent more people to cities. In countries with fragile economies, but also in middle income and wealthy countries, many cities that have been expanding in population have not grown economically. Consequently, urban settings offer diminishing levels of economic resources and opportunities to impoverished migrants. In its 2012/2013 report UN-HABITAT noted that "ill-fated development notions and policies have meant that instead of being the locus of opportunity and prosperity, cities all too often have become places of deprivation, inequality and exclusion" (UN-HABITAT 2012, 4).

This chapter argues against the policy objective of discouraging migration to cities. Instead, it advocates first, devoting attention, resources and expertise to reversing urban *and* rural environmental deterioration and patterns of violence; second, focusing on improved governance and economic opportunities in the urban informal settlements that exist inside and outside of the core municipal structures; and, third, supporting the efforts of formal municipalities to better serve residents, including crisis migrants, many of whom are squatters (UN-HABITAT 2012, 28–29). On a broader scale, central governments would do well to invest more, and more creatively, in the very rapidly expanding small and medium-sized urban spaces. These are potential venues for development initiatives, and promote closer connections between rural and urban life.

Urban growth and crisis migrants

The following pages will use case study material to illustrate the links between urban growth and crisis migration, and the particular challenges of adapting policies to take account of growing numbers of crisis migrants in urban settings. Most prominent among crisis migrants are those associated with displacement due to conflict and violence, and those displaced by environmentally related factors, including climate change and disasters. The examples here are taken from El Salvador, Liberia and South Sudan, but other country situations are also cited.

Vulnerable and politically charged groups are mixed together in cities of all sizes already home to growing numbers of rural migrants and unemployed young people whose economic incorporation poses major challenges. The various vulnerable populations in urban areas—from specific groups like refugees and internally displaced persons (IDPs), on the one hand, to broader categories like children, youth, handicapped and women's groups, on the other—have turned to national and international humanitarian agencies to supplement services and material needs unavailable to them through official channels. But, these agencies operate time-limited projects, almost all of which wind down early as internationally funded assistance ends.

Urban expansion as a consequence of conflict and violence

Forced displacement has been a major factor in the growth of cities in war-torn and violence-prone countries. The extremely rapid increase in city dwellers in Sub-Saharan Africa is directly tied to massive flight from the devastating effects of civil conflict and violence. Conflicts and violence in parts of Asia, the Middle East and Latin America, similarly, have added to bloated cities. Although refugees and IDPs generally flee with the intention of returning, large numbers of them ultimately cannot or will not do so. They may remain because the reasons for their flight persist: security threats, landmines, land seizure, institutional weakness, lack of services, weak or absent judicial mechanisms, and the disappearance of viable economies. Or, they may not have homes and property to which they can return even if there is peace (Fagen 2011a; Koser and Martin 2011). Small farmers who lose their land and property as a result of conflict are among those most likely to move to towns and cities and, lacking other options, to stay.

Physical environments change due to conflict, and so do the survivors. Refugees, IDPs and other war-affected populations change, sometimes dramatically, after years of disrupted lives and forced exile. In country after country formerly isolated small-scale farmers become effectively "urbanized." Even if they flee to a refugee or IDP camp, they will experience an urbanized lifestyle where agricultural options are very limited but education, healthcare and training are available. Thus, often as a result of experiences in exile, former farmers, and especially young people, choose not to "return to the past," and decide to abandon rural work and lifestyles. The empirical evidence in many countries is overwhelming: For similar reasons, war-affected populations in all categories go to and remain in large urban centers where neither they nor their families have previously established roots.

Conflicts in many parts of the world have created seemingly irreversible situations that, along with other factors, have undermined return and reintegration strategies and pushed people to remain in urban areas. In Iraq today, IDPs, refugees and returnees cannot live in their towns and cities of origin because these have become ethnic and/or sectarian enclaves and the return of minority populations would almost certainly provoke renewed violence.[2] The cities where they do live, inside and outside of Iraq, are overcrowded and at times dangerous. Afghanistan's major cities (especially Kabul) are unable to absorb the people who repatriated from Pakistan and Iran and went to the cities instead of returning to their villages (UNEP 2008, 22; Setchell 2009, 19).[3] Refugees who return to villages frequently find that livelihoods in their former rural homes are not viable; conflict is pervasive, schools are poor and restricted, and health facilities are severely limited. Further repatriations among the two million registered refugees still in Pakistan and Iran, few of whom own land or property, will exacerbate the already critical urban situation (see Fagen 2011a). The population of Kabul rose from about one million in 2001 to more than 4.5 million in 2010. As of 2009, an estimated 80 percent of Kabul's population consisted of returning refugees, IDPs and migrants (Setchell 2009; Metcalfe and Haysom 2012). Repairing urban material and social infrastructure in Kabul and other cities is understood to be a priority for both humanitarian and development agencies, but remains far behind urgent needs.

Perhaps the most dramatic example of an urbanized population ill equipped to return to a rural setting is the South Sudanese. Decades ago, millions of them fled ongoing brutal conflict by moving to Khartoum and other large cities inside and outside of Sudan. As IDPs and refugees, they mostly endured difficult lives of poverty and discrimination, with no good future prospects for their families. Since the end of conflict in 2005, and increasingly since Independence was declared in 2011, urbanized refugees and IDPs, with little or no knowledge of farming or awareness of conditions in their original homes, have been returning to the new South Sudan, some spontaneously, others under pressure. Spontaneous returnees from cities in Uganda and Kenya tend to go directly to Juba and other towns. But, humanitarian organizations have brought busloads of hope-filled South Sudanese IDPs and camp dwellers to villages where conditions are primitive, tribal based violence is widespread, and services are all but inexistent.[4] The unrealistic expectations that these unprepared and poorly served returnees would or could settle and earn livelihoods in such places paved the way for sizable secondary migration to towns and cities. South Sudanese cities, which are under resourced and have limited infrastructure, are utterly unprepared to absorb the newcomers (Martin and Mosel 2011, 3–4).

The 2005 population of Juba was about 250,000, having grown 450 percent since 1973. The population has probably doubled since the 2005 Peace Accords for the reasons stated above and also due to political/tribal conflicts in border locations and the interior. Rising land prices have made habitation in the city center prohibitive for nearly all new arrivals, so the most densely populated areas of Juba are in the peripheral zones where, as is the case in many cities across the globe, expansion is unregulated, services are absent, water is scarce, and employment is difficult to find. Planners in Juba, again similarly to other cases, tend to conceive of upgrading as a process of evicting the poor from existing slums and moving them still further away. This is easily achieved since few residents have land security (see Martin and Mosel 2011).

As of 2013, essential measures related to contested border area and oil revenues remain unresolved, giving rise to armed confrontations in, and renewed flight from, border locations. The inevitable conclusion at this time is that large-scale return movements to rural South Sudan are not only unsustainable in present circumstances, but have contributed to tribal conflict in parts of the new country.

The urban IDPs

Of the people displaced by conflict, the largest group populating cities is IDPs. Unless IDPs are able to live with family and friends, they are frequently housed in or near towns and cities, in camps and squatter settlements that are bereft of basic services. They lack the formal protections available to refugees under international auspices, but generally do receive short-term assistance to varying degrees. When armed confrontations end and international and/or national authorities deem it safe for IDPs to return home, urban-based IDP assistance and protection projects invariably wind down.[5] However, those who are supposed to return to their original homes may not do so, or they may again return to the cities after determining that rural life is not viable. A well-known but little researched fact is the large presence of *returned refugees* (as the Afghans noted above) who become urban IDPs after being repatriated to their original countries and communities. Finding meager survival prospects and continuing insecurity they uprooted again. Such people may or may not, strictly speaking, be defined as IDPs under the Guiding Principles on Internal Displacement (Guiding Principles) or by their respective governments. Nonetheless, the humanitarian problem inevitably becomes one of addressing urgent needs of vulnerable people whose lives have been disrupted by conflict, and who are now long-term and permanent urban residents.

Few cities better illustrate this phenomenon than Monrovia, the capital of Liberia. Today's Monrovia is a prime example of conflict driven urban growth, further exacerbated by rural deterioration and continuing ethnic tensions.[6] For most of the period from 1989 to 2003, the country was convulsed in conflict that produced a massive displacement of over a million refugees and IDPs. Moreover, it was very common for Liberians in flight to be IDPs or refugees at different times. The 2003 Accra Peace Agreement found the Liberian countryside in ruins. Villages were abandoned or nearly empty, rural and urban infrastructure destroyed and productive activity effectively halted. Refugees, IDPs and former combatants returning to villages in the highly conflictive center and southeast of Liberia frequently found their homes destroyed and their land appropriated by others, and had to leave again. Conflict over land ownership and loss, common in almost all protracted conflicts, has been especially vitriolic in Liberia due to ethnic hostilities and contradicting systems of jurisprudence.

Monrovia filled with displaced persons during the conflict, with smaller but significant numbers going to other Liberian town and cities. The UN High Commissioner for Refugees (UNHCR) operated urban camps until 2006, when it closed down its urban assistance programs and offered material and monetary packages to enable the IDPs to return to their villages. Although the Liberian government then ceased to recognize the people who had fled to Monrovia and other cities as IDPs, a large number of them remained, especially in Monrovia, for reasons related to security, land and livelihoods (UNHCR 2006). After the conflict, IDPs were removed from the public lands and buildings they had occupied. UNHCR returned 327,000 IDPs to their original villages and, with other donors, initiated community development projects (UNHCR 2006, 2007). The large residual population still in Monrovia is now settled in rapidly growing slums and shantytowns in and around the city. There, they have been joined by repatriated refugees who found conditions in their former villages to be bleak, and by demobilized former combatants and economic migrants. The population in Monrovia as of 2010 was variously estimated between 800,000 and 1,500,000, at least double its pre-conflict population of 400,000 to 600,000.[7] This growth is the more impressive taking into account that urban humanitarian assistance programs largely ended in the mid-2000s. Residents in the newer slums of Monrovia are mostly illegal squatters, occupying publicly or privately owned land, located in many instances on environmentally fragile swamp or waterfront areas.

Stated government priority has been, and remains, to relieve poverty across the board and restore rural productivity. Although the government of Liberia, with support from a variety of donors, has pursued a policy of trying to reduce the population in Monrovia with incentives and exhortations aimed at inducing people to return to their original homes, the results have been modest. Investments in agricultural improvement and community development do promote rural economic activity in some places and increase economies based on resource extraction, but relatively few people leave Monrovia. New arrivals continue. Despite the fact that Liberians—including the young—strongly identify with their tribal and regional origins even if they departed those regions a decade or more ago, Monrovia thus far shows no signs of losing population. This is especially the case among the young whose weak economic prospects and sometimes violent tendencies are of major concern (Sommers 2009, 9–10; Government Office for Science 2011, 155–156). The general disinclination among younger males (in Liberia and elsewhere) to consider a future in small-scale agriculture is an important factor thwarting agrarian recovery.[8]

Conditions do not bode well for the future of the uprooted poor in Monrovia. First, their extremely limited economic opportunities are almost entirely in the informal sector; second, they have created communities on land to which they have no formal claim and live with the constant threat of evictions; and third, they are vulnerable to physical harm, crime and gang violence. Monrovia is and always has been the chosen destination of the more entrepreneurial and better-educated Liberians and a mecca for Liberians seeking better opportunities. Nevertheless, far fewer would have chosen to migrate and live in crowded, dangerous, marginal slums, as so many now do, had they not been forced to flee during the fourteen-year conflict.

Former combatants and criminal gangs

A small but important group of post-conflict urban migrants consist of the combatants who are often blamed for having caused the flight of all the others. The majority of combatants in civil strife are not professional soldiers but rather conscripted young people. Nevertheless, even if they are not responsible for whatever has caused the conflicts in which they are fighting, they are major actors in carrying out the carnage that accompanies the fighting. Anecdotal evidence indicates that following brutal civil conflicts, former combatants have frequently rejected resuming former lives as subsistence farmers and/or have found themselves unwelcome in their original communities. Former combatants who have spent their early years living violently can find it difficult to reintegrate. African cities are plagued by violence carried out by armed former fighters without economic prospects.

Gang culture and criminal activity often attract former combatants, as well as marginalized young and unemployed people. Criminal life further weakens family and community ties overall, while placing both perpetrators and fellow citizens in grave danger.[9] The disruptive effect of gang culture is evident in El Salvador, and can be traced to the twelve years of civil conflict (1980 to 1992). The conflict depopulated many rural villages, displacing over a million within the country and internationally (today's population is approximately seven million). While the 1992 negotiated peace accord has held, persistent violence and impunity are lasting legacies of the conflict, as is continuing migration. Post-conflict land insecurity, adverse economic policies and poor rural conditions have discouraged the reinvigoration of rural productivity since the conflict, and rural El Salvador has stagnated economically.

Salvadorans obtained political asylum or extended temporary suspension of deportation from the United States, Canada and other countries on grounds of conflict and disaster (see below). At the same time, undocumented Salvadoran migration to the north grew apace. There have been no large-scale voluntary return movements. However, there have been numerous deportations that have impacted national security, particularly in urban areas. Deportees range from migrants caught without documents to criminals deported upon completing their sentences. The US Department of Homeland Security figures show deportations of Salvadorans to range from just over 8,000 to more than 20,000 per year from 2005 through 2010.[10] The figures identify about a quarter to a third of these deportees to have been in criminal status when deported, without specifying whether the criminal offenses were minor misdemeanors or violent crimes. It is certain that large numbers of deportees have had ties with criminal organizations and continue to pose social and economic risks.[11]

How many criminal deportees are in San Salvador and other cities is unknown, but it is reasonable to assume that they represent a significant number. The major gangs (*maras, pandillas*) were founded in Los Angeles and spread to Central America, primarily through deportees. While nearly all deportees are met at the airport and given minimal one-time assistance, there is no systematic follow-up on where they go from there. Insofar as today's criminal gangs support themselves largely by extortion from merchants and service providers, they are most profitably located in places with more population density. Accounts from nearly all sources attribute the present high levels of violence throughout the country, at least partially, to ties between criminal gangs in El Salvador and the US, presumably reinforced by gang members who move back and forth. Additionally, the conflict has left a legacy of plentiful arms and an appetite for violence.[12]

The recurring question as to why gang membership has attracted so many Salvadoran youth—who expect to die before they reach thirty and often do—produces depressingly predictable replies: if they are deportees, they arrive in the country lacking both social networks beyond other deportees and employment prospects; their role models and potential "employers" are apt to be economically successful gang members often encountered in prison either in El Salvador or the United States; their own families often are absent (perhaps in the US earning the remittances that sustain them), and the physical aspects of their gang membership—tattoos, manner of dress—make them unwelcome in wider family circles.

Refugees in host country cities

Victims of conflict who cross borders may attain refugee status, or be eligible for international protection and assistance from UNHCR or the government in the country they have entered. Millions of conflict refugees have been sheltered (warehoused may be a more accurate term) in designated host country camps and settlements, subjected to restrictions, limited in their freedom of movement, and unable to obtain rights and status under the 1951 Convention relating to the Status of Refugees. However, recognized and unrecognized refugees are increasingly avoiding long camp stays and moving to urban areas in their host countries. In the latter instance, they may obtain prima facie refugee recognition from UNHCR or a state entity and be protected from *refoulement*. Some do well after moving to cities; others face food insecurity, inadequate shelter and the loss of effective international protection.

The growing and increasingly visible presence of refugees in cities has attracted significant attention, especially from humanitarian agencies that have attended to refugees located in camps. Presently, only one-third of the persons of concern to UNHCR live in camps and over half live in cities (UNHCR 2009). Urban refugees' lack of status makes them especially vulnerable. Humanitarian assistance is impeded by the fact that they often are dispersed, living far from places where assistance is available. They usually are denied permission to work. In only a few instances are there host country mechanisms to grant legal status and make local integration a reality. Like other extremely poor people in urban slums, refugees contrive to survive with erratic poorly paid labor in the informal sector.[13]

International humanitarian organizations lack sufficient numbers of experienced staff to identify or protect the refugees in cities, although they have been retraining staff for work in urban settings and experimenting with different approaches, partners, and indicators of success.[14] New efforts notwithstanding, there is still a tendency to underserve refugees who have gone to large cities. UNHCR has only limited capacity to protect or assist urban refugees—a cause of growing concern and recently concerted corrective efforts within that organization. The High Commissioner for Refugees has confirmed his recognition that cities are becoming the main sites of humanitarian response for crisis migrants of all kinds (Guterres 2010). UNHCR published an institutionally groundbreaking position paper in September 2009, acknowledging its inadequate attention to urban refugees: "UNHCR Policy on Refugee Protection and Solutions in Urban Areas." Further UNHCR papers have elaborated response strategies in specific cases where urban refugees are a strong presence. The organization has recognized the need to expand its protection function in urban spaces and has elaborated strategies meant to achieve this goal. Nevertheless, critics accuse UNHCR of continuing to favor assistance projects in rural settings and camps (Refstie, Dolan and Okello 2010). A major problem in cities is to address the predictable resentment felt by local citizens living in the same or similar neighborhoods and receiving no assistance.

Environmental factors driving urban migration

Adding to the victims displaced by conflict are those displaced in the face of repeated acute environmental disasters from which they cannot recover, or slow-onset environmental degradation that undermines their health and livelihoods. These migrants are no less "forced" than migrants fleeing conflict, but do have a somewhat greater ability to make informed choices about destinations, family divisions of labor and economic alternatives. Survival strategies for coping with the consequences of climate change, for example, almost invariably include periods during which families or parts of families seek additional income in cities, where many remain.

Environmental deterioration and climate change affecting rural areas provoke sustained migration (see Government Office for Science 2011). When environmental factors are mixed with conflict conditions, acute environmental disasters, or both, rural livelihoods become virtually unsustainable, and the affected population migrates along previously established domestic or international routes (see the chapters by Lindley and Thomas, this volume—Chapters 8 and 3, respectively).

Traditional rural families still dispatch family members to urban settings, hoping for enough extra income to compensate for lower production and/or fewer fish, surviving livestock, etc.[15] After a time they too become more mobile. At some point, the moves initially intended to be temporary may become permanent.

An accelerated exodus to urban areas inevitably exacerbates resource and environmental problems already affecting the destination cities, and increases the impacts of acute environmental disasters and environmental deterioration in urban areas. Rebuilding after an urban disaster is especially challenging. Haiti's post-earthquake recovery after January 2010 should have been effective in view of the amount of funding pledged, but assistance was, and remains, impeded by institutional weakness, poor coordination, the absence of leadership and the physical difficulties posed by the utter devastation of the mega-city, Port-au-Prince. Even in the best-case scenarios, longer-term recovery falls behind because assistance declines after the emergency phase passes. Urban disasters are hardest on the poor and unemployed. Unable to afford repairs, they are more likely to piece together the scraps and to remain in even more miserable conditions than before.[16] But if the disasters are repeated, out migration is inevitable. Again, the case of El Salvador is illustrative.

The San Salvador Metropolitan Area encompassing fourteen autonomous municipal entities, previously separate from one another and from the city of San Salvador, is highly vulnerable to environmental disaster. Since the twelve-year conflict ended in 1992, its population has also expanded significantly. Although the central city has not greatly increased its population, the surrounding metropolitan area of San Salvador has grown by over a million since the time of the peace agreement in 1992.[17] As early as 1999, the Inter-American Development Bank (IDB) expressed concern that the metropolitan urban infrastructure and services would not grow adequately to keep up with urban expansion (IDB 1999). Government data for the early 2000s showed a roughly 60 percent to 40 percent population division favoring urban over rural areas (Ministry of Education 2008, 2).[18] Two decades prior, El Salvador had been predominantly rural. As of 2008, more than 32 percent of Salvadorans inside the country lived in the San Salvador Metropolitan region, whose population was then estimated at or near two million (Ministry of Education 2008, 2). The urban population growth here is less a function of the conflict per se than of its major causes—that is to say, the official economic policies before and after the conflict that favored (with poor results) *maquila*-type manufacturing based on cheap labor, and discouraged small agricultural production. The decline of Salvadoran coffee production after the conflict further weakened rural development.

Since 1992, El Salvador has suffered five natural disasters. Hurricane Mitch in 1998 left comparatively little damage in El Salvador; two destructive earthquakes in 2001 affected about a quarter of the population in rural and urban areas, and Tropical Storm Sam in 2005 flooded the western part of the country (Fagen 2008). In October 2010, the country experienced the most devastating flood in memory, cutting a wide swath that included San Salvador. Damaged infrastructure is far from fully repaired and more prone to landslides and floods than before. Only the urban core of San Salvador itself was sufficiently repaired so that most residents could remain in their homes. The already fragile slum areas in Metropolitan San Salvador's outlying neighborhoods received less attention and living conditions have deteriorated, but residents do not have viable alternative housing. These disasters caused extensive loss of lives and property, particularly among the poor. The state was neither able nor really willing to invest in rebuilding and redesigning the affected areas to be more disaster resistant. Weak investments in recovery have made it inevitable that more people leave the affected rural areas, while deteriorating urban conditions make it more difficult to absorb newcomers.

In November 2011 the IDB approved a $50 million project loan "to reduce the vulnerability of residents of slums" in the San Salvador Metropolitan Area (IDB 2011, 5). The project identified 514 slum neighborhoods lacking adequate access to electricity, sanitation and drinking water, and cited ninety-three of these as highly vulnerable to floods and landslides. Natural disasters already had increased these vulnerabilities and growing population density promised to worsen the problem. The IDB loan proposes to improve infrastructure so as to reduce flooding and landslides, and also to provide communities with water, sewage, drainage, sanitation and electricity. On a positive note in a bleak situation, the IDB project was the product of studies and consultation with the Council of Mayors representing the fourteen municipalities in the metropolitan area. Negotiations demonstrated the continuing commitment of San Salvador's metropolitan area local officials to improve their towns and districts.

Long-term prospects for urban and rural growth

Investments in agricultural technology and infrastructure are long overdue. Modernizing agriculture may in time revitalize rural areas, eventually making it more attractive for people to remain or return. But, the process is complicated and costly: The ravages of conflict have to be addressed; disputes over land and property must be resolved; disasters and environmental hazards need urgent attention including disaster risk reduction; more investment in small- and medium-scale commercial agriculture is needed; and stronger national institutions and local civil society must hold local governments accountable for what they do.

Even if rural living conditions can be vastly improved in the coming decades, cities—some of which have become mega cities—will still be crowded due to natural reproduction and migration. Land rights and environmental hazards need to be addressed just as urgently in cities as they do in war-torn and environmentally affected rural areas (see Zetter and Morrissey, Chapter 9 in this volume). New arrivals generally have no choice but to settle in densely populated, unregulated, informal slums. In these conditions environmental hazards multiply. As national and municipal leaders recognize the urgency of strengthening mechanisms of adaptation to cope with current and future population expansion, they need support for stronger, more reliable and protective municipal governance and more robust environmental risk reduction.

Most important, planning models that coordinate rural *and* urban development by taking advantage of increased mobility in both contexts are long overdue.

Rural–urban migration is not necessarily a one-way pattern. Newly arrived people in cities usually retain familial and cultural ties with their rural roots. When they are able to live in security and achieve a degree of economic stability, they are often able to bring positive benefits to their places of origin, even if they themselves do not return. By engaging in trade and facilitating education for family members, they can help to promote greater stability and agricultural productivity. With forward looking vision and innovation, planners, international donors, and the inhabitants themselves can build on rural–urban movements in ways that establish developmental links with the potential to benefit both the origin and destination contexts.

When municipal planners complain that rapidly increasing numbers of rural migrants are filling cities with ill-prepared residents, their typical response is to spearhead evictions and slum clearance. In Liberia, slum "improvements" and evictions throughout Monrovia seem inevitable for both environmental and economic reasons, but the present slum dwellers adamantly insist they have no place to go. Faced with current and future evictions, the Liberian Slum Dwellers Association, with support from UN-HABITAT, UN Development Programme (UNDP) and other entities, has sought to engage with the government by agreeing to resettle members of the communities in another part of the city or on land nearby. In exchange, they ask for decent housing, transportation and support for livelihood recovery.[19] This kind of approach would allow the dispossessed to be more productive, and therefore contribute more substantially to national development.

Municipal challenges in the face of rapid expansion

Having looked at urban expansion, crisis migrants and their vulnerabilities, and urban deterioration, it is important to consider the challenges facing national authorities, humanitarian agencies and development actors confronting present situations. Poorly constructed urban infrastructure has exacerbated the effects of floods and earthquakes, causing damage to private and public property, vulnerable geological sites and waterways. Water shortages, sanitation and health threats, toxic wastes and industrial pollution disproportionately affect the urban poor. In all poor and in many wealthier countries, municipal management is substandard and underfunded in the poorest areas. Strengthening urban capacities to absorb and integrate growing numbers means improving living conditions and addressing environmental shortcomings at the same time. For there to be progress on these fronts, municipal authorities must be technically prepared, proactive, and able to mobilize funds.

Not only is infrastructure deteriorating and often failing, the challenges of violence and drugs are overwhelming organized security agencies in cities across the globe, and young unemployed migrants are both prime recruits and major victims. Many poor urban neighborhoods have become extremely hazardous. Gender-based crimes multiply across both rural and urban settings, and reach epidemic proportions whenever soldiers and militia are able to prey upon civilians.

Women and girls come to cities sometimes to escape the threat and fear of assault and rape in their villages, and just as often because they already have suffered assault and rape, and no longer feel accepted in these tightly knit villages. A 2011 field study by the Women's Refugee Commission conducted in Kampala, Johannesburg and New Delhi underscored the vulnerable situations facing growing number of female inhabitants in urban settings, and the limited protection they are able to access (Krause-Vilmar 2011).

Local authorities need funding to be able to enact reforms, reinforce security and cope with the environmental consequences of rapid expansion. While a strong argument can be made for channeling national and international funding directly to the municipal level, doing so is fraught with difficulty on all fronts. Often, the most serious opposition to municipal improvements comes from national governments, who resist allocating budgetary power to local leaders they perceive as challenges to their own authority. Local officials often misuse funds to enhance personal wealth or to back warlords, criminals, and specific ethnic groups. Enhancing local financial power by no means ensures that benefits will go to the most vulnerable. For this outcome, the beneficiaries need to be engaged in the process and the process must be free of corruption. Organizing beneficiary participation is difficult in diverse, often multi-ethnic, and disconnected urban settings. Nevertheless, while there is no doubt that structures of governance and service delivery in some towns and cities are beyond repair due to corruption and their leaders' affiliations with perpetrators of violence, leaders in many municipalities would use—and have used—added resources to improve the services they offer to their citizens.

Colombia is an important case example of almost all the problems outlined above. In addition to enormously powerful drug cartels and widespread corruption at all levels, the country has between four and five million IDPs (depending how they are counted), who are found in nearly every municipality. The smaller municipalities are especially heavily impacted, as they lack the resources and viable institutions to cope with an internally displaced population that sometimes outnumbers original residents. The most impacted municipalities, as one would expect, have been deeply involved in the country's long-standing civil conflict and coca production (see Ibáñez 2008; USAID 2011). The US-supported security stabilization initiatives in Colombia feature combining humanitarian assistance with strong oversight over governance practices. The country has adopted a comprehensive system to register, assist and deliver public services to IDPs, but municipal authorities complain that the mechanisms and funding are far too centralized to be effective. With more funding and more control over how funding is channeled, some municipal authorities, including some that are in conflict areas, would effectively use additional resources for the benefit of the vulnerable population. Others would not (see Vidal Lopez 2011). To avoid funding corrupt leaders who lack political will, guidelines issued by the Constitutional Court in December 2010 require that municipal budgets for IDP subsidies be vetted by the national government. Funding is not to be made available if municipalities fail to channel allocated resources to IDP needs and services.[20]

In El Salvador, associations of mayors at the regional and national levels meet to collectively address issues of mutual interest. A national level Fund for Social Investment for Local Development (FISDL) also channels funding to municipalities for local project implementation and lobbies for further decentralization and municipal autonomy. Nevertheless, funding is minimal and the smaller municipalities are progressively poorer. Both in Colombia and El Salvador, medium and large municipalities are confronted with levels of crime and drug trafficking with which they cannot cope.

Adapting humanitarian assistance to urban settings

The major international humanitarian organizations have long played essential roles in helping refugees, IDPs, former combatants, those who are displaced by acute disasters among others, in rural communities. They are now preparing to devote more attention to the same vulnerable groups who have left these communities for the urban areas.

Historically, humanitarian agencies that address the needs of crisis migrants have worked in settings outside of cities, and especially in small communities and camps. Now, to be effective, they have to better understand diverse social organizations and the highly mobile character of the urban displaced. Victims of conflict, disasters, and environmentally induced economic ruin seek stability and safety in cities where they are strangers; the challenges of integrating them in unstable, crime ridden and impoverished urban spaces are daunting. As noted above, UNHCR is partnering more often with other development and humanitarian agencies in the context of the UN Cluster system and the oversight coordinating body, Inter-Agency Standing Committee (IASC).

Referring to the reality that urban residents represent over half the global population and therefore humanitarian agencies are required to adapt to this reality, the IASC developed a strategy paper to help members in "Meeting Humanitarian Challenges in Urban Areas" (IASC 2011). The paper addressed the importance of developing agency expertise and capacities, creating partnerships (with one another, with local and national government, etc.), developing new strategies, and identifying good practices. The IASC targets urban dwellers who are especially vulnerable (to evictions, epidemics, urban violence, lack of shelter, food insecurity, etc.), and acknowledges the difficulties facing the humanitarians identifying, accessing and targeting humanitarian assistance in crowded urban contexts where it is extremely difficult to distinguish chronic needs of poor households from those of crisis migrants.

Humanitarian agencies are accustomed to working with beneficiaries who are largely separated from the general population, which is far from the case in urban contexts. For reasons elaborated above, partnering with municipal authorities can prove frustrating, partnering with development agencies can seem thankless, and even maintaining contact with the target population poses difficulties. Humanitarian programs and projects need to be available to populations defined by needs rather than by the origin of their presence in the city. Services are most sustainable, and best accepted, if located within established urban institutions serving the wider public. For example, reproductive health services or psychosocial attention can be located in a municipal or state operated clinic in an area where former IDPs have concentrated. Humanitarian organizations, international and national, can work with community inhabitants to improve sanitation and water use. They can bring their vast experience in mobilizing community participation to urban settings, taking care to be inclusive in appropriate ways, but recognizing the special needs of their traditional target populations—crisis migrants. The goal is to establish mechanisms that support these populations, while simultaneously encouraging broader community participation and ownership.

On the development side, national urban planners justly want to upgrade infrastructure, remove people living in environmentally fragile areas, encourage formal employment, and enhance investment opportunities. Achieving these goals by removing people from their fragile habitats without engaging them in the process and offering viable economic alternatives has produced tragic results. There are better alternatives. The authors of an Overseas Development Institute study of urbanization in Sudan make a compelling case that the IDP returnees from northern cities, whose past experiences make them unfit to be farmers in rural Southern Sudan, as described above, have potentially valuable skills for growth and development in Juba and other southern cities. Hence, they should be taken account of, not excluded, in urban planning processes (Martin and Mosel 2011). As in other cases, new migrants to Juba often maintain rural links that may facilitate coordinated rural and urban development. At the level of national development policy, coordinated funding and strategies for urban and rural development contribute to comprehensive development overall.

Present dichotomies between rural and urban development planning have the added disadvantage of neglecting economic strategies and funding for medium and small cities and market towns. Yet, the smaller urban settlements potentially are able to bridge the separation between rural village life and life in a mega city. As noted, towns and small cities receive growing numbers of destitute people from farms and villages, but are largely unable to absorb them productively.[21] For example, Liberian cities such as Ganta, Gbargna and Buchanan are expanding but are not yet as impacted as Monrovia. Should investment in these secondary cities and in the larger market towns increase, economic opportunities, urban type employment, schools and health facilities may be able to expand as well. Secondary cities potentially would then become more attractive alternatives than Monrovia among residents who neither want to remain in that increasingly difficult metropolis nor return to village life.[22]

Conclusions

Crisis migrants are not new to cities, but unprecedented numbers of such migrants are moving to cities due to the combined effects of conflicts, environmental degradation and disasters, and economic models that have undermined rural economies. To a greater or lesser degree, Asian, African and Latin American cities are characterized by expanding unregulated and often violence-prone slums, inadequate informal employment and failing infrastructure (water, sanitation, electricity). These conditions invite or exacerbate urban-based disasters and deterioration, epidemics, violence, and declining quality in education, health and other public services.

On the positive side, experts and policy makers are aware that urban spaces are major venues for addressing poverty and for providing services and offering economic opportunities. Present awareness, however, also requires reversing widespread negative assumptions prevalent among national authorities, donors, international organizations, and humanitarian agencies about expanding cities. The widely shared but highly questionable mantra has been: "Cities are bad places for rural migrants and rural migrants are bad for urban prosperity." It is essential to target actions aimed both at preventing and managing crises that give rise to displacement *and* at addressing the crises in urban destination locations, improving protection mechanisms in both. A disconnect between the humanitarian and development communities continues to impede progress in this area.

There are several promising options for future paths, not mutually exclusive, but all bringing their own challenges for implementation.

- **Agricultural revitalization:** Across the planet, rural–urban flows are vastly accelerated by the virtual destruction of rural economies. Subsistence farmers who have been forcibly displaced, whatever the cause, have rejected returning to subsistence agriculture, even if rural conditions are relatively secure. Yet, potentially, greater investment in production units near cities that grow food for urban residents, (which, in turn, requires reform and regularization of land and properties) could draw people from *both* rural *and* urban contexts. Rural revitalization in poor countries must incorporate modernized commercial agriculture, but in ways that benefit small holders as well as large. Local food security is as important a goal as sales in distant markets. Social services and judicial access also are *sine qua non* for rural peace and stability.
- **Measures to accommodate inevitable migration:** Ultimately, the urban core and its densely inhabited unregulated periphery need to be upgraded, with land legally accounted for and registered so as to benefit recent migrants as well as long-standing residents. The impacts of natural or industrial disasters and epidemics are exacerbated by large scale unplanned migration. They also are indisputably linked to inadequate physical and social infrastructure, as well as to weak local institutions, corruption, and unheeded regulations.
- **Protection:** Urban planning often ignores the needs of new arrivals and the especially vulnerable crisis migrants defined in these pages. To strengthen protection for all vulnerable sectors, new arrivals and longer-term residents, the best option is to promote effective governance at the local level, supported at the national level. Judicial mechanisms (police, laws, courts, prisons, etc.),

sensitive to gender-based violence, child abuse and exploitation, should be accessible when needed.

- **Development and resettlement:** Urban modernization and reforms that include slum clearance are valid development tools. Unfortunately, because crisis migrants and refugees are generally unwanted, they are likely not taken into account when local authorities put into action their urban reform plans. The poor in marginal areas are also likely to be the first evicted when the urban landscape is upgraded and under more solid environmental control. To evict a population recently displaced by conflict, or to oblige displaced persons to reside in remote settlements lacking services or employment possibilities, is surely contrary to the Guiding Principles and is unacceptable even in the name of development. Resettlement guidelines, such as those long issued by the World Bank, should be used by governments undertaking forced urban resettlement.

- **Investment in small and medium-sized cities:** Development plans and planners tend to dichotomize rural and urban environments, without recognizing the extent to which rural populations and urban migrants move back and forth for purposes of trade, investment, family and cultural ties. In this regard, small cities and towns that link rural and urban interests are well placed to promote more even development. At present, the smaller urban entities in predominantly agricultural or mineral rich areas are faced with migration pressures beyond their meager resources. The population is already in place but trade, commerce and productive enterprises are constrained by multiple factors. Increased employment opportunities would allow people, especially younger people who now reject rural life or are forced off their land, to retain their former rural ties and bring funds to their rural families, while enjoying urban amenities, such as schools, healthcare and recreation. There are promising indications of greater investment in smaller urban spaces that are close to agricultural production and prospects for generating more private investment in such activities are realistic. Nevertheless, small cities with potential for commerce often lack banking and administrative facilities, regularization of property ownership, and mechanisms for metropolitan income generation. The many national governments that advocate decentralization need to improve budgeting practices and invest in the kinds of institutions that would enable small cities to function more effectively.

- **Changing roles of humanitarian organizations:** Having accepted that long-term migrants to the city fall within its responsibility, the humanitarian community is now moving more decisively to address the needs of urban-based victims of conflict, disasters and environmental degradation. Humanitarian agencies working outside of cities have played fundamental roles in these regards for many decades. Advocates in urban areas, until recently, have largely focused on helping forcibly displaced people to return to small communities. Or, they have initiated projects on behalf of specific segments of the urban population, e.g. street children, trafficked women. It is difficult to overstate the challenges facing

UNHCR and numerous non-governmental organizations (NGOs) in reorienting their staff and deploying their resources to cities. Nevertheless, it is especially
important that humanitarian agencies work in closer partnerships with development actors and government officials than has been the case historically.

- **Expanding priorities of development actors:** Urban planners in most
places are very well aware of the severity of the problems they face as a result
of rapid growth. They seem less aware either of the dimensions of the problems that are producing such rapid urbanization, or of the positive and negative implications of vastly expanded mobility in even traditional segments of
their societies. Development actors too often, and mistakenly, consider crisis
migration as a temporary phenomenon and primarily a humanitarian problem.
As has become abundantly clear, people forced to flee and to move to cities
more often than not remain there for long or indefinite periods. Municipal and
national authorities now need to find ways to integrate them.
- **National-level crisis management:** The management of incipient conflict,
enactment of land and property reforms, and support for disaster risk reduction must include commitments by ministries, including planning, justice and
human rights, at the national level. Equally important are mechanisms for consultation and participation of, and support for, counterparts in departments and
municipalities. This kind of approach might help to prevent situations that are
likely to cause forced migration in the first place.

Notes

1 For a comprehensive overview of urban transformations, especially in underdeveloped
countries, see National Research Council (2003) and UN-HABITAT (2010, 3). For an
overview of the social, political and ecological dangers of the rapid growth of cities, see
Liotta with Miskel (2012).
2 Contrary to most crisis migration situations, the Iraqi flight consisted mainly of urban
dwellers who went to other cities.
3 Nordland (2012) describes the tragic toll of winter on underfed, poorly clothed and
housed Afghan IDP children in "refugee camps" in Kabul.
4 The deadly conflict that broke out in 2013 has vastly exacerbated rural flight.
5 In Colombia, uniquely, the Guiding Principles have been incorporated into national
law, and buttressed with further legislatively enacted measures for the protection and
assistance of IDPs. Colombian cities across the nation host millions of IDPs. However,
the assistance is very uneven, and Colombia is still experiencing conflict and flight.
6 This section summarizes findings in Fagen (2011b), and draws on several sources, including Isser, Lubkemann and N'Tow (2009), Butman (2009) and Williams 2011.
7 Estimates vary widely from both national and international sources. On post-conflict
growth in Monrovia, see Williams (2011) and Ngafuan (2010). The latter emphasizes
that not only are the numbers unreliable, but so are the formal distinctions between
urban and rural spaces.
8 This generalization does not necessarily hold for women, especially those who have lost
male providers and family members.
9 See Pavanello and Metcalfe with Martin (2012). Sommers (2010, 5–6) questions that
conflict is the primary factor in African urban growth, and emphasizes instead what he
calls a "youth bulge in African cities." He affirms that young people who have not been
displaced by conflict are showing the same urban preference as the displaced.

10 Yearly from 2005 to 2010: 8,305,11,050, 20,045, 20,849, 19,809. Deportations are similarly high among Hondurans and Guatemalans, with similar consequences as well.

11 Material on Salvadoran gangs, including criminal deportees, is summarized from unpublished data obtained by Thomas Bruneau in connection with his work with the Salvadoran National Police Intelligence Services, received February 2012.

12 The Washington-based Washington Office on Latin America has published several recent reports on drug culture, the human cost of the drug traffic and policy alternatives to current practice (see www.wola.org).

13 Landau and Duponchel (2011) contend that refugees in host country cities in Africa are often able to overcome vulnerabilities, despite lack of status and difficulties with security agents, thanks to their own networks.

14 An entire issue of *Forced Migration Review* (February 2010, Issue 34) is devoted to articles on and about "Adapting to Urban Displacement." The various articles describe the often miserable conditions and lack of security refugees experience in cities, and offer guidelines related to various sectors of humanitarian operations. An ongoing research project undertaken by the Overseas Development Institute, Internal Displacement Monitoring Centre, International Committee of the Red Cross, and Tufts University has been publishing studies of refugees in cities in Africa, Latin America and Asia. The International Rescue Committee and Women's Refugee Commission have published a comprehensive analysis of how humanitarian agencies can improve work with refugees in urban settings (Lyytinen and Kullenberg 2013).

15 Ballesteros and Jaramillo (2011), reporting on response to natural disasters in Colombia (plagues ruining crops, diseases killing animals, floods) found little in the way of preventive action, but a notable increase in the number of individuals looking for wage labor. The authors found that victims were more likely to move to smaller dwellings and share space with other relatives than to move away.

16 Government Office for Science 2011, 155–160, 198; Commins 2011, 3, 5. The World Bank, UNDP and UN-HABITAT have initiated projects in several countries, but these often fail to reach peripheral areas.

17 Metropolitan San Salvador population in 2007: 1,566,629; in 1992: 521,855.

18 The total population of the country at that time was about nine million. The publication was prepared with census and survey data to support efforts aimed at expanding education and social services in the country.

19 The proposal of the slum dwellers is elaborated in a declaration of the Slum Dwellers Association of Liberia, inc. Monrovia, written by Bestman D. Toe, Slum Dweller Association President, October 2010.

20 Fagen (2011b, 31–34) elaborates conditions in two cities where corruption has been reduced and the political will is present to integrate large IDP populations, but doing so depends on more resources and the ability to manage them locally.

21 Colombia, with 4 to 5 million IDPs, living in cities of all sizes, has made efforts to integrate them, assuming correctly that they will probably remain. The impacts have been strong, with some small cities (e.g. Quibdo, Florencia, Sincelejo) reporting that the IDPs represent from 33 to 46 percent of the total population (see Fagen 2011b; USAID 2011). The results are decidedly mixed but worthwhile examining.

22 Buttressing resources in these cities is also one of Ngafuan's (2010) core recommendations.

References

Ballesteros, M.C. and Jaramillo, C.R. (2011) "Choques adversos a los hogares y sus reacciones," in *Colombia en Movimiento. Un análisis descriptivo basado en la Encuesta Longitudinal Colombiana de la Universidad de los Andes ELCA*, Bogotá: Universidad de los Andes.

Butman, M.A. (2009) "Urbanization vs Rural Return in Post-Conflict Liberia," MA thesis, George Washington University.

Commins, S. (2011) "Urban Fragility in Africa," *Africa Security Brief*, 12.

Disasters (2012) "Humanitarian Action in Urban Areas," Special Issue, London: ODI.

Fagen, P.W. (2008) "El Salvador: A Case Study in the Role of the Affected State in Humanitarian Action," HPG Working Paper.

Fagen, P.W. (2011a) "Refugees and IDPs after Conflict. Why They Do Not Go Home," Special Report, US Institute for Peace.

Fagen, P.W. (2011b) *Uprooted and not Unrestored: A Comparative Review of Durable Solutions for People Displaced by Conflict in Colombia and Liberia*, UNHCR, PDES.

Government Office for Science (2011) *Migration and Global Environmental Change: Future Challenges and Opportunities*, Final Project Report *Foresight Migration and Global Environmental Change*, London.

Gozdziak, E.M. and Walter, A. (2012) Urban Refugees in Cairo, ISIM, available online at: http://issuu.com/georgetownsfs/docs/urban_refugees_in_cairo/1.

Guterres, A. (2010) "Protection Challenges for Persons of Concern in Urban Settings," *Forced Migration Review*, February: 8.

IASC (2011) "Meeting Humanitarian Challenges in Urban Areas," available online at: http://www.humanitarianinfo.org/iasc/pageloader.aspx?page=content-subsidi-common-default&sb=74.

Ibáñez, A.M. (2008) "Public Policies to Assist Internally Displaced Persons: The Role of the Municipalities," Brookings Institution Occasional Paper, December, available online at: http://www.brookings.edu/~/media/research/files/reports/2008/12/03%20colombia%20ibanez/1203_colombia_ibanez.pdf.

Inter-American Development Bank (1999) "El Salvador: Los Retos del Desarrollo Sostenible y Equitativo," discussion document of May 12, Banco Interamericano de Desarrollo, Departmento Regional de Operaciones II.

Inter-American Development Bank (2011) "El Salvador: Reduction of Vulnerability in Informal Urban Neighborhoods in the San Salvador Metropolitan Area" (ES-L1016), available online at: http://www.iadb.org/en/projects/project-description-title,1303.html?id=es-l1016.

International Federation of Red Cross and Red Crescent Societies (2012) "Forced Migration in an Urban Context: Relocating the Humanitarian Agenda," *World Disaster Report* for 2012.

International Organization for Migration (2010) "Total Returns to South Sudan, Post CPA to December 2009," Tracking of Returns Project, available online at: http://www.iom.int/jahia/webdav/shared/shared/mainsite/activities/countries/docs/tracking_returns_annual_report.pdf.

Isser, D.H., Lubkemann, S.C. and N'Tow, S. (2009) *Looking for Justice: Liberian Experiences with and Perceptions of Local Justice Options*, US Institute for Peace, *Peaceworks* No. 63.

Koser, K. and Martin, S. (2011) "Introduction," in Koser, K. and Martin, S. (eds.), *The Migration-displacement Nexus: Patterns Processes, and Policies*, New York: Berghahn Books.

Krause-Vilmar, J. (2011) *Dawn in the City: Guidance for Achieving Self-Reliance for Urban Refugees*, New York: Women's Refugee Commission.

Landau, L.B. and Duponchel, M. (2011) "Laws, Policies, or Social Position? Capabilities and the Determinants of Effective Protection in Four African Cities," *Journal of Refugee Studies*, 24 (1): 1–22.

Lindley, A. (2011) "Between a Protracted and a Crisis Situation: Policy Responses to Somalis in Kenya," *Refugee Survey Quarterly*, 30 (4): 14–49.

Liotta, P.H. with Miskel, J. (2012) *The Real Population Bomb*, Stirling, VA: Potomac Books.

Lyytinen, E. and Kullenberg, J. (2013) "Urban Refugee Research and Social Capital: A Roundtable Report and Literature Review," February 19, International Rescue Committee and Women's Refugee Commission.

Martin, E. and Mosel, I. (2011) *City Limits: Urbanisation and Vulnerability in Sudan: Juba Case Study*, London: Overseas Development Institute, Humanitarian Policy Group.

Metcalfe, V. and Haysom, S. (2012) "Sanctuary in the City: Urban Displacement in Kabul," London: Overseas Development Institute, Humanitarian Policy Group.

Ministry of Education (2008) *Informe Nacional Sobre el Desarrollo y el Estado de la Cuestión Sobre el Aprendizaje de Adultos en Preparación de la Confintea VI*, San Salvador: Ministry of Education.

National Research Council (2003) *Cities Transformed: Demographic Change and its Implications in the Developing World* (Panel on Urban Population Dynamics, Mark R. Montgomery, Richard Stren, Barney Cohen, and Holly E. Reed, eds.), Washington, DC: National Academies Press.

Ngafuan, R.F. (2010) "The Overcrowding of Monrovia and its Link to Rural–urban Migration in Liberia: Causes, Consequences and Solutions," *The Perspective*, Atlanta, Georgia, June 13, available online at: http://www.theperspective.org/.

Nordland, R. (2012) "Afghan Camps Receive Winter Aid, but Officials Say it Isn't Enough," *New York Times*, December 30.

Pavanello, S., Elhawary, S. and Pantuliano, S. (2010) "Hidden and Exposed: Urban Refugees in Nairobi, Kenya," Overseas Development Institute, available online at: http://www.odi.org.uk/publications/4786-urban-refugees-nairobi-kenya.

Pavanello, S. and Metcalfe, V. with Martin, E. (2012) "Survival in the City: Youth, Displacement and Violence in Urban Settings," HPG Policy Brief 44, available online at: http://www.odi.org.uk/publications/6400-youth-urban-gang-violence-refugee-idp.

Refstie, H., Dolan, C. and Okello, M.C. (2010) "Urban IDPs in Uganda: Victims of Institutional Convenience," *Forced Migration Review*, 34, 32–33.

Setchell, C.A. (2009) "Urban Displacement and Growth Amidst Humanitarian Crisis: New Realities require a new strategy in Kabul," *Monday Developments*, USAID.

Sommers, M. (2009) *Africa's Young Urbanites: Challenging Realities in a Changing Region*, UNICEF/ADAP, available online at: http://www.unicef.org/adolescence/files/ADAP_Learning_Series-5_Africas_Young_Urbanites.pdf.

Sommers, M. (2010) "Urban Youth in Africa," *Environment & Urbanization*, 22 (2): 5–6.

UNEP (2008) "Afghanistan's Environment, 2008," available online at: http://reliefweb.int/report/afghanistan/afghanistans-environment-2008.

UN-HABITAT (2008) *The State of African Cities: A Framework for Addressing Urban Challenges in Africa*, Nairobi, UN-HABITAT.

UN-HABITAT (2010) Annual Report HS/036/11.

UN-HABITAT (2011) Annual Report.

UN-HABITAT (2012) *State of the World's Cities, 2012/2013: Prosperity of Cities*, UN-HABITAT.

UNHCR (2006) "Liberia, IDP Camp Closure Assessment Report," June.

UNHCR (2007) "Real-time Evaluation of UNHCR IDP Operation in Liberia," PDES/2007/02-RTE, prepared by Wright, N., Savage, E. and Tennant, V.

UNHCR (2009) "Trying to Get By in the City," available online at: http://www.unhcr.org/pages/4b0e4cba6.html.

USAID (2011) "Casos de Ciudades Intermedias Altamente Receptoras," prepared by Forrero, E. Oficina de poblaciones vulnerables, Unidad de Población dezplazada, January 2.

US Committee for Refugees (2008) *World Refugee Survey 2007*.

Vidal Lopez, R. (2011) *The Effects of Internal Displacement on Host Communities*, Brookings Institution-London School of Economics Project on Displacement, Bogotá.

Williams, R. (2011) *Beyond Squatters' Rights: Durable Solutions and Development-induced Displacement in Monrovia, Liberia*, Norwegian Refugee Council Report.

PART IV

Governance

17

THE GLOBAL GOVERNANCE OF CRISIS MIGRATION

Alexander Betts

Introduction

Crisis migration represents an umbrella term (Martin, Weerasinghe and Taylor 2014, Chapter 1 in this volume). It highlights a range of emerging migration challenges that arise in the context of humanitarian crisis: displacement, including that which falls outside existing protection frameworks (Betts 2013a), trapped or stranded populations (Dowd 2008; Collyer 2010), mixed migration (Van Hear, Brubaker and Bessa 2009; Koser and Martin 2011), and anticipatory movements. Many of these challenges are relatively newly recognized and so do not map neatly onto the mandates of existing international organizations (IOs) or onto the coverage of existing international norms. There is no single, coherent, and unified "crisis migration regime," for either the overarching concept or its constitutive elements. As the other chapters in this volume highlight, there are therefore significant protection gaps for different groups of vulnerable migrants affected by crisis.

But what does this mean for global governance? Does it necessarily imply that new international institutions are required to address these gaps? Or, alternatively, to what extent can existing institutions adapt to meet the various challenges that comprise crisis migration? Is it realistic to believe that existing norms and IOs might adapt or "stretch" to fill these gaps and address the emerging challenges, without the need for root and branch reform? And, if so, are there things that international public policy makers can do to facilitate a gradualist approach to institutional adaptation?

As with all areas of migration, the absence of a formal regime for crisis migration does not mean that there are not relevant existing governance structures (Betts 2011). Crisis migration highlights a diverse set of governance challenges that straddle numerous issue-areas and institutional mandates, and responses depend upon the interplay of the humanitarian and migration regimes, in particular, but also the potential role of other regimes such as human rights, international security and development. The purpose of this chapter is to trace what already exists in terms of norms and organizations, and how they relate to crisis migration in both theory and practice, in order to identify gaps, and areas in which incremental adaptation might yield more effective and efficient responses.

In order to make sense of the global governance of crisis migration, the chapter draws upon two concepts within the wider global governance literature that have particular relevance for considering the dynamic ways in which "old" international institutions may adapt to "new" challenges: "regime complexity" (Raustiala and Victor 2004; Alter and Meunier 2009; Young 2010; Orsini, Morin and Young 2013) and "regime stretching" (Betts 2013a). Put simply, the former relates to the horizontal relationship across different international institutions; the latter relates to the vertical relationship between the global level and national and local practices. Crucially, both levels entail immanent opportunities for gradualist institutional adaptation and for potentially filling gaps based on "making existing institutions work better."

Regime complexity is a concept that has been introduced to make sense of the way in which international institutions have proliferated since the Second World War to create a dense tapestry of overlapping, parallel and nested institutions. Regime complexity can be defined as "the presence of nested, partially overlapping and parallel international regimes that are not hierarchically ordered" (Alter and Meunier 2009, 13). The idea of "not hierarchically ordered" regimes highlights the way in which an issue may be governed by a disparate range of institutions within and across issue-areas, even in areas where there is no obvious referent institution or organization (Raustiala and Victor 2004). Regime complexity is therefore a particularly useful concept for understanding the governance of policy fields—like crisis migration—which lack a referent institution but fall under the actual or potential purview of numerous international regimes. The concept has been usefully applied to look at climate change (Keohane and Victor 2010), human rights (Hafner Burton 2009), and trade (Davis 2009), for example, and also issues relating to refugees and migration (Betts 2009, 2011, 2013b; Trachtmann 2011).

Regime stretching relates to the "degree to which the scope of a regime at the national or local level takes on tasks that deviate from those prescribed at the global level" (Betts 2013a, 30). It highlights how international institutions should not simply be understood as existing in abstraction in New York or Geneva. Instead, they are subject to translation processes from global to national to local levels, such that what they do at the level of implementation in one country context may be quite different from the way they are generically defined within international agreements. The concept is especially relevant to thinking about the constitutive elements of crisis migration, in which a range of international regimes—not least the norms and organizations of the global refugee regime and the humanitarian regime—may sometimes "stretch" at implementation to fill gaps (Betts 2013a). The concepts of regime complexity and regime stretching may be considered to be related insofar as regime stretching in relation to crisis migration is likely to draw upon a full range of institutions from the crisis migration regime complex in order to fill protection gaps.

The chapter divides into three broad sections: theory, empirics, and policy. First, the chapter maps out the crisis migration regime complex, highlighting the range of norms and IOs with relevance to the global governance of crisis migration, and explores the scope for regime stretching to fill gaps within the regime complex. Second, the chapter examines how and to what extent the complex works in practice, by examining three case studies relating to the four constitutive elements of the crisis migration umbrella: (1) cross-border displacement; (2) trapped populations; (3) mixed migration; and (4) anticipatory movements. Third, I explore what all of this means for making global governance more effective and efficient in its responses to different elements of crisis migration.

Theory

The concept of "global governance" is often over-used and poorly defined. In terms of process, it implies all actions directed toward organizing collective action between states and other transnational actors. In terms of substance, it relates to the norms and IOs that regulate the behavior of states and transnational actors within and across particular policy fields. In the aftermath of World War II, global governance was relatively straightforward. It generally involved a set of multilateral treaties and IOs created to oversee those treaties or to provide services to states within clearly delineated and distinct policy fields. Today, the character of global governance is very different. There has been a proliferation of institutions at the multilateral, regional, bilateral, and transnational levels, both formal and informal. New trans-boundary challenges have emerged that defy the boundaries of existing institutions, requiring forms of adaptation and coordination.

Crisis migration represents one such emerging new challenge, which falls outside the post-World War II framework of global governance but which lies embedded within, and across the purview of a variety of existing institutional frameworks and issue-areas. This section traces the variety of norms and IOs that comprise the regime complex for crisis migration, and the extent to which the different institutions within the complex might "stretch" to fill potential protection gaps. It does so in order to identify what exists, and trace sources of actual and potential complementary or contradictory overlap across those relevant sets of institutions.

The regime complex for crisis migration

The concept of a regime is often defined by its "consensus definition" of "principles, norms, rules, and decision-making procedures around which actor expectations converge in a given issue-area" (Krasner 1982, 3). Yet the definition is unwieldy and almost impossible to operationalize. It makes far more sense to see regimes as having just two core elements: norms and IOs (which can both be subsumed under the notion of "international institutions"). The traditional premise of regime theory—and of the concept's emphasis on "in a given issue-area"—is the assumption that regimes are defined by discrete sets of norms and organizations in a particular policy field.

In recent years, though, regime complexity has emerged as a concept based on the recognition that with institutional proliferation over time, there is an increasingly dense tapestry of institutions. Regime complexity refers to the way in which two or more institutions intersect in terms of their scope and purpose. This literature sets out three concepts that describe different aspects of complexity. First, institutions may be *nested*—regional or issue-specific institutions may be part of wider multilateral framework. Second, they may be *parallel*—obligations in similar areas may or may not contradict one another. Third, they may be *overlapping*—multiple institutions may have authority over the same issue (Alter and Meunier 2009).

The existence of nesting, parallel and overlapping institutions may be complementary or contradictory in its implications for a given regime. The research agenda on complexity has tried to explore how it emerges as a dependent variable and its role as an independent variable in influencing the behavior of states and IOs. It is conceptually most useful for analyzing issue-areas in which there is no clear referent institution and a range of institutions are organized in a non-hierarchical way (Raustiala and Victor 2004). Global migration governance in general is characterized by regime complexity (Betts 2011), but one might also identify complexes for specific areas of migration such as crisis migration.

The regime complex for crisis migration can be illustrated in two ways—by a table or a Venn diagram. The former (Table 17.1) allows the norms and IOs involved in each area of crisis migration to be clearly highlighted. The latter (Figure 17.1) allows the overlaps across different regimes, and the way in which particular activities straddle extant regimes, to be graphically depicted.

The normative and organizational terrain governing the different areas of crisis migration is complex and challenging (see McAdam 2014, Chapter 2 in this volume).

In terms of displacement, there are well-established norms and clear division of organizational responsibility for both internal displacement and refugees. The Guiding Principles on Internal Displacement consolidate International Human Rights Law (IHRL) and International Humanitarian Law (IHL) norms within an authoritative soft law document on the protection of internally displaced persons (IDPs), whether they are displaced by conflict or man-made or natural disaster. The "cluster approach" creates a clear basis for coordinating organizational responses to internal displacement. Meanwhile the refugee regime sets out clear legal norms based on International Refugee Law (IRL) and the UN High Commissioner for Refugees (UNHCR) has responsibility for refugee protection.

The main gap for displacement relates to what I have described as survival migration (Betts 2013a)—in other words, cross-border displacement caused by serious human rights deprivations that falls outside the dominant interpretation of who is a "refugee." Here, IHRL norms exist and should create certain obligations on host states to refrain from *non-refoulement*. However, there is an operational gap, and so

TABLE 17.1 Regime complex for crisis migration

	Norms	International organizations
Displacement		
– Internal displacement	IHRL/IHL "Guiding Principles"	UN "cluster approach"
– Refugees	IRL/IHRL	UNHCR
– Survival migration	IHRL UNFCCC Article 14f.	Ad hoc responses "Nansen Initiative"
Trapped populations	IHRL/IHL IML Principles R2P	Ad hoc responses UN Security Council IOM's Operational Framework
Mixed migration	IRL/IHRL UNCLOS	Ad hoc responses "UNHCR 10-point plan" "Regional Mixed Migration Secretariat" International Maritime Organization
Anticipatory movement	IHRL Development norms R2P	Disaster Risk Reduction (e.g. World Bank/UNDP) Relocation (World Bank) UN Security Council

Notes: IRL = International Refugee Law; IHRL = International Human Rights Law; IHL = International Humanitarian Law; UNCLOS = United Nations Convention on the Law of the Sea; R2P = Responsibility to Protect; UNFCCC = United Nations Framework Convention on Climate Change.

responses to survival migration are subject to the ad hoc response of IOs and non-governmental organizations (NGOs). Sometimes, some combination of UNHCR, the International Organization for Migration (IOM), the International Federation of Red Cross and Red Crescent Societies (IFRC), or NGOs such as Médecins Sans Frontières will fill protection gaps. But responses are not consistent and there is a need for greater clarity of both normative obligations and the organizational division of responsibility. Hence, the Nansen Initiative began in late 2012 as an informal intergovernmental process to consider protection gaps relating to cross-border displacement in the context of natural disasters.

In terms of so-called "trapped" or "involuntarily immobile" populations, there are similarly normative principles, which can be found in both IHRL and International Migration Law (IML) more broadly (Cholewinski, Perruchoud and Mac-Donald 2007). But on an organizational level, responses are generally subject to ad hoc IO and NGO responses, which depend on selective political will and the availability of funding. The inconsistency in response is illustrated by the stark variation between responses in Libya, where UNHCR and IOM coordinated their efforts in humanitarian evacuation of stranded migrants, and in Côte d'Ivoire, where there was almost no international response to parallel dynamics during 2011. IOM has recently developed an Operational Framework in the context of its work on "Migration Crisis" to enable it to respond better to such situations in future.

Responses to anticipatory movements represent a very different area of governance since they take us into the realm of prevention. It is an area that therefore cuts across a range of regimes specific to crisis, involving, for example, development, governance and actors such as the United Nations Development Program (UNDP) and the World Bank, through Disaster Risk Reduction (DRR) and relocation. It also potentially involves the interplay of the humanitarian and migration regimes with the governance of international security and the environment.

Mixed migration is a concept that is challenging to define, not least because it can refer to either mixed populations or mixed motives (Van Hear *et al.* 2009). Generally, though, it is used to refer to the separation of persons in need of international protection from broader migratory flows. IRL and IHRL both offer a clear normative framework on the people who should be in need of international protection. However, operational responses have, again, often been ad hoc. For example, in East Africa and the Horn of Africa, UNHCR, IOM and the Danish Refugee Council have coordinated on the development of the Regional Mixed Migration Secretariat, based in Nairobi. Elsewhere, UNHCR has tried to operationalize its "10-point plan" on mixed migration in ways that have often attempted to reach across the migration regime by developing partnerships with IOM. For mixed migration across territorial waters, the United Nations Convention on the Law of the Sea (UNCLOS) and the role of the International Maritime Organization are highly relevant.

Drawing upon the normative and organizational structures explained above, Figure 17.1 is intended to highlight the interplay between the humanitarian and migration regimes and the ways in which they intersect in relation to crisis migration. The diagram depicts the different regimes by institutional function to highlight areas in which functions overlap across the two different regimes. It does not map on specific organizations because, although UNHCR is the referent organization for the refugee regime and IOM for the migration regime, in practice both are now "itinerant actors" who work beyond the boundaries of "their" regime (Betts 2010). For example, IOM plays a humanitarian role as part of the "cluster approach" and UNHCR's work is not reducible to the refugee regime per se. One might also have added other relevant regimes, including the development and security regimes, as having relevant overlaps with core areas of crisis migration.

Figure 17.1 helps to illustrate how, when a humanitarian crisis occurs, there are clear zones of intersection across the regimes, which draw in different sets of norms and actors. For example, depending on the nature of the crisis—whether it links to asylum or a humanitarian emergency—the refugee regime or the cluster approach may be triggered from the humanitarian governance side. However, core functions of the migration regime such as assisted voluntary return may be drawn in. In other

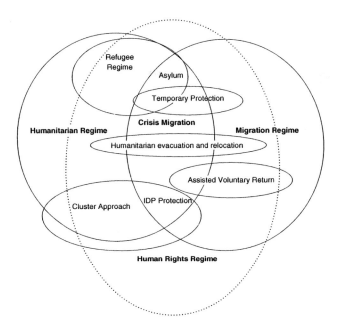

FIGURE 17.1   The regime complex for crisis migration

areas, there will be overlaps in function, requiring mechanisms of coordination to maximize complementarities and minimize contradictions. In addition to this depiction of the humanitarian–migration interplay, it is important to recognize that there are also significant interplays with international security and development. Furthermore, the entire regime complex is also nested within the human rights regime, with crisis migration needing to be understood within a human rights based framework. More specific aspects of the complex emerge in starker relief when explored through case studies.

Identifying regime complexity has been highlighted to have a host of international public policy implications (Orsini *et al.* 2013). First, it gives rise to the recognition of *implicit forms of governance* even in emerging areas (Betts 2011). Second, it highlights how some policy areas may be simultaneously governed by multiple regimes in ways that may lead to either *overlaps or gaps* (Alter and Meunier 2009). Third, where there are gaps or overlaps these may create a *case for improved coordination mechanisms* (Morin and Orsini 2013). Fourth, complexity is likely to create opportunities for states to engage in *forum-shopping* or a range of strategies that potentially undermine regimes (Abbott and Snidal 2001; Helfer 2004). Fifth, complexity may in turn place *traditional international organizations in a competitive environment* within which states can select between IOs as alternative forums, venues and service providers (Betts 2013b; Margulis 2013). All of these challenges are likely to characterize the governance of crisis migration within which—with the exception of refugees—no IO or regime enjoys de facto monopoly status.

Prospects for regime stretching

Many of the constitutive areas of crisis migration are organizationally character-ized by "ad hoc IO and NGO responses," within which there are protection gaps because of a lack of predictability and consistency. For example, UNHCR some-times works beyond the boundaries of its core refugee protection mandate. IOM sometimes plays a protection role across a number of areas of crisis migration. IFRC and NGOs sometimes fill protection gaps. Occasionally, there are also ad hoc coor-dination mechanisms at the field level. However, beyond refugee protection for those fleeing state persecution, there is enormous inconsistency on responses to crisis migration.

This is where the second concept used by this chapter becomes important. "Regime stretching" refers to the *degree to which a regime at the national or local level takes on tasks that deviate from those prescribed at the global level*. First, such stretching may be regime-consistent (taking on tasks that are complementary to the under-lying purpose of the regime) or regime-inconsistent (contradicting the underly-ing purpose of the regime). Second, it may be based on adjustment of the refer-ent regime or of other regimes adjusting to fill gaps. Third, it may be the result of normative adjustment, organizational adjustment or both. The concept high-lights the way in which a regime may adapt at the national level at implementa-tion, even in the absence of adaptation at the levels of international bargaining or institutionalization.

This is a particularly important concept in the context of a world in which new problems and challenges are emerging but new formal institutions are created at a much slower pace, and there is often a need for "old global institutions" to adapt to "new national challenges." As problems emerge that were not within the scope of a regime at its creation, the norms and organizations may adapt, even without formal renegotiation. Dan Drezner (2007) has written about the "viscosity of global governance." He argues that one of the tragedies of global governance is the ease with which states can fluidly create new international institutions, and so engage in forum shopping or regime shifting. However, he notes that the costs of forum shopping are likely to vary with issue-areas, implying different degrees of "viscos-ity" (or resistance). The concept of regime stretching adds an additional dimension to that of the viscosity of an international regime. Where Drezner (2007) implies that adaptation to new problems takes place through movement to new or alter-native institutions, regime stretching highlights an alternative method of adapta-tion—within the regime itself. In other words, rather than (1) *creating a new institu-tion* (institutional proliferation) or (2) *moving to another institution* (forum shopping or regime shifting) (Alter and Meunier 2009), states may also (3) *adapt an existing institution*—not only through international bargaining or institutionalization, but also at the level of implementation through "regime stretching."

Furthermore, regime stretching adds a spatial dimension to how we think about regime adaptation. Comparative politics has explored temporal explanations of institutional change (Lieberman 2002; Pierson 2004; Hall and Thelen 2009). James

Mahoney and Kathleen Thelen's (2010) work in particular shows the range of causal mechanisms (displacement, layering, drift, convergence) through which an institution changes between time period t1 and time period t2. However, exploring the relationship between the global and the local opens up a spatial dimension to the question of institutional change, highlighting how the same global regime can have different national manifestations (in states a, b and c) at the same time period (whether t1 or t2).

The degree to which norms stretch can be assessed in relation to the benchmark of the formal global norms. In cases where a formal treaty, such as the 1951 Convention relating to the Status of Refugees (1951 Refugee Convention), exists, this offers a basis for identifying the aims and scope of the regime at the global level. In the case of the refugee regime, Article 1a of that Treaty defines "who is a refugee," limiting it to people who face a well-founded fear of persecution because of race, religion, nationality or membership of a social group. The rest of the treaty ascribes certain rights to people who fall within that category. This benchmark means that, in a given national context, if one can identify activity that falls either side of that line—for example, the inclusion or exclusion of additional groups of people—that might be regarded as "regime stretching." In the case of the refugee regime, that might be measured by looking at the *type* and *number* of additional people included within the refugee framework, and the *degree of rights* that they receive.

In my wider work on survival migration, for example, I have argued that there is considerable variation in regime stretching by the refugee regime to protect people who flee serious human rights deprivations that fall outside the dominant interpretation of persecution (Betts 2013a). In some host states, survival migrants are protected as refugees. In others, they face round-up, detention and deportation. Whether or not the refugee regime stretches at the national level to protect other categories of cross-border displaced populations depends largely on politics. In the absence of legal and normative precision, domestic and international incentives for elites in host state governments have determined whether the regime has stretched beyond its core mandate. These dynamics of regime stretching illustrate that even without formal renegotiation of an international regime, what regimes do in practice can vary—and be influenced—at the level of implementation. In other words, regimes—individually or through collective coordination—can sometimes be made to stretch to address new challenges at the national and local levels. These insights have potential relevance across the whole regime complex for crisis migration.

Empirics

While the regime complex can be mapped out abstractly at the global level, the way it plays out in particular contexts will vary. This section therefore explores what the complex does in practice when it encounters particular crisis migration scenarios. It takes three contrasting cases: displacement by the Horn of Africa drought and famine of 2011; trapped populations in the Libya crisis in 2011; mixed and anticipatory movement from Zimbabwe 2006–11. They offer a useful comparison insofar as the cases touch upon the global governance of each of the four areas of migration under the crisis migration umbrella. They reveal different aspects of how the regime complex works, and the conditions under which existing institutions are or are not able to adapt or stretch to meet the emerging challenges of crisis migration. They highlight the conditions under which the existing refugee regime and humanitarian cluster approach have or have not adapted to address the humanitarian consequences of crisis migration.

In the Horn of Africa, the refugee regime was able to stretch to play a role because of a "link" between crisis migration and the national asylum systems. The refugee regime thereby stretched to fill the gap. In contrast, this was not possible in the Libyan context, and alternative ad hoc coordination between existing IOs was required. In the Zimbabwean case, formal governance stretched only a limited extent to protect vulnerable people within mixed or anticipatory migration flows. Instead, informal protection structures based on NGOs, church or community-based self-protection filled some of the protection gaps. The three cases thereby illustrate different modes through which protection gaps resulting from different elements of crisis migration have been filled until now.

Where the refugee regime and cluster approach were unable to stretch to meet the challenges of crisis, ad hoc arrangements were required, which have necessitated the flexible and creative coordination mechanisms across existing institutions, or alternatively, the mobilization of informal protection structures.

Displacement: drought and famine in the Horn of Africa

The 2011 drought and famine in the Horn of Africa precipitated large-scale displacement (Betts 2013a). Worst affected was Somalia, where the complex interaction of conflict, state fragility and food insecurity exacerbated the impact on the population and limited the prospects for mitigating the crisis in country. Responses to the IDP crisis within South-Central Somalia were limited by insecurity and access constraints. During the year around 289,000 Somalis fled to Ethiopia, Kenya, Djibouti and Yemen, including 44,000 in the first three weeks of July alone.

Within the country of origin, responses were highly constrained—both to addressing the underlying causes of the crisis and to meeting the humanitarian assistance and protection needs of IDPs. Fighting between Al-Shabaab and the internationally backed Transitional Federal Government, alongside fragile governance, had meant that the UN system's response to Somalia had for a long time been predominantly coordinated from Nairobi. The cluster approach was applied to address in-country needs, coordinated through the UN Country Team for Somalia. The national protection cluster coordinated an immediate tripartite response based on collaboration between the World Food Program (WFP), United Nations Children's Fund (UNICEF) and UNHCR, within which UNHCR provided emergency assistance packages, including food and non-food items, and engaged in protection monitoring. All of the organizations used a network of local implementing partners to deliver and administer basic assistance. Yet the in-country response was notably slow and cumbersome in both planning and implementation, due to humanitarian space constraints and limited access and capacity within the country.

Across the border, the humanitarian response in neighboring countries was largely based on stretching of the refugee regime. The ability to "link" crisis migration to the refugee regime because of the nature of refugee legislation and policy within Kenya and Ethiopia meant that crisis migration could fall within the mandate of UNHCR and so all fleeing Somalis were able to receive protection as though they were refugees. In particular, the 1969 OAU Convention, incorporated in both countries' national refugee frameworks, enabled all Somalis to be recognized on a "prima facie" basis. That in turn placed primary responsibility for protection with UNHCR by connecting crisis migration to a refugee protection mandate.

In Ethiopia, the 2004 Refugee Proclamation allows prima facie recognition under both the 1951 Refugee Convention and the 1969 Organization of African Unity (OAU) Convention Governing the Specific Aspects of Refugee Problems in Africa (1969 OAU Convention). The Ethiopian Administration for Refugee and Returnee Affairs (ARRA) jointly coordinates a Standing Task Force on refugee affairs with UNHCR, with twenty-three implementing partners in place. This enabled a coordinated "refugee regime" response to the influx with agency cooperation in the Dollo Ado refugee camp, in particular, where new arrivals came. For example, UNHCR established functional partnerships with UNICEF on nutrition and water, sanitation and health (WASH) and child protection, and with the United Nations Population Fund (UNFPA) and WHO on sexual and gender-based violence (SGBV) and health respectively.

In Kenya, the 2006 Kenyan Refugee Act also allows prima facie status under the 1951 Refugee Convention and the 1969 OAU Convention. As in Ethiopia, UNHCR has had a long-standing and established range of implementing partners in the Dadaab refugee camps that took most of the 2011 influx. While prima facie recognition for all Somalis enabled the refugee regime to stretch to address those fleeing the drought and famine, it has, however, presented a range of challenges, and strained the refugee regime almost to breaking point (Lindley 2014, Chapter 3, this volume). UNHCR faced considerable funding challenges and even by November 2011 had only 60 percent funding for its revised emergency budget. Similarly, the scale of the influx led to a political shift within Kenya in which the then Minister of Home Affairs, George Saitoti, and the Defense Minister, Yousef Haji, in particular, sought to question the generosity of the Kenyan refugee system, even proposing the creation of "safe havens" within Somalia as an internal flight alternative.

Trapped populations: the response in Libya

Between March 2011 and October 2011, the North Atlantic Treaty Organization (NATO) invasion in Libya in support of the National Transition Council's attempt to overthrow Gaddafi, triggered a complex array of types of crisis migration. Koser (2014, Chapter 13, this volume) highlights five categories of migrants resulting from political instability: evacuated migrant workers, Libyan nationals moving into

Tunisia and Egypt, "boat people" moving to Europe, IDPs, and asylum seekers and refugees. These complex migration dynamics highlight gaps and challenges for the existing structures of global migration governance at three levels: in-country, in neighboring states and in third countries.

First, within Libya, over 200,000 Libyans were internally displaced in Eastern Libya, around 58,000 of whom were in spontaneous settlements (i.e. settled without international assistance) by mid-2011 (UNHCR 2011b). UNHCR, working together with the Libyan Committee for Humanitarian Aid and Relief (LCHR) and organizations such as WFP, provided shelter, food and non-food items to IDPs in and around Benghazi. However, there was very limited access to large areas of the country and UN international staff were absent from Tripoli for most of the crisis. This made implementation of a national cluster response extremely challenging

The situation of stranded migrant workers posed a challenge to governance that fell largely outside of established institutional responses. Given Libya's long-standing role as both a migrant receiving and transit country, around 2–2.5 million migrant workers (notably from Egypt, Tunisia, Sub-Saharan Africa and Bangladesh) were in Libya when violence broke out. The majority of these, especially those from Sub-Saharan Africa, were trapped within the country unable to get home, and faced human rights violations from both rebels and Gaddafi loyalists, due in part to the fact that some had been portrayed as "African mercenaries." However, one of the remarkable successes of the crisis response was the unprecedented UNHCR–IOM cooperation on developing a joint Humanitarian Evacuation Cell in Geneva. As of mid-May around 60,000 people had been evacuated on about 300 flights. Although based on an ad hoc response, its successful implementation represents a compelling source of future best practice for addressing the humanitarian needs of stranded migrant workers.

Second, the crisis highlights challenges for addressing acute cross-border displacement. By the end of the crisis over 900,000 Libyans and foreign migrants had fled repression and violence, mainly across the border with Egypt and Tunisia. From there, many have sought assistance and protection from IOs; many foreign nationals have been supported to return to their home countries in emergency evacuation missions; and others have spontaneously transited. The mass influx required a significant emergency response at the Salloum border crossing with Egypt and the Dehiba border crossing with Tunisia. Fleeing political instability, their exact status was ambiguous in the absence of refugee status determination. However, they were registered and given a de facto form of temporary protection at the border, with UNHCR taking the lead in providing protection.

Third, in addition to the effect on neighboring states, a relatively limited number of people attempted to cross the Mediterranean into Europe as a result of the conflict. By May 18, 2011, however, out of 37,111 total new arrivals to Europe from North Africa, only 14,000 people had arrived in Italy and Malta by boat from Libya (UNHCR 2011b). The so-called "boat people" movements have led to drownings at sea—one boat sinking with around 600 people on board (UNHCR 2011a).

These movements have triggered political debate and intra-European rifts, notably between France and Italy. However, a fraction (less than 2 percent) of the total number displaced by the Libya crisis is crossing to Europe.

In theory, regional governance in Europe should have offered a means to support temporary protection and related burden-sharing to protect people fleeing Libya. However, in practice, political divisions made implementation impossible. At the meeting of the European Union's (EU) Justice and Home Affairs ministers in Luxembourg on April 1, Italy and Malta called on the EU to activate the 2001 EU Directive on Temporary Protection to grant temporary protection to migrants in cases of "mass influx" and to share the burden of absorbing the newcomers. However, this appeal was rejected by other European states as "premature," with claims that the influx to Europe was far less numerically significant than the mass influx into Germany from the Balkans in the 1990s (*The Economist* 2011). The influx also led to internal political tension over burden sharing within the EU and challenges to the Schengen Agreement on freedom of movement within the EU. This was exacerbated by Italy granting temporary Schengen travel permits to newly arrived Tunisians, enabling this group to move across the border to France. In response, France stepped up identity checks in areas near the border with Italy and some Tunisians were expelled back to Italy.

Mixed migration: the Zimbabwean case

Since the start of the millennium, Zimbabwe has gone from being one of the most developed countries in Africa to being mired in economic and political crisis. After the government of Robert Mugabe initiated a wave of land invasions to transfer white-owned farms to its political supporters in 2000, the resulting international sanctions, capital flight, declining agricultural productivity, and hyperinflation have conspired to plunge living standards to a level that ranks the country alongside the most fragile and failed states in the world. Simply in order to survive or to provide basic subsistence for their families, millions of people have been forced to migrate within or beyond the borders of Zimbabwe, with nearly one quarter of the population going into exile in neighboring countries. The resulting movement has been described as "The largest migration event in the region's recent history" (Polzer, Kiwanuka and Takavirwa 2010, 30). Although there are no accurate statistics available, it is commonly suggested that around two million Zimbabweans crossed into neighboring countries between 2000 and 2012 (Polzer 2008; Crisp and Kiragu 2010; Solidarity Peace Trust 2012), making it by far the largest mass influx situation anywhere in the world since the start of the twenty-first century.

However, despite the fact that the Zimbabweans have left a desperate humanitarian situation in which their most fundamental human rights cannot be guaranteed, the overwhelming majority have not been recognized as refugees. This is because governments and UNHCR have consistently argued that—with very few exceptions—the Zimbabweans could not be regarded as refugees given that most were not fleeing individualized persecution. In the words of one South African human rights advocate, "most have been escaping the economic consequences of the political situation" rather than political persecution per se.[1] Hence UNHCR described most Zimbabweans as in a "neither/nor" situation, being neither refugees under the 1951 Refugee Convention, nor voluntary, economic migrants (Betts 2013a).

The Zimbabwean influx has been characterized in a number of ways. Polzer (2008) has argued that it represents a form of "mixed migration" insofar as the movements involved a complex array of motives and circumstances, and have also included a significant number of anticipatory movements, with people leaving in large numbers prior to the elections in 2008, for example, in anticipation of significant violence. Others, such as Betts (2013a), have suggested that a significant proportion of people within those mixed migratory movements have been "survival migrants," falling outside the framework of the 1951 Refugee Convention but still fleeing serious human rights deprivations and being in need of international protection.

The overwhelming majority of Zimbabweans have fled to neighboring South Africa (Polzer 2008; Crush and Tevera 2010). The response of the South African government to the influx can be characterized as ad hoc. Because of a quirk in the asylum and immigration system, all Zimbabweans have been allowed access to South African territory through so-called "asylum-seeker permits," allowing them to self-settle and have the right to work, pending assessment of their asylum claim.

Until April 2009, the main policy toward Zimbabweans was "arrest, detain and deport" for all those outside formal asylum or labor migration channels. Furthermore, once Refugee Status Determination was complete, the Zimbabweans were no longer able to receive an Asylum Seeker Permit and were liable for arrest, detention and deportation. Until 2009, the refugee recognition rate for Zimbabweans was extremely low. In the early 2000s, the Department of Home Affairs' informal practice was to reject all Zimbabwean applications (Polzer *et al.* 2010). Even at the peak of the crisis in Zimbabwe in 2008 and 2009, the refugee recognition rate was only around 10 percent of Zimbabweans, with refugee status only being made available to people individually persecuted because of direct political links to the opposition Movement for Democratic Change (Betts 2013a). According to the Department of Home Affairs' own statistics, around 150,000 people were deported each year between 2001 and 2003, increasing to 175,000 in 2004, 200,000 in 2005, 250,000 in 2006, and 300,000 in 2007 and 2008 (Crisp and Kiragu 2010; Vigneswaran *et al.* 2010, 466).

After April 2009, there were some attempts to adapt policy and the application of existing legislation. The possibility of applying the broader refugee definition contained in the 1969 OAU Convention and covering events "seriously disturbing public order" in the country of origin was discussed (Polzer 2008). Indeed, South Africa's Refugee Act incorporates both the 1951 Refugee and 1969 OAU Conventions. However, both UNHCR and the Government resisted this on the grounds that this clause within the 1969 OAU Convention lacks "doctrinal clarity" (Crisp and Kiragu 2010, 21). Beyond that, as refugee rights advocates—and UNHCR—began lobbying the government to grant Zimbabweans temporary residence permits under Section 31(2)(b), the Home Affairs Minister Nosiviwe Mapisa-Nqakula announced a moratorium on the deportation of Zimbabweans. This ran only from May 2009 until October 2011.

The protection of Zimbabweans in South Africa has fallen between cracks of different IOs' mandates. UNHCR's role in relation to the Zimbabweans in South Africa has, by its own admission, "been a subtle and arguably ambiguous one" (Crisp and Kiragu 2010, 21). It has consistently regarded most Zimbabweans as not being refugees. Only the quirk of the South African asylum system—of granting "asylum seeker permits" to all who request them—has designated Zimbabweans as people who fall within the purview of UNHCR's mandate. This "asylum seeker" link to UNHCR's mandate has meant that UNHCR has played a practical role in trying to ensure and oversee access to the asylum system at the border and at Refugee Reception Offices. However, due to its limited staff and presence in both contexts, its effectiveness in this role has been called into question; one member of staff at the Department of Home Affairs described UNHCR as "largely invisible." According to UNHCR, other members of the UN Country Team were "relatively inactive" because "they perceive it as a 'UNHCR problem'" and "irregular migrants constitute a 'grey zone' in the UN system" (Crisp and Kiragu 2010, 22).

Consequently, the most relevant sources of protection for many Zimbabweans have been informal sources. Local NGOs, church organizations, Zimbabwean diaspora organizations, and community-based self-protection strategies have filled some of the gaps left by the absence of adequate international or national-level responses. In Johannesburg, for example, the Central Methodist Church, working under the leadership of Bishop Paul Verryn, has played an important role in offering sanctuary and protection. Meanwhile Zimbabwean groups such as the range of NGOs working under the auspices of the Global Zimbabwe Forum have played an important humanitarian role within urban centers such as Johannesburg and Cape Town (Betts 2013a).

Policy

The above cases highlight the variability in the extent to which existing institutions are or are not fit for purpose in relation to the challenges posed by different aspects of crisis migration. In some areas, existing governance structures adequately address the humanitarian challenge. In other areas, structures exist in theory, but problems of implementation exist in practice. In still other areas, there are gaps that need to be filled. The refugee regime and the humanitarian cluster approach, in particular, offer the main sources of response to cross-border and internal crisis migration. However, in some contexts crisis migration either goes beyond their intended scope or there are challenges to implementation. This begs the question of how, where there are gaps, they should be addressed and on what basis.

The emergence of crisis migration represents a particular case of a broader phenomenon in world politics. While the core of global governance was created during a particular historical juncture, new challenges have emerged over time. The question is, how should existing institutions adapt given that few old institutions die and it is difficult to create new formal multilateral institutions? An important analytical feature of many "new" trans-boundary problems that emerge and require international cooperation is that they are implicitly embedded in existing structures of governance. In other words, they relate to and touch upon the purview of a set of norms and organizations that already exist, even if the relationship is not explicit from the title or mandate of the norms and organizations. Crisis migration is one such area that is implicitly embedded within a pre-existing set of institutions.

In such a situation, rather than reinventing the wheel by assuming the need for a new organization or treaty, it makes sense to begin with a principle of making existing institutions work better. It also makes sense to work from the "bottom up" and consider institutional change at three levels: implementation, institutionaliza-tion, and international agreements (Betts 2013a).

At the level of *implementation*, a range of norms and organizational structures exist, have been signed and ratified by states, but are not always fully implemented. Examples include the ways in which signatory states of the 1969 OAU Conven-tion do not always implement protection for those fleeing "serious disturbances to public order," while the EU has rarely implemented its directive on temporary protection. Furthermore, there are a range of normative frameworks that have implications for states' obligations toward crisis migration that falls outside the 1951 Refugee Convention or the Guiding Principles on Internal Displacement. Most notably, states have signed up to and ratified human rights norms, which have sig-nificant implications for how they should respond to crisis migration.

At the level of *institutionalization*, there are ways in which existing norms or practices might be better incorporated within legal and policy frameworks. Many states in Africa have not institutionalized the 1969 OAU Convention. The UN Convention on the Rights of All Migrant Workers and their Families has poten-tial implications for the rights of stranded migrant workers in the context of crisis. Where practices are developed—such as the joint UNHCR–IOM Humanitarian Evacuation Program in Libya—these might be documented in ways that enable them to become informally institutionalized as "good practice."

At the level of *international agreements*, it makes sense only to consider the need for new structures once the possibility to improve implementation or institution-alization of existing institutions is exhausted. Even then, reforming international agreements need not imply the creation of new treaties or organizations. Instead, it may involve processes of consolidation in relation to existing norms and proc-esses of coordination in relation to existing institutions. When new challenges are "embedded" in a broad set of existing norms of organizations, as is the case with crisis migration, "soft law" frameworks may offer a means to provide an authorita-tive and applied consolidation of existing legal and normative standards within a single document. Similarly, when issue-areas are embedded within organizational frameworks, creating improved coordination structures may help fill gaps. The "cluster approach" offers a starting point; however, it is exclusively focused on humanitarian issues, and even there, lacks a focus on migrant protection. It is therefore important to look beyond a single policy field and understand the intersections across, for example, migration, environmental, security, development, and human rights governance.

Conclusion

Even though crisis migration is a recently recognized umbrella term, there are significant areas of global governance of actual and potential relevance. The existing global governance framework for crisis migration can be understood as a regime complex, characterized by a range of overlapping, nested and parallel institutions that are not hierarchically ordered. That complex exists in abstraction at the global level—in terms of the range of norms and IOs of actual and potential relevance to addressing crisis migration. However, it also exists at the level of practice, where the implementation of the complex may have different "on the ground" manifestations in relation to different crises in different places.

Looking at the regime complex for crisis migration both in theory and practice draws attention to the interplay between the humanitarian and migration regimes that lies at the core of responses to crisis migration. It sheds light on areas in which there are complementarities, contradictions, and also underexploited opportunities for improved coordination, not least between UNHCR and IOM. In some areas existing institutions are working to address crisis migration. The Horn of Africa case shows how, when there is a "link" to national refugee legislation, the refugee regime may stretch to cover gaps. In contrast, Libya highlights how the challenge of trapped and stranded migrant workers has required new and creative ad hoc responses. Meanwhile, the Zimbabwean case shows how, when existing institutions have largely failed to adapt to complex mixed migratory movements, a range of informal structures and community-based self-protection mechanisms have filled some of the gaps.

Where gaps exist, these are likely to be most effectively addressed through strategies that exist at both the national and global levels. At the national level, improved implementation of existing structures can sometimes fill significant gaps. At the global level, it is crucial to be aware that the global governance challenge posed by crisis migration is not analytically unique. It is similar to the challenge posed by the emergence of many "new" issues against the backdrop of "old" institutions. Crisis migration is implicitly embedded within a dense tapestry of existing norms and organizations. Rather than requiring new treaties or organizations, it requires the coherent consolidation of existing norms and practices and the better coordination of existing organizations across all levels of governance.

Note

1   Interview with Kajaal Ramjathan-Keogh, Head of Refugee and Migrant Rights Programme, Lawyers for Human Rights (LHR), Johannesburg, March 18, 2009.

References

Abbott, K.W. and Snidal, D. (2001) "International 'Standards' and International Governance," *Journal of European Public Policy*, 8(3): 345–370.

Alter, K.J. and Meunier, S. (2009) "The Politics of International Regime Complexity," *Perspectives on Politics*, 7(1): 13–24.

Betts, A. (2009), "Institutional Proliferation and the Refugee Regime," *Perspectives on Politics*, 7(1): 53–58.

Betts, A. (2010) "The Refugee Regime Complex," *Refugee Survey Quarterly*, 29(2): 12–37.

Betts, A. (ed.) (2011) *Global Migration Governance*, Oxford: Oxford University Press.

Betts, A. (2013a) *Survival Migration: Failed Governance and the Crisis of Displacement*, Ithaca: Cornell University Press.

Betts, A. (2013b), "Regime Complexity and International Organizations: UNHCR as a Challenged Institution," *Global Governance*, 19(1): 69–81.

Cholewinski, R., Perruchoud, R. and MacDonald, E. (eds.) (2007) *International Migration Law: Developing Paradigms and Key Challenges*, The Hague: T.M.C. Asser Press.

Collyer, M. (2010) "Stranded Migrants and the Fragmented Journey," *Journal of Refugee Studies*, 23(3): 273–293.

Crisp, J. and Kiragu, E. (2010) "Refugee Protection and International Migration: A Review of UNHCR's Role in Malawi, Mozambique and South Africa," Geneva: UNHCR Policy Development and Evaluation Service (PDES).

Crush, J. and Tevera, D.S. (eds.) (2010) *Zimbabwe's Exodus: Crisis, Migration, Survival*, Cape Town: Southern African Migration Project/Ottawa: International Development Research Centre.

Davis, C. (2009), "Overlapping Institutions in Trade Policy," *Perspectives on Politics*, 7(1): 25–31.

Dowd, R. (2008) "Trapped in Transit: the Plight and Human Rights of Stranded Migrants," *New Issues in Refugee Research*, 156, Geneva: UNHCR.

Drezner, D.W. (2007) *All Politics is Global: Explaining International Regulatory Regimes*, Princeton, NJ: Oxford: Princeton University Press.

Economist, The (2011) "The Next European Crisis: Boat People," *The Economist Online*, April 11, available online at: http://www.economist.com/blogs/charlemagne/2011/04/north_african_migration.

Hafner Burton, E.M. (2009) *Forced to Be Good: Why Trade Agreements Boost Human Rights*, London: Cornell University Press.

Hall, P.A. and Thelen, K.A. (2009) "Institutional Change in Varieties of Capitalism," *Socio-Economic Review Special Issue: Changing Institutions in Developed Democracies: Economics, Politics and Welfare*, 7(1): 7–34.

Helfer, L. (2004) "Regime Shifting: The TRIPS Agreement and New Dynamics of International Intellectual Property Lawmaking," *Yale Journal of International Law*, 29: 1.

Keohane, R.O. and Victor, D.G. (2010) "The Regime Complex for Climate Change," Discussion Paper 10–33, Cambridge: Harvard Project on International Climate Agreements.

Koser, K. and Martin, S. (eds.) (2011) *The Migration–displacement Nexus: Patterns, Processes, and Policies*, Oxford: Berghahn Books.

Krasner, S.D. (1982) "Structural Causes and Regime Consequences: Regimes as Intervening Variables," *International Organization*, 36(2): 185–205.

Lieberman, R.C. (2002) "Ideas, Institutions, and Political Order: Explaining Political Change," *American Political Science Review*, 96(4): 697–712.

Mahoney, J. and Thelen, K.A. (eds.) (2010) *Explaining Institutional Change: Ambiguity, Agency and Power*, Cambridge: Cambridge University Press.

Margulis, M. (2013) "The Regime Complex for Food Security: Implications for the Global Hunger Challenge," *Global Governance: A Review of Multilateralism and International Organizations*, 19(1): 53–67.

Morin, J.-F. and Orsini, A. (2013) "Regime Complexity and Policy Coherency: Introducing a Co-adjustments Model," *Global Governance*, 19(1): 41–51.

Orsini, A., Morin, J.F. and Young, O. (2013) "Regime Complexes: A Buzz, A Boom or a Boost for Global Governance?" *Global Governance*, 19(1): 27–39.

Pierson, P. (2004) *Politics in Time: History, Institutions, and Social Analysis*, Princeton: Princeton University Press.

Polzer, T. (2008) "Responding to Zimbabwean Migration in South Africa: Evaluating Options," *South African Journal of International Affairs*, 15(1): 1–15.

Polzer, T., Kiwanuka, M. and Takavirwa, K. (2010) "Regional Responses to Zimbabwean Migration, 2000–2010," *Open Space: On the Move: Dynamics of Migration in Southern Africa*, 3: 30–34.

Raustiala, K. and Victor, D.G. (2004) "The Regime Complex for Plant Genetic Resources," *International Organization*, 58(2): 277–309.

Solidarity Peace Trust (2012) *Perils and Pitfalls: Migrants and Deportation in South Africa*, Johannesburg: Solidarity Peace Trust, available online at: www.solidaritypeacetrust. org/perils-and-pitfalls.

Trachtmann, J. (2011) "Coherence and the Regime Complex for International Economic Migration," in Kunz, R., Lavenex, S. and Panizzon, M. (eds.), *Multilayered Migration Governance*, London: Routledge.

UNHCR (2011a) "Hundreds Risk Return to Libya in Bid to Reach Europe by Boat," May 17, available online at: http://www.unhcr.org/4dd27eea9.html.

UNHCR (2011b) "Update No. 25 on the Humanitarian Situation in Libya," *UNHCR*, May 18, available online at: http://www.unhcr.org/4dd61ebc9.html.

Van Hear, N., Brubaker, R. and Bessa, T. (2009) *Managing Mobility for Human Development: The Growing Salience of Mixed Migration*, UNDP Human Development Research Paper 2009/20, New York: UNDP.

Vigneswaran, D., Araia, T., Hoag, C. and Tshabalala, X. (2010) "Criminality or Monopoly? Informal Immigration Enforcement in South Africa," *Journal of Southern African Studies*, 36(2): 465–485.

Young, O. (2010) *Institutional Dynamics: Emergent Patterns in International Environmental Governance*, Cambridge: MIT Press.

INDEX